The Microbiological Quality of Food

Related Titles

Foodborne Parasites in the Food Supply Web
(ISBN: 978-1-78242-332-4)

Improving and Tailoring Enzymes for Food Quality and Functionality
(ISBN: 978-1-78242-285-3)

Advances in Food Traceability Techniques and Technologies
(ISBN: 978-0-08-100310-7)

The Microbiological Quality of Food

Foodborne Spoilers

Edited by

Antonio Bevilacqua, Maria Rosaria Corbo and Milena Sinigaglia

Department of the Science of Agriculture
Food and Environment (SAFE)
University of Foggia
Foggia, Italy

ELSEVIER

AMSTERDAM • BOSTON • HEIDELBERG • LONDON
NEW YORK • OXFORD • PARIS • SAN DIEGO
SAN FRANCISCO • SINGAPORE • SYDNEY • TOKYO
Woodhead Publishing is an imprint of Elsevier

WP
WOODHEAD PUBLISHING

Woodhead Publishing is an imprint of Elsevier
The Officers' Mess Business Centre, Royston Road, Duxford, CB22 4QH, United Kingdom
50 Hampshire Street, 5th Floor, Cambridge, MA 02139, United States
The Boulevard, Langford Lane, Kidlington, OX5 1GB, United Kingdom

British Library Cataloguing-in-Publication Data
A catalogue record for this book is available from the British Library

Library of Congress Cataloging-in-Publication Data
A catalog record for this book is available from the Library of Congress

ISBN: 978-0-08-100502-6 (print)
ISBN: 978-0-08-100503-3 (online)

For information on all Woodhead Publishing
visit our website at https://www.elsevier.com

Publisher: Nikki Levy
Acquisition Editor: Rob Sykes
Editorial Project Manager: Karen Miller
Production Project Manager: Julia Haynes
Designer: Maria Inês Cruz

Typeset by MPS Limited, Chennai, India

Contents

Part I
The Spoilers

2. *Pseudomonas* and Related Genera

Nadia Andrea Andreani and Luca Fasolato

3. Specific Spoilage Organisms (SSOs) in Fish

Ioannis S. Boziaris and Foteini F. Parlapani

4. Spore-Forming Bacteria

Fabienne Remize

5. Yeasts

Marianne Perricone, Mariangela Gallo, Maria Rosaria Corbo, Milena Sinigaglia and Antonio Bevilacqua

List of Contributors

Nadia Andrea Andreani University of Padova, Padova, PD, Italy

Francisco Noé Arroyo-López Instituto de la Grasa (CSIC), University Campus Pablo de Olavide, Seville, Spain

Antonio Bevilacqua University of Foggia, Foggia, FG, Italy

Ioannis S. Boziaris School of Agricultural Sciences, University of Thessaly, Volos, Greece

Daniela Campaniello University of Foggia, Foggia, FG, Italy

Altieri Clelia University of Foggia, Foggia, FG, Italy

Giuseppe Comi University of Udine, Udine, UD, Italy

Maria Rosaria Corbo University of Foggia, Foggia, FG, Italy

Luca Fasolato University of Padova, Padova, PD, Italy

Mariangela Gallo University of Foggia, Foggia, FG, Italy

Antonio Garrido-Fernández Instituto de la Grasa (CSIC), University Campus Pablo de Olavide, Seville, Spain

Michael Lu University of Maryland, College Park, MD, United States

Eduardo Medina-Pradas Instituto de la Grasa (CSIC), University Campus Pablo de Olavide, Seville, Spain

Chioma Nwakanma Michael Okpara University of Agriculture, Abia, Nigeria

Foteini F. Parlapani School of Agricultural Sciences, University of Thessaly, Volos, Greece

Ilenys M. Pérez-Díaz North Carolina State University, Raleigh, NC, United States

Marianne Perricone University of Foggia, Foggia, FG, Italy

Leonardo Petruzzi University of Foggia, Foggia, FG, Italy

Fabienne Remize Université de La Réunion, ESIROI, UMR QualiSud, Saint-Denis, Réunion

Milena Sinigaglia University of Foggia, Foggia, FG, Italy

Barbara Speranza University of Foggia, Foggia, FG, Italy

Marian Unachukwu Godfrey Okoye University, Enugu, Nigeria

Nam Sun Wang University of Maryland, College Park, MD, United States

Woodhead Publishing Series in Food Science, Technology and Nutrition

Chilled foods: A comprehensive guide
Edited by C. Dennis and M. Stringer

Yoghurt: Science and technology
A. Y. Tamime and R. K. Robinson

Food processing technology: Principles and practice
P. J. Fellows

Bender's dictionary of nutrition and food technology Sixth edition
D. A. Bender

Determination of veterinary residues in food
Edited by N. T. Crosby

Food contaminants: Sources and surveillance
Edited by C. Creaser and R. Purchase

Nitrates and nitrites in food and water
Edited by M. J. Hill

Pesticide chemistry and bioscience: The food-environment challenge
Edited by G. T. Brooks and T. Roberts

Pesticides: Developments, impacts and controls
Edited by G. A. Best and A. D. Ruthven

Dietary fibre: Chemical and biological aspects
Edited by D. A. T. Southgate, K. W. Waldron, I. T. Johnson and G. R. Fenwick

Vitamins and minerals in health and nutrition
M. Tolonen

Technology of biscuits, crackers and cookies Second edition
D. Manley

Instrumentation and sensors for the food industry
Edited by E. Kress-Rogers

Food and cancer prevention: Chemical and biological aspects
Edited by K. W. Waldron, I. T. Johnson and G. R. Fenwick

Food colloids: Proteins, lipids and polysaccharides
Edited by E. Dickinson and B. Bergenstahl

Food emulsions and foams
Edited by E. Dickinson

Maillard reactions in chemistry, food and health
Edited by T. P. Labuza, V. Monnier, J. Baynes and J. O'Brien

The Maillard reaction in foods and medicine
Edited by J. O'Brien, H. E. Nursten, M. J. Crabbe and J. M. Ames

Encapsulation and controlled release
Edited by D. R. Karsa and R. A. Stephenson

Flavours and fragrances
Edited by A. D. Swift

Feta and related cheeses
Edited by A. Y. Tamime and R. K. Robinson

Biochemistry of milk products
Edited by A. T. Andrews and J. R. Varley

Physical properties of foods and food processing systems
M. J. Lewis

Food irradiation: A reference guide
V. M. Wilkinson and G. Gould

Kent's technology of cereals: An introduction for students of food science and agriculture Fourth edition
N. L. Kent and A. D. Evers

Biosensors for food analysis
Edited by A. O. Scott

Separation processes in the food and biotechnology industries: Principles and applications
Edited by A. S. Grandison and M. J. Lewis

Handbook of indices of food quality and authenticity
R. S. Singhal, P. K. Kulkarni and D. V. Rege

Principles and practices for the safe processing of foods
D. A. Shapton and N. F. Shapton

Biscuit, cookie and cracker manufacturing manuals Volume 1: Ingredients
D. Manley

Biscuit, cookie and cracker manufacturing manuals Volume 2: Biscuit doughs
D. Manley

Biscuit, cookie and cracker manufacturing manuals Volume 3: Biscuit dough piece forming
D. Manley

Biscuit, cookie and cracker manufacturing manuals Volume 4: Baking and cooling of biscuits
D. Manley

Biscuit, cookie and cracker manufacturing manuals Volume 5: Secondary processing in biscuit manufacturing
D. Manley

Biscuit, cookie and cracker manufacturing manuals Volume 6: Biscuit packaging and storage
D. Manley

Practical dehydration Second edition
M. Greensmith

Lawrie's meat science Sixth edition
R. A. Lawrie

Yoghurt: Science and technology Second edition
A. Y. Tamime and R. K. Robinson

New ingredients in food processing: Biochemistry and agriculture
G. Linden and D. Lorient

Benders' dictionary of nutrition and food technology Seventh edition
D. A. Bender and A. E. Bender

Technology of biscuits, crackers and cookies Third edition
D. Manley

Food processing technology: Principles and practice Second edition
P. J. Fellows

Managing frozen foods
Edited by C. J. Kennedy

Handbook of hydrocolloids
Edited by G. O. Phillips and P. A. Williams

Food labelling
Edited by J. R. Blanchfield

Cereal biotechnology
Edited by P. C. Morris and J. H. Bryce

Food intolerance and the food industry
Edited by T. Dean

The stability and shelf-life of food
Edited by D. Kilcast and P. Subramaniam

Functional foods: Concept to product
Edited by G. R. Gibson and C. M. Williams

Chilled foods: A comprehensive guide Second edition
Edited by M. Stringer and C. Dennis

HACCP in the meat industry
Edited by M. Brown

Biscuit, cracker and cookie recipes for the food industry
D. Manley

Cereals processing technology
Edited by G. Owens

Baking problems solved
S. P. Cauvain and L. S. Young

Thermal technologies in food processing
Edited by P. Richardson

Frying: Improving quality
Edited by J. B. Rossell

Food chemical safety Volume 1: Contaminants
Edited by D. Watson

Making the most of HACCP: Learning from others' experience
Edited by T. Mayes and S. Mortimore

Food process modelling
Edited by L. M. M. Tijskens, M. L. A. T. M. Hertog and B. M. Nicolaï

EU food law: A practical guide
Edited by K. Goodburn

Extrusion cooking: Technologies and applications
Edited by R. Guy

Auditing in the food industry: From safety and quality to environmental and other audits
Edited by M. Dillon and C. Griffith

Handbook of herbs and spices Volume 1
Edited by K. V. Peter

Food product development: Maximising success
M. Earle, R. Earle and A. Anderson

Instrumentation and sensors for the food industry Second edition
Edited by E. Kress-Rogers and C. J. B. Brimelow

Food chemical safety Volume 2: Additives
Edited by D. Watson

Fruit and vegetable biotechnology
Edited by V. Valpuesta

Foodborne pathogens: Hazards, risk analysis and control
Edited by C. de W. Blackburn and P. J. McClure

Meat refrigeration
S. J. James and C. James

Lockhart and Wiseman's crop husbandry Eighth edition
H. J. S. Finch, A. M. Samuel and G. P. F. Lane

Safety and quality issues in fish processing
Edited by H. A. Bremner

Minimal processing technologies in the food industries
Edited by T. Ohlsson and N. Bengtsson

Fruit and vegetable processing: Improving quality
Edited by W. Jongen

The nutrition handbook for food processors
Edited by C. J. K. Henry and C. Chapman

Colour in food: Improving quality
Edited by D. MacDougall

Meat processing: Improving quality
Edited by J. P. Kerry, J. F. Kerry and D. A. Ledward

Microbiological risk assessment in food processing
Edited by M. Brown and M. Stringer

Performance functional foods
Edited by D. Watson

Functional dairy products Volume 1
Edited by T. Mattila-Sandholm and M. Saarela

Taints and off-flavours in foods
Edited by B. Baigrie

Yeasts in food
Edited by T. Boekhout and V. Robert

Phytochemical functional foods
Edited by I. T. Johnson and G. Williamson

Novel food packaging techniques
Edited by R. Ahvenainen

Detecting pathogens in food
Edited by T. A. McMeekin

Natural antimicrobials for the minimal processing of foods
Edited by S. Roller

Texture in food Volume 1: Semi-solid foods
Edited by B. M. McKenna

Dairy processing: Improving quality
Edited by G. Smit

Hygiene in food processing: Principles and practice
Edited by H. L. M. Lelieveld, M. A. Mostert, B. White and J. Holah

Rapid and on-line instrumentation for food quality assurance
Edited by I. Tothill

Sausage manufacture: Principles and practice
E. Essien

Environmentally-friendly food processing
Edited by B. Mattsson and U. Sonesson

Bread making: Improving quality
Edited by S. P. Cauvain

Food preservation techniques
Edited by P. Zeuthen and L. Bøgh-Sørensen

Food authenticity and traceability
Edited by M. Lees

Analytical methods for food additives
R. Wood, L. Foster, A. Damant and P. Key

Handbook of herbs and spices Volume 2
Edited by K. V. Peter

Texture in food Volume 2: Solid foods
Edited by D. Kilcast

Proteins in food processing
Edited by R. Yada

Detecting foreign bodies in food
Edited by M. Edwards

Understanding and measuring the shelf-life of food
Edited by R. Steele

Poultry meat processing and quality
Edited by G. Mead

Functional foods, ageing and degenerative disease
Edited by C. Remacle and B. Reusens

Mycotoxins in food: Detection and control
Edited by N. Magan and M. Olsen

Improving the thermal processing of foods
Edited by P. Richardson

Pesticide, veterinary and other residues in food
Edited by D. Watson

Starch in food: Structure, functions and applications
Edited by A.-C. Eliasson

Functional foods, cardiovascular disease and diabetes
Edited by A. Arnoldi

Brewing: Science and practice
D. E. Briggs, P. A. Brookes, R. Stevens and C. A. Boulton

Using cereal science and technology for the benefit of consumers: Proceedings of the 12P^{thP} International ICC Cereal and Bread Congress, 24–26P^{thP} May, 2004, Harrogate, UK
Edited by S. P. Cauvain, L. S. Young and S. Salmon

Improving the safety of fresh meat
Edited by J. Sofos

Understanding pathogen behaviour: Virulence, stress response and resistance
Edited by M. Griffiths

The microwave processing of foods
Edited by H. Schubert and M. Regier

Food safety control in the poultry industry
Edited by G. Mead

Improving the safety of fresh fruit and vegetables
Edited by W. Jongen

Food, diet and obesity
Edited by D. Mela

Handbook of hygiene control in the food industry
Edited by H. L. M. Lelieveld, M. A. Mostert and J. Holah

Detecting allergens in food
Edited by S. Koppelman and S. Hefle

Improving the fat content of foods
Edited by C. Williams and J. Buttriss

Improving traceability in food processing and distribution
Edited by I. Smith and A. Furness

Flavour in food
Edited by A. Voilley and P. Etievant

The Chorleywood bread process
S. P. Cauvain and L. S. Young

Food spoilage microorganisms
Edited by C. de W. Blackburn

Emerging foodborne pathogens
Edited by Y. Motarjemi and M. Adams

Benders' dictionary of nutrition and food technology Eighth edition
D. A. Bender

Optimising sweet taste in foods
Edited by W. J. Spillane

Brewing: New technologies
Edited by C. Bamforth

Handbook of herbs and spices Volume 3
Edited by K. V. Peter

Lawrie's meat science Seventh edition
R. A. Lawrie in collaboration with D. A. Ledward

Modifying lipids for use in food
Edited by F. Gunstone

Meat products handbook: Practical science and technology
G. Feiner

Food consumption and disease risk: Consumer–pathogen interactions
Edited by M. Potter

Acrylamide and other hazardous compounds in heat-treated foods
Edited by K. Skog and J. Alexander

Managing allergens in food
Edited by C. Mills, H. Wichers and K. Hoffman-Sommergruber

Microbiological analysis of red meat, poultry and eggs
Edited by G. Mead

Maximising the value of marine by-products
Edited by F. Shahidi

Chemical migration and food contact materials
Edited by K. Barnes, R. Sinclair and D. Watson

Understanding consumers of food products
Edited by L. Frewer and H. van Trijp

Reducing salt in foods: Practical strategies
Edited by D. Kilcast and F. Angus

Modelling microorganisms in food
Edited by S. Brul, S. Van Gerwen and M. Zwietering

Tamime and Robinson's Yoghurt: Science and technology Third edition
A. Y. Tamime and R. K. Robinson

Handbook of waste management and co-product recovery in food processing Volume 1
Edited by K. W. Waldron

Improving the flavour of cheese
Edited by B. Weimer

Novel food ingredients for weight control
Edited by C. J. K. Henry

Consumer-led food product development
Edited by H. MacFie

Functional dairy products Volume 2
Edited by M. Saarela

Modifying flavour in food
Edited by A. J. Taylor and J. Hort

Cheese problems solved
Edited by P. L. H. McSweeney

Handbook of organic food safety and quality
Edited by J. Cooper, C. Leifert and U. Niggli

Understanding and controlling the microstructure of complex foods
Edited by D. J. McClements

Novel enzyme technology for food applications
Edited by R. Rastall

Food preservation by pulsed electric fields: From research to application
Edited by H. L. M. Lelieveld and S. W. H. de Haan

Technology of functional cereal products
Edited by B. R. Hamaker

Case studies in food product development
Edited by M. Earle and R. Earle

Delivery and controlled release of bioactives in foods and nutraceuticals
Edited by N. Garti

Fruit and vegetable flavour: Recent advances and future prospects
Edited by B. Brückner and S. G. Wyllie

Food fortification and supplementation: Technological, safety and regulatory aspects
Edited by P. Berry Ottaway

Improving the health-promoting properties of fruit and vegetable products
Edited by F. A. Tomás-Barberán and M. I. Gil

Improving seafood products for the consumer
Edited by T. Børresen

In-pack processed foods: Improving quality
Edited by P. Richardson

Handbook of water and energy management in food processing
Edited by J. Klemeš, R.. Smith and J.-K. Kim

Environmentally compatible food packaging
Edited by E. Chiellini

Improving farmed fish quality and safety
Edited by Ø. Lie

Carbohydrate-active enzymes
Edited by K.-H. Park

Chilled foods: A comprehensive guide Third edition
Edited by M. Brown

Food for the ageing population
Edited by M. M. Raats, C. P. G. M. de Groot and W. A Van Staveren

Improving the sensory and nutritional quality of fresh meat
Edited by J. P. Kerry and D. A. Ledward

Shellfish safety and quality
Edited by S. E. Shumway and G. E. Rodrick

Functional and speciality beverage technology
Edited by P. Paquin

Functional foods: Principles and technology
M. Guo

Endocrine-disrupting chemicals in food
Edited by I. Shaw

Meals in science and practice: Interdisciplinary research and business applications
Edited by H. L. Meiselman

Food constituents and oral health: Current status and future prospects
Edited by M. Wilson

Handbook of hydrocolloids Second edition
Edited by G. O. Phillips and P. A. Williams

Food processing technology: Principles and practice Third edition
P. J. Fellows

Science and technology of enrobed and filled chocolate, confectionery and bakery products
Edited by G. Talbot

Foodborne pathogens: Hazards, risk analysis and control Second edition
Edited by C. de W. Blackburn and P. J. McClure

Designing functional foods: Measuring and controlling food structure breakdown and absorption
Edited by D. J. McClements and E. A. Decker

New technologies in aquaculture: Improving production efficiency, quality and environmental management
Edited by G. Burnell and G. Allan

More baking problems solved
S. P. Cauvain and L. S. Young

Soft drink and fruit juice problems solved
P. Ashurst and R. Hargitt

Biofilms in the food and beverage industries
Edited by P. M. Fratamico, B. A. Annous and N. W. Gunther

Dairy-derived ingredients: Food and neutraceutical uses
Edited by M. Corredig

Handbook of waste management and co-product recovery in food processing Volume 2
Edited by K. W. Waldron

Innovations in food labelling
Edited by J. Albert

Delivering performance in food supply chains
Edited by C. Mena and G. Stevens

Chemical deterioration and physical instability of food and beverages
Edited by L. H. Skibsted, J. Risbo and M. L. Andersen

Managing wine quality Volume 1: Viticulture and wine quality
Edited by A. G. Reynolds

Improving the safety and quality of milk Volume 1: Milk production and processing
Edited by M. Griffiths

Improving the safety and quality of milk Volume 2: Improving quality in milk products
Edited by M. Griffiths

Cereal grains: Assessing and managing quality
Edited by C. Wrigley and I. Batey

Sensory analysis for food and beverage quality control: A practical guide
Edited by D. Kilcast

Managing wine quality Volume 2: Oenology and wine quality
Edited by A. G. Reynolds

Winemaking problems solved
Edited by C. E. Butzke

Environmental assessment and management in the food industry
Edited by U. Sonesson, J. Berlin and F. Ziegler

Consumer-driven innovation in food and personal care products
Edited by S. R. Jaeger and H. MacFie

Tracing pathogens in the food chain
Edited by S. Brul, P. M. Fratamico and T. A. McMeekin

Case studies in novel food processing technologies: Innovations in processing, packaging, and predictive modelling
Edited by C. J. Doona, K. Kustin and F. E. Feeherry

Freeze-drying of pharmaceutical and food products
T.-C. Hua, B.-L. Liu and H. Zhang

Oxidation in foods and beverages and antioxidant applications Volume 1: Understanding mechanisms of oxidation and antioxidant activity
Edited by E. A. Decker, R. J. Elias and D. J. McClements

Oxidation in foods and beverages and antioxidant applications Volume 2: Management in different industry sectors
Edited by E. A. Decker, R. J. Elias and D. J. McClements

Protective cultures, antimicrobial metabolites and bacteriophages for food and beverage biopreservation
Edited by C. Lacroix

Separation, extraction and concentration processes in the food, beverage and nutraceutical industries
Edited by S. S. H. Rizvi

Determining mycotoxins and mycotoxigenic fungi in food and feed
Edited by S. De Saeger

Developing children's food products
Edited by D. Kilcast and F. Angus

Functional foods: Concept to product Second edition
Edited by M. Saarela

Postharvest biology and technology of tropical and subtropical fruits Volume 1: Fundamental issues
Edited by E. M. Yahia

Postharvest biology and technology of tropical and subtropical fruits Volume 2: Açai to citrus
Edited by E. M. Yahia

Postharvest biology and technology of tropical and subtropical fruits Volume 3: Cocona to mango
Edited by E. M. Yahia

Postharvest biology and technology of tropical and subtropical fruits Volume 4: Mangosteen to white sapote
Edited by E. M. Yahia

Food and beverage stability and shelf life
Edited by D. Kilcast and P. Subramaniam

Processed Meats: Improving safety, nutrition and quality
Edited by J. P. Kerry and J. F. Kerry

Food chain integrity: A holistic approach to food traceability, safety, quality and authenticity
Edited by J. Hoorfar, K. Jordan, F. Butler and R. Prugger

Improving the safety and quality of eggs and egg products Volume 1
Edited by Y. Nys, M. Bain and F. Van Immerseel

Improving the safety and quality of eggs and egg products Volume 2
Edited by F. Van Immerseel, Y. Nys and M. Bain

Animal feed contamination: Effects on livestock and food safety
Edited by J. Fink-Gremmels

Hygienic design of food factories
Edited by J. Holah and H. L. M. Lelieveld

Manley's technology of biscuits, crackers and cookies Fourth edition
Edited by D. Manley

Nanotechnology in the food, beverage and nutraceutical industries
Edited by Q. Huang

Rice quality: A guide to rice properties and analysis
K. R. Bhattacharya

Advances in meat, poultry and seafood packaging
Edited by J. P. Kerry

Reducing saturated fats in foods
Edited by G. Talbot

Handbook of food proteins
Edited by G. O. Phillips and P. A. Williams

Lifetime nutritional influences on cognition, behaviour and psychiatric illness
Edited by D. Benton

Food machinery for the production of cereal foods, snack foods and confectionery
L.-M. Cheng

Alcoholic beverages: Sensory evaluation and consumer research
Edited by J. Piggott

Extrusion problems solved: Food, pet food and feed
M. N. Riaz and G. J. Rokey

Handbook of herbs and spices Second edition Volume 1
Edited by K. V. Peter

Handbook of herbs and spices Second edition Volume 2
Edited by K. V. Peter

Breadmaking: Improving quality Second edition
Edited by S. P. Cauvain

Emerging food packaging technologies: Principles and practice
Edited by K. L. Yam and D. S. Lee

Infectious disease in aquaculture: Prevention and control
Edited by B. Austin

Diet, immunity and inflammation
Edited by P. C. Calder and P. Yaqoob

Natural food additives, ingredients and flavourings
Edited by D. Baines and R. Seal

Microbial decontamination in the food industry: Novel methods and applications
Edited by A. Demirci and M.O. Ngadi

Chemical contaminants and residues in foods
Edited by D. Schrenk

Robotics and automation in the food industry: Current and future technologies
Edited by D. G. Caldwell

Fibre-rich and wholegrain foods: Improving quality
Edited by J. A. Delcour and K. Poutanen

Computer vision technology in the food and beverage industries
Edited by D.-W. Sun

Encapsulation technologies and delivery systems for food ingredients and nutraceuticals
Edited by N. Garti and D. J. McClements

Case studies in food safety and authenticity
Edited by J. Hoorfar

Heat treatment for insect control: Developments and applications
D. Hammond

Advances in aquaculture hatchery technology
Edited by G. Allan and G. Burnell

Open innovation in the food and beverage industry
Edited by M. Garcia Martinez

Trends in packaging of food, beverages and other fast-moving consumer goods (FMCG)
Edited by N. Farmer

New analytical approaches for verifying the origin of food
Edited by P. Brereton

Microbial production of food ingredients, enzymes and nutraceuticals
Edited by B. McNeil, D. Archer, I. Giavasis and L. Harvey

Persistent organic pollutants and toxic metals in foods
Edited by M. Rose and A. Fernandes

Cereal grains for the food and beverage industries
E. Arendt and E. Zannini

Viruses in food and water: Risks, surveillance and control
Edited by N. Cook

Improving the safety and quality of nuts
Edited by L. J. Harris

Metabolomics in food and nutrition
Edited by B. C. Weimer and C. Slupsky

Food enrichment with omega-3 fatty acids
Edited by C. Jacobsen, N. S. Nielsen, A. F. Horn and A.-D. M. Sørensen

Instrumental assessment of food sensory quality: A practical guide
Edited by D. Kilcast

Food microstructures: Microscopy, measurement and modelling
Edited by V. J. Morris and K. Groves

Handbook of food powders: Processes and properties
Edited by B. R. Bhandari, N. Bansal, M. Zhang and P. Schuck

Functional ingredients from algae for foods and nutraceuticals
Edited by H. Domínguez

Satiation, satiety and the control of food intake: Theory and practice
Edited by J. E. Blundell and F. Bellisle

Hygiene in food processing: Principles and practice Second edition
Edited by H. L. M. Lelieveld, J. Holah and D. Napper

Advances in microbial food safety Volume 1
Edited by J. Sofos

Global safety of fresh produce: A handbook of best practice, innovative commercial solutions and case studies
Edited by J. Hoorfar

Human milk biochemistry and infant formula manufacturing technology
Edited by M. Guo

High throughput screening for food safety assessment: Biosensor technologies, hyperspectral imaging and practical applications
Edited by A. K. Bhunia, M. S. Kim and C. R. Taitt

Foods, nutrients and food ingredients with authorised EU health claims: Volume 1
Edited by M. J. Sadler

Handbook of food allergen detection and control
Edited by S. Flanagan

Advances in fermented foods and beverages: Improving quality, technologies and health benefits
Edited by W. Holzapfel

Metabolomics as a tool in nutrition research
Edited by J.-L. Sébédio and L. Brennan

Dietary supplements: Safety, efficacy and quality
Edited by K. Berginc and S. Kreft

Grapevine breeding programs for the wine industry
Edited by A. G. Reynolds

Handbook of antimicrobials for food safety and quality
Edited by T. M. Taylor

Managing and preventing obesity: Behavioural factors and dietary interventions
Edited by T. P. Gill

Electron beam pasteurization and complementary food processing technologies
Edited by S. D. Pillai and S. Shayanfar

Advances in food and beverage labelling: Information and regulations
Edited by P. Berryman

Flavour development, analysis and perception in food and beverages
Edited by J. K. Parker, S. Elmore and L. Methven

Rapid sensory profiling techniques and related methods: Applications in new product development and consumer research
Edited by J. Delarue, J. B. Lawlor and M. Rogeaux

Advances in microbial food safety: Volume 2
Edited by J. Sofos

Handbook of antioxidants for food preservation
Edited by F. Shahidi

Lockhart and Wiseman's crop husbandry including grassland: Ninth edition
H. J. S. Finch, A. M. Samuel and G. P. F. Lane

Global legislation for food contact materials
Edited by J. S. Baughan

Colour additives for food and beverages
Edited by M. Scotter

A complete course in canning and related processes 14th Edition: Volume 1
Revised by S. Featherstone

A complete course in canning and related processes 14th Edition: Volume 2
Revised by S. Featherstone

A complete course in canning and related processes 14th Edition: Volume 3
Revised by S. Featherstone

Modifying food texture: Volume 1: Novel ingredients and processing techniques
Edited by J. Chen and A. Rosenthal

Modifying food texture: Volume 2: Sensory analysis, consumer requirements and preferences
Edited by J. Chen and A. Rosenthal

Modeling food processing operations
Edited by S. Bakalis, K. Knoerzer and P. J. Fryer

Foods, nutrients and food ingredients with authorised EU health claims Volume 2
Edited by M. J. Sadler

Feed and feeding practices in aquaculture
Edited by D. Allen Davis

Foodborne parasites in the food supply web: Occurrence and control
Edited by A. Gajadhar

Brewing microbiology: design and technology applications for spoilage management, sensory quality and waste valorisation
Edited by A. E. Hill

Specialty oils and fats in food and nutrition: Properties, processing and applications
Edited by G. Talbot

Improving and tailoring enzymes for food quality and functionality
Edited by R. Yada

Emerging technologies for promoting food security: Overcoming the world food crisis
Edited by C. Madramootoo

Innovation and future trends in food manufacturing and supply chain technologies
Edited by C. E. Leadley

Functional dietary lipids: Food formulation, consumer issues and innovation for health
Edited by T. Sanders

Handbook on natural pigments in food and beverages: Industrial applications for improving color
Edited by R. Carle and R. M. Schweiggert

Integrating the packaging and product experience in food and beverages: A road-map to consumer satisfaction
Edited by P. Burgess

The stability and shelf life of food Second edition
Edited by Persis Subramaniam and Peter Wareing

Multisensory flavor perception: From fundamental neuroscience through to the marketplace
Edited by Betina Piqueras-Fiszman and Charles Spence

Flavor: from food to behaviors, wellbeing and health
Edited by Andrée Voilley, Christian Salles, Elisabeth Guichard and Patrick Etiévant

Developing food products for consumers with specific dietary needs
Edited by Wayne Morley and Steve Osborn

Advances in food traceability techniques and technologies: Improving quality throughout the food chain
Edited by Montserrat Espiñeira and J. Francisco Santaclara

Innovative food processing technologies: Extraction, separation, component modification and process intensification
Edited by Kai Knoerzer, Pablo Juliano, and Geoffrey Smithers

Steamed breads: Ingredients, process and quality
Sidi Huang and Diane Miskelly

Handbook of hygiene control in the food industry Second edition
Edited by Huub Lelieveld, Domagoj Gabrić, and John Holah (Editors)

Handbook for sensory and consumer-driven new product development: Innovative technologies for the food and beverage industry
Maurice O'Sullivan

Early nutrition and long-term health: Mechanisms, consequences and opportunities
Edited by Jose M. Saavedra and Anne Dattilo

Baking problems solved Second edition
S. P. Cauvain

Non-equilibrium states and glass transitions in food
Edited by Bhesh Bhandari and Yrjö Roos

New aspects of meat quality: From genes to ethics
Edited by Peter Purslow

Advances in food authenticity testing
Edited by Gerald Downey

Food for the ageing population Second edition
Edited by Monique Raats, Lisette De Groot, Dieneke van Asselt

Microbiological quality of food
Edited by Antonio Bevilacqua, Maria Rosaria Corbo and Milena Sinigaglia

Food Spoilage: A New Threat?

Food safety represents a primary goal all over the world and national regulations are more and more restrictive. However, food quality is a challenge for producers as any loss of flavor and color, and any undesirable change, can result in significant economic loss. What is spoilage and why address this topic in a book?

One of the most commonly used definitions of food spoilage is: "*Food spoilage can be considered as any change which renders a product unacceptable for human consumption; it may arise from insect damage, physical damage (bruising, freezing, drying, etc.), indigenous enzyme activity in animal or plant tissues, and chemical changes (usually involving oxygen). Therefore, spoilage is a complex phenomenon, involving physical, chemical, microbiological, and biochemical changes.*"

This book covers the microbiological spoilage of foods, with a focus on the spoilers (the characters), the foods themselves (the stage), and the signs of spoilage (the plot).

THE CHARACTERS

Many microorganisms can act as spoilers (bacteria, yeasts, and filamentous fungi). The book addresses some traditional spoilers (filamentous fungi, spore forming bacteria, yeasts, SSO in fish), as well as some emerging spoilers (*Pseudomonas*), now recognized as primary targets in some foods due to some recent spoiling phenomena (e.g., blue mozzarella cheese in Europe).

THE STAGE

The second section of the book is a brief overview of some important foods (vegetables, milk and dairy products, meat, and fish), and addresses the practical implications of spoilage. This section offers detail on the signs of spoilage, how to prevent spoilers, and the methods of detecting spoilage and spoilage microorganisms in food.

THE PLOT

Microbial spoilage of food can occur in different ways and the first chapter offers an overview of the most important phenomena connected to food spoilage. However, the book covers and addresses an elephant in the room in Food Microbiology: the impact of biofilm on the quality of many foods, and the implications of biofilm for spoilage and shelf life.

Another hot topic covered by this book is the possible link between food spoilage and food safety as a consequence of antibiotic resistance spreading all over the world. This is an ongoing and unexplored area; we have few evidences, but we also know that the situation can exert a strong effect on future trends in food science and microbiology.

This book is the result of a collaboration with colleagues from different countries (Spain, France, Italy, the United States, Greece, and Nigeria), and it is a link between various expertise and research trends, to address this difficult task in food science: microbiological spoilage. The importance of this topic can be inferred by the increasing number of references and published papers on WoS (Web of Science) and Scopus, as well as through the continued efforts to search for and design new technologies and quantification methods.

Who are the readers of this book? Scientists, students, stakeholders, and teachers, because food spoilage is “a topic in progress.”

We have started with a question: Is spoilage a new threat? We strongly believe so!

Chapter 1

Microbial Spoilage of Foods: Fundamentals

Leonardo Petruzzi, Maria Rosaria Corbo, Milena Sinigaglia and Antonio Bevilacqua
University of Foggia, Foggia, FG, Italy

1.1 INTRODUCTION

Food spoilage results when microbiological, chemical, or physical changes occur, rendering the food product unacceptable to the consumer. Microbiological food spoilage is caused by the growth of microorganisms which produce enzymes that lead to objectionable by-products in the food. Chemical food spoilage occurs when different components in the food react with each other or with some added component which alter the food's sensory characteristics. Examples of this include: oxidation; enzymatic browning; and nonenzymatic browning. Physical food spoilage results when moist foods are excessively dehydrated or dried foods absorb excessive moisture (Benner, 2014).

For years, safety (i.e., the exclusion or elimination of pathogens from food) has been studied separately from the prevention of spoilage. In most countries the legislation has tended to reinforce this concept. However, from a microbiological-ecological point of view the two areas cannot be distinguished. In spite of considerable efforts, microbiological safety assurance seems as remote as ever, even in advanced countries. Death, suffering, economic losses, and civil claims on behalf of victims of foodborne diseases are matched by the economic losses caused by food spoilage (Di Renzo et al., 2015).

Although exact figures of the total economic losses due to food spoilage are unknown, it is clear that it constitutes an enormous financial burden (Blackburn, 2006) estimated at 1.3 billion tons per year by FAO (Cichello, 2015). According to the USDA Economic Research Service estimates, about 96 billion pounds of food, or more than a quarter of the 356 billion pounds of edible food available for human consumption in the United States, were lost to human use at three marketing stages: retail, foodservice, and consumers. Fresh fruits and vegetables (19.6%), fluid milk (18.1%), grain products

The Microbiological Quality of Food. DOI: http://dx.doi.org/10.1016/B978-0-08-100502-6.00002-9

(15.2%), and sweeteners (12.4%), mostly sugar and high-fructose corn syrup, accounted for two-thirds of these losses (Nguyen Van Long et al., 2016).

A significant proportion of the loss is due to spoilage by microorganisms, resulting in final products with an inadequate shape or appearance (Dousset et al., 2016). It has been estimated that about 25% of all foods produced globally are lost due to microbial spoilage (Bondi et al., 2014).

The food industry is continually creating new microbial habitats, either by developing new products and reformulating traditional ones, or by chance, as a result of the composition of raw materials or in production (Argyri et al., 2014). Also, the modern consumers' preference for fresh foods with extended shelf life and products that are free of chemical pesticides leave foods more vulnerable to spoilage, as well as increasing the diversity of spoilage species (Blackburn, 2006; Brandelli, 2015).

Understanding how different properties of a food, its environment and "history," can influence the microbiota that develops in products, is an important first step toward control quality and safety (Argyri et al., 2014).

The present chapter discusses the characteristics of microbial spoilage of foods with a focus on the major spoilage microorganisms and how they can be detected and monitored.

1.2 MICROBIAL FOOD SPOILAGE: OVERVIEW

The range of spoilage microorganisms is wide. Bacteria are responsible for some of the most rapid and evident spoilage events of proteinaceous foods such as meat, poultry, fish, shellfish, milk, and some dairy products. The growth of yeasts and molds is generally slower than that of bacteria, but the wide variety of ecological niches they can exploit, the ability to utilize a variety of substrates, and tolerance of more extreme conditions than (vegetative) bacteria makes them formidable spoilage agents (Blackburn, 2006).

The spoilage flora of perishable products is comprised on the one hand of the natural flora from the product itself, which is mainly influenced by its environmental conditions. For example, the natural microflora of fish is influenced by the habitat, the geography (tropical or arctic water), and the fishing season, whereas the natural flora of fruits and vegetables is influenced by the soil, air, irrigation water, insects, and animals. On the other hand, microorganisms will be transferred to food by cross contamination during or after processing via surfaces, machines, humans, or the surrounding atmosphere (Kreyenschmidt and Ibald, 2012).

The microbiological flora is not static, so the initial flora differs from the flora at the end of shelf life. Furthermore, only a small number of microorganisms are often responsible for the loss of quality based on food characteristics, environmental conditions, and interactions between the microorganisms. Those organisms mainly responsible for spoilage are known as specific spoilage organisms (SSOs) (Kreyenschmidt and Ibald, 2012).

The many kinds of microorganisms that can grow on food have evolved biochemical mechanisms to digest components of the food, thereby providing energy sources for their own growth (Sperber, 2009). Transformation of the available compounds gives a wide range of end products that affect the food's sensory, chemical, and physical properties (Howell, 2016). The range of possible compounds produced is dependent on not only the genetic capability of the microorganism in question but also the intrinsic characteristics of the product and the extrinsic characteristics of handling and the storage environment that can vastly change biochemical pathways (Benner, 2014; Howell, 2016).

1.3 FACTORS AFFECTING MICROBIAL SPOILAGE

1.3.1 Food Structure

Raw food is often protected from attack by microorganisms by means of structures that are not easily degradable (fruit skin, nut shells, bran, collagen muscle tissue, egg shells, etc.). Only microbial species with appropriate enzymes (cellulase, pectinase, protease, etc.) can attack these structures. As a result, raw food prior to harvest is often only contaminated on the surface. However, as soon as the processing stage begins, this protection ceases and the food becomes more vulnerable to the spread of microorganisms (Baron and Gautier, 2016).

In general, fluid foods spoil rapidly because the organisms can easily spread throughout the food by means of their own motility or by convection currents. Semi-solid foods such as meat stews, soup, and tinned fruits can spoil as rapidly as fluid foods. Solid foods tend to spoil from their outside surfaces inwards, these being the first surfaces to become contaminated (Modi, 2009).

1.3.2 Food Composition

The nutrient composition of food also influences the growth of the most suitable species of microorganisms. Protein foods such as meat, fish, and eggs are liable to be attacked by proteolytic organisms; "carbohydrate" foods such as bread, flour, pasta, syrups, and jams are more liable to attack by fermentative organisms; fats are liable to be attacked by lipolytic organisms (Modi, 2009).

Lipids can be degraded enzymatically to produce free fatty acids that have rancid and unpleasant off-aromas. Breakdown of lipids is important in meat, dairy, and olive oil systems (Howell, 2016).

Some foods also have inhibitory substances or naturally occurring antimicrobials present (Davidson and Critzer, 2012). Such substances can inhibit or slow the growth of some microorganisms (Kreyenschmidt and Ibald, 2012).

For example, egg albumin contains the enzyme lysozyme, which disrupts the cell walls of Gram-positive bacteria by hydrolyzing the glycosidic bonds of *N*-acetylmuramic acid and *N*-acetylglucosamine in the peptidoglycan layer, avidin, which ties up the vitamin biotin, conalbumin, which ties up iron, and protease inhibitors, which inhibit protein degradation. All of these compounds act in concert to inhibit microbial growth. Raw milk also contains lysozyme and the lactoperoxidase system, which requires the interaction of the enzyme lactoperoxidase, thiocyanate, and hydrogen peroxide to produce the antimicrobial hypothiocyanate (Davidson and Critzer, 2012).

Lactoferrin is also present, which sequesters the iron necessary for microbial growth similar to conalbumin found in eggs (Davidson and Critzer, 2012).

In general, Gram-positive bacteria are sensitive to many molecules such as citrate, nisin, butylated hydroxyanisol or butylated hydroxytoluene, as well as molecules known for their antifungal activity, such as sorbates or benzoates. Gram-negative bacteria are more resistant than Gram-positive bacteria, but are still susceptible to a broad spectrum of additives as well as SO_2 (Baron and Gautier, 2016).

There are some yeast species that are tolerant to ethanol at high concentrations, although their metabolism may be affected, and these include *Zygosaccharomyces*, *Dekkera*, *Pichia*, and *Saccharomycodes*. Many yeast species have evolved resistance to weak organic acids and include *Zygosaccharomyces*, along with some strains of *Candida krusei* and *Pichia membranifaciens* (Howell, 2016).

Foods with low sugar content, such as juices, jellies, and jams, contain acids, sweeteners, and hydrocolloids. These additives are needed for sensory purposes. Aspartame is among the most used sweeteners, though the use of the natural sweetener stevia is increasing in recent years. The presence of mentioned additives affects the growth of *Zygosaccharomyces bailii* (Campos et al., 2015).

Some yeasts are able to grow in media or substrates with high salt or sugar concentrations. For example, *Debaryomyces hansenii* is very salt tolerant, and some strains can tolerate up to 24% (w/v) NaCl. *Saccharomyces cerevisiae* also exhibits significant salt tolerance, but this is strain-dependent and is also influenced by pH of the food or beverage in question (Howell, 2016). Some species such as *Hemimysis anomala* and *Candida pseudotropicalis* may grow at NaCl up to 11% (Campos et al., 2015). *Zygosaccharomyces rouxii* is exceptionally tolerant to high sugar concentrations (up to 70% w/v sucrose) (Howell, 2016). Similarly, molds are tolerant to high concentrations of sugar or salt (Modi, 2009).

1.3.3 Water Activity

The water activity (a_w), which is the ratio of the vapor pressure of the food to the vapor pressure of pure water, has an important influence on the growth of the microorganisms since an aqueous phase is a main requirement for their metabolic activity. The optimum a_w for most microorganisms is in the range 0.995–0.980. Lowering the a_w (e.g., due to salt, sugar, drying, or curing) inhibits microbiological growth or results in an increased lag phase and a decreased growth rate and thus leads to a longer shelf life. Since microorganisms have different demands with regard to water activity, the composition of the microflora is influenced by the prevalent a_w (Kreyenschmidt and Ibald, 2012).

Bacteria require a high a_w and cannot therefore contribute to the spoilage of dry foods while they remain dry (Modi, 2009). With a few exceptions, the bacterial growth continues up to a $\geq$0.88 (Kreyenschmidt and Ibald, 2012). Some bacteria produce molecules called osmoprotectants (ectoine produced by *Brevibacterium linens*) that can keep water inside the cytoplasm (Baron and Gautier, 2016).

Generally, yeasts must have an a_w above 0.87–0.94, but some osmotolerant species grow in an a_w as low as 0.60. Molds can grow in foods with the lowest water activity, 0.70–0.80; *Xeromyces* is one species of mold that can grow in an a_w as low as 0.60 (Davidson and Critzer, 2012).

1.3.4 pH

The pH value of a food limits the range of organisms which it can support (Modi, 2009). Most microorganisms develop around neutral pH, which corresponds to the pH of bacterial cytoplasm and is optimal for bacterial enzyme activities (Baron and Gautier, 2016). However, there are bacteria that can tolerate a pH outside these ranges. For example, lower pH ranges can be tolerated by *Lactobacillus* (3.0–4.4) and *Acetobacter* (2.8) (Davidson and Critzer, 2012).

Yeasts exhibit best growth in the range of pH 4.5–7.0. The lower end of the pH scale is preferred by many yeast species, and many species readily grow at pH 2.0–2.5. More rarely, *Hanseniaspora uvarum* may grow at pH 1.5 (Modi, 2009). For most yeasts, tolerance to a_w is decreased at extreme pH values. Some yeasts such as *Z. bailii* and *P. membranifaciens* are more tolerant to acidic pH while others such as *D. hansenii* and *S. cerevisiae* are more tolerant to pH values within 5.0–7.0. (Campos et al., 2015). Stronger growth of yeasts at low pH means that they are successful at competing with and dominating other microbial partners, including bacteria (Howell, 2016).

In general, microorganisms have established response mechanisms to deal with stress, enabling them to survive in certain pH conditions. With *Salmonella* Typhimurium, there is a two-step process corresponding to two systems that are activated at different levels of acidity: tolerance response (between pH 4.5 and 6) and resistance response to more acidic pH (Baron and Gautier, 2016). Some proteolytic bacteria can grow at a higher maximum pH because during growth they produce amines to buffer the high pH (Davidson and Critzer, 2012).

Oxidative or film yeasts oxidize sugars, organic acids, and alcohol, and in their growth raise the pH; thus, they tend to grow on the surface of liquors forming a skin or film (Modi, 2009).

1.3.5 Temperature

Temperature conditions during processing, transport, and storage are one of the most important factors that influence the spoilage process. Increasing temperature conditions lead to a decrease in the length of the lag phase and generation times and thus to an increase in the growth rate. They also influence protein synthesis, enzyme activity, solute uptake, and thus the length of shelf life (Kreyenschmidt and Ibald, 2012).

Every type of organism is able to grow within a certain temperature range, maximum growth occurring at and around the optimum temperature. So organisms are divided on the basis of their temperature preferences into rough categories: psychrophiles, mesophiles, and thermophiles (Modi, 2009). Psychrophiles are organisms that have a growth range of −5°C to 35°C. Their minimum growth range is −5°C to 5°C, optimum 12–15°C, and maximum 15–20°C. Examples include the foodborne pathogens *Yersinia enterocolitica* and *Listeria monocytogenes*, and the spoilage bacterium *Pseudomonas*. The category in which most human and animal pathogens and most food spoilage microorganisms are found is the mesophiles: their minimum growth range is 5–15°C, optimum 30–45°C, and maximum 35–47°C. Examples of these pathogens are *Escherichia coli*, *Salmonella*, *Clostridium botulinum*, and *Staphylococcus aureus*. For thermophiles, minimum growth range is 40–45°C, optimum 55–75°C, and maximum 60–90°C. Examples include *Clostridium thermosaccharolyticum* and *Geobacillus stearothermophilus* (Davidson and Critzer, 2012).

Molds and yeasts tend to grow best at room temperature and below, and therefore assume greater importance in the spoilage of foods held at cool and chilled temperatures (Modi, 2009). Additionally, some yeasts of the species *Candida*, *Cryptococcus*, and *Rhodotorula* can grow at temperatures <0°C and can spoil frozen foods, although this is quite uncommon (Howell, 2016).

Bacterial spores are among the most resistant, as well as certain fungal spores that can survive treatments lasting several hours at temperatures above 100°C (Baron and Gautier, 2016).

In many microorganisms, sublethal heat treatment is accompanied by an increase in thermotolerance. Generally, the induction of thermotolerance is related to the concomitant synthesis of heat shock proteins. On another level, most microorganisms are more resistant to freezing when the process is rapidly carried out. This is because of the formation of intercellular microcrystals that do not alter the cells as much as large crystals formed during a slow freezing process. Whatever the freezing method, cell structures suffer deformation, damage, and destruction, which causes cell death at a rate of 1 Log. In general, Gram-positive bacteria are more resistant than Gram-negative bacteria (Baron and Gautier, 2016).

1.3.6 Gaseous Conditions

The oxygen tension and the oxidation-reduction (O-R) potential of a food influences the type of organisms which can grow in it. Spoilage by aerobic organisms occurs at the surfaces of foods; most fresh plant and animal foods have a low O-R throughout and they are aerobic at the surface only. Facultative organisms grow both on the surface of the foods and within them, as shown when cans of food are spoiled by members of the genus *Bacillus*. Anaerobic organisms grow within foods held under anaerobic conditions such as inside cans; similarly vacuum-packs are liable to fermentative spoilage by bacteria and yeasts (Modi, 2009). Yeasts grow both in aerobic and anaerobic conditions (Modi, 2009). Molds are aerobic, since spoilage generally begins at the surface, although the mycelium later penetrates deep into the food (Modi, 2009).

1.3.7 Interaction Phenomena

Each of these physicochemical parameters have a specific influence, but they also interact and it is necessary to consider their combined action with regard to the growth of microorganisms. These interactions can be positive or negative depending on the target microorganism, the food itself and the permissible levels of food additives (Baron and Gautier, 2016).

When bacteria, yeasts and molds are present in a food which is capable of supporting the growth of all three it is most likely that the bacteria will become dominant first. Mold or yeast spoilage may occur at a later stage if the conditions in the food at that time permit. The waste products that the dominant organisms produce may either stimulate or inhibit the growth of other organisms present. For example, some molds of the *Penicillium* species may produce antibiotics in their growth which are inhibitory to other organisms (Modi, 2009). For bacteria, examples include the competitive advantage of *Pseudomonas* spp. due to the production of iron-chelating siderophores; the generation of substrates for spoilage reactions by one organism from another organism (so-called metabiosis); and the "switching on" of spoilage metabolism through cell-to-cell communication (Blackburn, 2006).

In some cases, spoilage microflora (or a portion of it) is able to inhibit, suppress, or eliminate a microbial pathogen in a food product, and consequently that food product has less chance of causing consumer illness. Foodborne bacterial pathogens that can be inhibited by spoilage microorganisms include *Aeromonas sobria*, *Bacillus cereus*, *Campylobacter jejuni*,

C. botulinum, *E. coli* O157:H7, *L. monocytogenes*, *Salmonella* spp., *S. aureus*, and *Y. enterocolitica*. The most extensively studied spoilage microorganisms that exhibit the ability to inhibit bacterial pathogens are *Pseudomonas* spp. and lactic acid bacteria (LAB). Other naturally occurring bacteria, yeasts, and molds have also been shown to exert inhibitory activity toward pathogens. Some *Pseudomonas* spp. have also been shown to inhibit other spoilage microorganisms including other pseudomonads and *Shewanella putrefaciens* (Benner, 2014).

1.4 MAJOR MICROORGANISMS INVOLVED IN FOOD SPOILAGE

1.4.1 Bacteria

LAB, encompassing *Lactococcus*, *Lactobacillus*, *Leuconostoc*, *Weissella*, and *Carnobacteria* species, as well as *Enterococcus* are frequently associated with spoilage. Although LAB are generally beneficial for food and are used for the fermentation of a variety of food and raw materials, where they contribute to flavor, texture, and shelf life, some species can play a significant role in food spoilage and decay (Remenant et al., 2015). Undesirable changes caused by LAB include greening of meat and gas formation in cheeses (blowing), pickles (bloater damage), and canned or packaged meat and vegetables. Off-flavors described as mousy, cheesy, malty, acidic, buttery or liver-like may be detected in wine, meats, milk, or juices spoiled by these bacteria. LAB may also produce large amounts of an exopolysaccharide that causes slime on meats and ropy spoilage in some beverages (Rawat, 2015).

In addition to LAB species, other Gram-positive bacteria can play a significant role in food spoilage. One of the most prominent is the psychrotrophic species *Brochothrix thermosphacta*, known as an important spoiler bacterium of various food matrixes. Enterobacteriaceae can also play a key role in food spoilage due to their ability to metabolize amino acids to malodorous volatile compounds, such as foul-smelling diamines and sulfuric compounds. Gram-negative bacteria with species belonging to the genera *Serratia*, *Hafnia*, and *Pseudomonas* have also often been incriminated. *Pseudomonas* spp., particularly *P. fluorescens*, *P. putida*, and *P. fragi*, also contribute to a large extent to the spoilage process of food. These are the predominant spoilers of foods stored under aerobic refrigerated conditions, especially aerobically chill-stored beef, seafood, poultry, and milk (Remenant et al., 2015).

There are several sporeformers bacteria which can produce significant economic spoilage of foodstuff (Brown, 2000). Some species can be psychrotrophic which can spoil refrigerated dairy products; others such as *Bacillus subtilis* are mesophilic and can spoil bakery products, whereas others such as *G. stearothermophilus* are thermophilic and spoil foods that are canned or in hermetically sealed packages. Clostridia generally spoil foods of low oxygen/reduction potential such as canned or vacuum-packaged foods (Setlow and Johnson, 2013). Several psychrophilic and psychrotolerant species including *Clostridium algidicarnis*, *Clostridium algidixylanolyticum*, *Clostridium estertheticum*, *Clostridium frigidicarnis*, and *Clostridium gasigenes* have been implicated in red meat spoilage (softening of the meat, large amounts of drip, offensive odor). Of these species, *C. estertheticum* (subspecies *estertheticum* and *laramiense*) and *C. gasigenes* additionally produce gas and have been recognized as causative agents of "blown pack" spoilage of vacuum-packed chilled meats at normal storage temperatures (−1.5 to 2°C) (Heyndrickx, 2011). In addition, the acid tolerant species *Clostridium barati* and *Clostridium butyricum* have been reported as the causative agents of the spoilage of canned pasteurized mung bean sprouts, stored under acidic conditions (Remenant et al., 2015). Spores of *Clostridium tyrobutyricum* can cause spoilage of semihard cheeses with long ripening times such as Gouda and Emmenthaler (Heyndrickx, 2011). Spoilage from *Clostridium sporogenes* produces typically blown or burst packs with a strong putrefactive odor. (Brown, 2000). Spoilage from *C. thermosaccharolyticum* manifests itself by blown or burst packs with a strong butyric or cheesy odor (Brown, 2000).

Alicyclobacillus species have been recognized to spoil acidic products. Spoilage is sometimes associated with a slight increase in turbidity and white sediment at the bottom of packages, but the most important fault is taint of strong medicinal or antiseptic flavor caused mainly by the production of guaiacol, but also by 2,6-dibromophenol and 2,6-dichlorophenol. Affected products are pasteurized fruit juices (mainly apple and orange) and fruit juice blends, but there have also been reports of spoiled carbonated fruit drinks, berry juice containing iced tea, and diced canned tomatoes (Heyndrickx, 2011).

The polysaccharides produced by microorganisms during their growth can also cause food spoilage. An example is the contamination of certain fermented beverages (cider or wine) by *Leuconostoc* species giving the product an oily consistency (Baron and Gautier, 2016).

For the most part, the spoilage organisms discussed have been harmless bacteria that pose no threat to human health. However, in certain situations, the bacteria responsible for food spoilage may, in fact, be pathogenic. The pathogenic species, *B. cereus*, has been linked to spoilage and human illness in fluid milk products. *Clostridium perfringens* can comprise a large portion of the spoilage flora under anaerobic storage conditions. In addition, Enterobacteriaceae, including pathogenic *Salmonella* spp. and *E. coli*, can dominate the spoilage microflora under aerobic and anaerobic storage conditions. *Vibrio* spp. are commonly present on seafood products harvested from warm water climates. These organisms tend to dominate the microflora of seafood products that are abused at temperatures more than 20°C. In particular, the human pathogen *Vibrio parahaemolyticus* has been shown to be the predominant spoilage organism in shrimp held at temperatures of 24°C or greater (Benner, 2014). Moreover, biogenic amines can be produced in meat and fish by several members of this group while others produce off-odors or colors in beer (*Obesumbacterium*), bacon and other cured meats (*Proteus, Serratia*), cheeses (several genera), cole slaw (*Klebsiella*), and shell eggs (*Proteus, Enterobacter, Serratia*) (Rawat, 2015).

1.4.2 Yeasts

Species of *Zygosaccharomyces* and related genera are usually the yeasts that colonize and spoil high sugar and high salt products (Blackburn, 2006). The *Zygosaccharomyces* genus consists of six species, of which *Z. bailii*, *Z. bisporus*, and *Z. rouxii* are the most relevant to the spoilage of foods and beverages. These yeasts are characterized by fermentative spoilage of products such as fruit juices, fruit concentrates, syrups, sauces, alcoholic beverages, honey, jams, and confectionary. The fermentative nature of *Zygosaccharomyces* metabolism produces carbon dioxide that is of concern if the food product is in a sealed container, as the increase in gas can cause the container to leak or, in extreme cases, explode. Haze or sediment can occur in beverages as cellular biomass accumulates, and in some cases, surface biofilms can form. Taste modification can occur by the production of secondary metabolites with sensorial impacts on the food or beverage and include acetic acid, esters, and higher alcohols (Howell, 2016).

Saccharomyces are best known for their positive contributions to food and beverage production, but they also can have deleterious effects including circumstances when the same species spoil the very commodities they produce (Blackburn, 2006). In particular, end products such as CO_2 and ethanol, as well as minor compounds like acids, esters, ketones, aldehydes, alcohols, and sulfur compounds drastically change the aroma and flavor of a spoiled substrate (Howell, 2016).

Candida species make up one-quarter of all known yeasts and their heterogeneity means that they are responsible for the spoilage of a wide range of foods (Blackburn, 2006). *Dekkera* produces distinctive off-taints with secondary metabolism, where phenolic and mousy flavors can spoil the aroma and desirability of the wine (Howell, 2016). Several yeasts such as *H. uvarum* and *C. krusei* are able to produce significant amounts of acetic acid during spoilage of beverages, particularly wine (Howell, 2016). *Yarrowia lipolytica* is associated with surface discoloration of cheese, where catabolism of the amino acid tyrosine gives the pigmented compound melanin. Surface spoilage of uncooked and cooked meats (such as chicken or turkey) has been also observed, where the extracellular proteolytic and lipolytic activities result in spoilage aromas (Howell, 2016). Of particular importance is the growth of *D. hansenii* in fermented dairy products with the removal of lactic acid (Howell, 2016). The growth of *Schizosaccharomyces pombe* removes malic acid from fruit juices, which increases the pH and dramatically affects the taste and stability of the juice (Howell, 2016).

Meat products are rarely degraded by proteolysis by spoilage yeasts, as most do not produce extracellular proteases. There are some notable exceptions, e.g., *Y. lipolytica*, *Rhodotorula*, and *Cryptococcus* spp. in meat products cured with nitrates, spoilage with *Pichia* and *Cryptococcus* spp. can lead to metabolism of sodium nitrate and sodium nitrite (Howell, 2016). In sauerkraut, a pink discoloration is induced by *Rhodotorula* spp. (Campos et al., 2015).

1.4.3 Molds

Although the range of molds is immense there are a specific and rather limited number of genera and species that are spoilage hazards for each kind of food. The *Zygomycetes*, popularly known as the "pin molds" are often seen as rapid growers following a "hit and run" strategy for the foods that they spoil. The Penicillia and Aspergilli are common spoilage molds with the latter generally growing more rapidly and at higher temperatures or lower water activities than the former. Other types of molds are significant in food spoilage, but many molds may be present on, or isolated from, foods in which they never or rarely cause spoilage (Blackburn, 2006).

However, molds are also responsible for the formation of undesired mycotoxins that are small (molecular weight of ~700 DA) secondary metabolites produced by several fungi belonging mainly to the genera *Aspergillus*, *Penicillium*, *Fusarium*, and *Alternaria*. Some mycotoxins appear to be produced in response to environmental changes, usually due to the onset of stress conditions. The most significant mycotoxins, from both public health and agronomic perspectives, include aflatoxins, trichothesenes, fumonisins, ochratoxin A, patulin, tremogenic toxins, and ergot alkolids (Miescher Schwenninger et al., 2011).

1.5 DETECTION AND QUANTIFICATION OF FOOD-SPOILAGE MICROORGANISMS

1.5.1 Classical and Innovative Methods

A great number of different methodologies have been applied to evaluate food spoilage. Among these, microbiological methods have been used almost exclusively in the actual evaluation of spoilage (Nychas and Panagou, 2011). Classical microbiological methods require multiple culturing steps for microbes to be isolated and identified, and they may not provide results quickly enough for appropriate interventions to be made. Further, these methods are not always appropriate since some spoilage microorganisms fail to grow in culture media (Linares et al., 2014). However, the recent development of high-throughput sequencing has led to a deeper analysis of food microbial ecosystems. These approaches, mostly focused on 16S rDNA, have confirmed previous studies performed with classical cultural methods about the nature of bacteria found in foods. In addition, the data obtained by direct analysis of DNA extracted from food matrixes have enabled a better description of the ecosystems at genus or even species level, and have also revealed that the involvement of some species in food spoilage might have been underestimated (Remenant et al., 2015).

Actually, some methodologies such as Enzyme-Linked Immuno-Sorbent Assay or Polymerase Chain Reaction (PCR) have already become classic identification methods. The development of PCR protocols coupled with sequencing tools has provided a large amount of information that has already been deposited into public databases and is freely available. Conversely, the development of rapid and high sensitivity techniques, such as real-time PCR (qPCR), DNA microarrays and biosensors, have led to the replacement of traditional culturing methods in the field of bacterial identification for clinical diagnostics and the food sector. Furthermore, Fourier transform infrared spectrometry (Quintela-Baluja et al., 2014) and Raman spectroscopy have been also investigated for their potential on evaluating food spoilage (Argyri et al., 2014). At the same time, proteomic tools, such as mass spectrometry, were introduced for the identification of microorganisms (Quintela-Baluja et al., 2014).

1.5.2 Microbial Metabolites as Spoilage Indicators

Numerous attempts have been made to associate certain metabolites with microbial spoilage (Nychas and Panagou, 2011). The microbial volatile organic compounds (MVOCs) are generated in this process as well, and the amount of MVOCs can reflect the condition of microbial growth and food spoilage directly (Wang et al., 2016).

In a recent review, Wang et al. (2016) reported the following compounds as potential indicators of spoilage: (1) volatile amines as an indicator of poultry meat spoilage in air, vacuum and two modified atmosphere; (2) ethanol and 2,3-butanediol as indicator of microbial spoilage in Vietnamese *Pangasius hypophthalmus* fillets; (3) alcohols such as ethyl hexanoate, ethyl octanoate, ethyl nonanoate, ethyl decanoate, and isoamyl acetate for beef samples contaminated with *Serratia proteamaculans* 42M; (4) ketones (octan-3-one, pentadecan-2-one), alcohols (2-methylpropan-1-ol, hexan-1-ol), and phenols (2,6-bis(1,1 dimethylethyl)-4-ethyl, myrcene) for coffee beverage inoculated with *Penicillium brevicompactum*; (5) ethyl acetate and acetic acid for *S. aureus* and *E. coli* in milk; (6) 1,3-Pentadiene and styrene for *Penicillium* spp., *Aspergillus niger* VH10, *Emericella* sp. VH12, and *Trichoderma* sp. VS20 inoculated in strawberry jam.

Unfortunately, correlating the production of molecules responsible for or associated with spoilage appearance to the functions of spoilage microorganisms is not always possible. There are several reasons for this limitation;

spoilage may result from a large variety of processes. It can be the consequence of a complex succession of enzymatic reactions, potentially associated with nonenzymatic reactions, such as meat discoloration, or with enzymatic reactions originating from both the spoilage organisms and the food matrix (like lytic activities of enzymes from muscle cells). Another example is metabiosis that results from enzymatic reactions successively carried out by different microorganisms (Remenant et al., 2015). Spoilage can also result from reactions catalyzed by enzymes that are not well identified. Finally, some molecules responsible for spoilage can be produced by many different enzymes and identifying those that produce them in food may be a challenge (Remenant et al., 2015).

1.5.3 Limits and Perspectives of Detection/Quantification Methods

Although the majority of detection methods show a consistent analytical sensitivity, their universality is far from complete. For this reason, each food should be subjected to multiple analyses; this unfortunately involves long analytical periods and high costs. Microorganism detection and unequivocal identification is an important issue for the food industry: a rapid assessment of a potential microbial risk can predict and enhance the shelf life of foodstuffs, avoiding health hazards and economic losses, through the application of preventive measures. For all these reasons, inexpensive and reliable molecular identification methods are necessary to detect and identify food microorganisms associated with infections and diseases (Galimberti et al., 2016). A possible approach relies upon the possibility of using the genomics of spoilage microorganisms and the metagenomics of food ecosystems combined with other "omics," particularly metabolo mics, to determine the volatile compounds or molecules putatively responsible for spoilage to identify the metabolic pathways involved in the appearance of spoilage (Remenant et al., 2015).

1.5.4 Statistical Tools and Predictive Microbiology

The enormous amount of information provided by the mentioned methods makes the data produced unmanageable. Thus, the application of advanced statistical methods (discriminant function analysis, clustering algorithms, chemo-metrics) and intelligent approaches (neural networks, fuzzy logic, evolutionary algorithms, and genetic programming) may be used as qualitative rather than quantitative indices since their primary aim is to distinguish objects, groups or populations. This is an unsupervised learning method. Nowadays, modern machine learning procedures are based on supervised learning algorithms. The last mentioned approach, together with the development of artificial neural networks, has been proposed for evaluation of food spoilage (Nychas and Panagou, 2011).

Similarly, the development of predictive models for specific food spoilage flora is a complementary approach to ensure safe and stable products. Predictive models are based on the premise that responses of microbial populations to environmental factors are reproducible and predictable. In predictive microbiology, a reductionist approach is generally adopted and microbial responses are measured under defined and controlled conditions. The results are summarized in mathematical equations which, by interpolation, can predict probable behavior in novel conditions, i.e., those not actually tested (Braun and Sutherland, 2006). In the early stages of predictive microbiology, most studies were related to pathogenic bacteria causing foodborne diseases. In the last two decades, the usefulness of predictive models to monitor food spoilage has been recognized and studies have emerged with respect to the modeling of spoilage organisms, mainly yeasts (Van Derlinden et al., 2013). Conversely, several software systems and associated databases have been developed to predict the growth of a range of relevant microorganisms under specific conditions (Kreyenschmidt and Ibald, 2012).

1.6 PREVENTING MICROBIOLOGICAL SPOILAGE OF FOODS

Various processes and methods have been applied to prevent the microbiological spoilage of foods (Nychas and Panagou, 2011). The major preservation methods are based upon the reduction of microbial growth due to unfavorable environmental conditions such as temperature reduction, lowering of pH and a_w and denaturation due to heat treatment. However, due to consumer pressure, the trend in recent years has been to use less severe, milder preservation methods and modified atmosphere packaging (MAP) (Forsythe, 2000).

1.6.1 Traditional Approaches

Traditional approaches include among others such methods as heating, dehydration, fermentation, refrigeration, freezing, and preservatives (Rawat, 2015; Fu et al., 2016).

In general, thermal processing is based on time-temperature combinations that are required to inactivate cells of pathogenic microorganisms in a significant number of food products, in addition to reducing spoilage microorganisms. Pasteurization is commonly used for high-acid food products ($pH < 4.6$) to inactivate target pathogenic bacteria and to extend product shelf life for a few weeks. It is also utilized for low-acid foods followed by refrigeration. The most common pasteurization method, high temperature short time, uses temperatures around 72°C for 15 seconds in the case of milk. Sterilization is a stronger thermal treatment used for low-acid food products ($pH > 4.6$); it inactivates spores, extends the shelf life of the product for months at a time, and uses temperatures around 121°C for several minutes (e.g., 15 minutes). Nevertheless, some microbial growth can be observed during food storage, e.g., in pasteurized products (i.e., milk) (Barbosa-Cánovas and Bermúdez-Aguirre, 2011). In this respect, practical control of sporeformers bacteria causing spoilage includes monitoring of raw foods entering the cannery, particularly sugars, starches, spices, onions, mushrooms, and dried foods, to limit the initial spore load in a food product, adequate thermal processing depending on subsequent storage and distribution conditions, rapid cooling of products, chlorination of cooling water, and implementing and maintaining good manufacturing practices within the food plant (Setlow and Johnson, 2013).

Some heat-resistant fungi can spoil acidic foods, particularly fruit products. While most filamentous fungi and yeasts are killed by heating for a few minutes at 60–75°C, heat-resistant fungi produce thick-walled ascospores that survive heating at ca. 85°C for 5 minutes. The most common genera of heat-resistant fungi causing spoilage are *Byssochlamys*, *Neosartorya*, *Talaromyces*, and *Eupenicillium*. Certain heat-resistant fungi also produce mycotoxins. To prevent spoilage of heat-treated foods, raw materials should be screened for heat-resistant fungi and strict food manufacturing practices and sanitation programs should be followed during processing (Setlow and Johnson, 2013).

Dehydration, or reducing the a_w of foods, is another way of inhibiting microbial growth in foods. Dehydration is when the water content is reduced 75–92%. This will generally result in a very stable food product. However, the dehydration process, even using heat, may allow some pathogens to survive. For example, *Salmonella* has been found to survive in spray-dried foods, and *Cronobacter* (*Enterobacter*) *sakazakii* survives in powdered infant formula. Ingredients commonly used to decrease water activity include salt, sucrose, glycerol, fructose, galactose, hydrocollids, high-fructose corn syrup, and polyols (xylitol, mannitol, sorbitol) (Davidson and Critzer, 2012).

Fermentation is one of the oldest forms of preservation and is dependent upon the activity of microorganisms such as LAB to limit the growth of undesirable microorganisms from foods (Fu et al., 2016). Preservation by fermentation relies on the fact that the end-products of oxidation of carbohydrates (e.g., acids, alcohol, and CO_2) can control the growth of food spoilage microorganisms (Galimberti et al., 2016). This process also produces antimicrobial compounds such as bacteriocins to protect the food from spoilage after the fermentation process is complete (Fu et al., 2016). Many types of food products undergo fermentation for preservation including dairy products such as cheese, buttermilk, and yogurt, which are products of milk fermentation. These types of fermentations are carried out by LAB like *Lactococcus* and *Lactobacillus*. Vegetable products like sauerkraut, pickles, and olives use *Leuconostoc* and *Lactobacillus* species during fermentation (Davidson and Critzer, 2012).

Meats that undergo fermentation include summer sausage, salami, and pepperoni. *Pediococcus* is one microorganism that is commonly used for these fermentations. Cereal products like bread, crackers, and pretzels can be fermented by *Saccharomyces* (Davidson and Critzer, 2012).

Storage temperature is one of the most important factors in the spoilage processes. As an example, the shelf life of fresh pork is decreased by nearly 25% if the storage temperature is 4°C instead of 2°C. Also, short temperature abuses can significantly increase the growth rate of SSOs. Shelf life reductions of up to 20% were observed for fresh poultry and fresh pork when the periods with out-of-range temperatures were less than 5% of the total storage time and the temperature shift was 11°C (Kreyenschmidt and Ibald, 2012).

Antimicrobial food preservatives can be incorporated into foods to delay spoilage or inhibit the growth of foodborne pathogens (Davidson and Critzer, 2012). Many chemical compounds have a long history of safety use and exhibit different mechanisms of action depending on their chemical structure (Campos et al., 2015). Commonly used preservatives include common salt, sugar, dextrose, spices, vinegar, ascorbic acid, benzoic acid and its salt, SO_2 and the salts of sulphuric acid, nitrates, sorbic acid and its salts, propionic acid and its salts, lactic acid and its salts (Rawat, 2015). Their effectiveness is markedly influenced by food composition and by environmental factors. In general, a_w depression and the solute used modified the antimicrobial action of preservatives (Campos et al., 2015). However, as mentioned above, some fungi and bacteria are able to adapt to the presence of certain preservatives (Brandelli, 2015).

Although chemical treatments are effective in reducing surface microbial counts, they do not sanitize tissue crevices. Low-temperature is highly effective in conjunction with chemical treatment in reducing metabolic activity and microbial proliferation, but is not a good defence against psychrotrophic microorganisms such as *L. monocytogenes*, *Aeromonas*, etc., which tend to cause foodborne illness (Lu and Xia, 2012).

Overall, these features are an important aspect to consider when discussing the need for new preservation methods to inhibit the growth of spoilage microorganisms, and this is motivating the food industry toward a focus on different preservation approaches.

1.6.2 Alternative Approaches

The main alternative approaches to prevent microbiological spoilage of foods include: (1) nonthermal processing technologies; (2) biopreservation; (3) MAP; (4) antimicrobial packaging; and (5) hurdle technology (Gálvez et al., 2012; Campos et al., 2015; Rahman, 2015; Fu et al., 2016; Mauriello, 2016).

Nonthermal technologies describe those processing technologies where alternative form of source of energy is used as the main lethal agent (Martínez-Monteagudo and Balasubramaniam, 2016). The main novel technologies that are being assayed are high hydrostatic pressure, irradiation, pulsed electric fields, and ultrasound with temperature (thermosonication) (Campos et al., 2015).

Derived from material science, high hydrostatic pressure processing (HPP) is the technology by which a product is treated at or above 100 MPa. In contrast to heat, HPP does not disrupt covalent bonds, thus maintaining the primary structure of proteins and thereby retaining appearance, flavor, texture, and nutritional qualities of the unprocessed product. HPP has been applied to raw bivalve shellfish, fruit juices, cider, jams, and jellies, drinkable yoghurt, smoothies, avocado products, chopped onions, and ready-to-eat meat products (Lu and Xia, 2012). In order to inactivate spores, heat (70–80°C) and pressure must be utilized together (Davidson and Critzer, 2012).

Irradiation is another process employed to inactivate foodborne pathogens and spoilage microorganisms. Two categories of irradiation are employed by the food industry, ultraviolet (UV) radiation (nonionizing) and ionizing radiation (Davidson and Critzer, 2012). The UV-C-irradiation has germicidal action since it breaks down the DNA of microorganisms causing cell death. Since UV irradiation has low penetration effect, it is used for the surface sterilization of foods by the application of intense short pulses of near infrared region wavelengths (Campos et al., 2015). Application of gamma-radiation is a well-known method used to eliminate/inactivate the spoilage-causing and pathogenic microorganisms with no adverse effects on nutritional and sensory quality of foods (Lu and Xia, 2012). Other common types of irradiation that could be utilized in commercial food processing include electron beam and X-rays (Davidson and Critzer, 2012).

The application of pulsed electric fields to microorganisms mainly produces deformation of the cells, apparent fusion of cells, and the formation of pores. This technology had been successfully assayed as an alternative system of microbiological control in wineries. Its use does not cause sensory changes in food since it is a cold pasteurization (Campos et al., 2015).

In thermosonication, cellular death mainly occurs due to erosion and disruption of the cellular membrane, formation of orifices on the surface, and release of intracellular contents by lysis of cells (Campos et al., 2015).

Biopreservation or biocontrol refers to the use of natural or controlled microbiota, or its antibacterial products to extend the shelf life and enhance the safety of foods (Gálvez et al., 2014). LAB are particularly useful due to their important role in many food fermentations. In addition, they produce a variety of antimicrobial compounds, e.g., the pH-reducing fermentation products lactic and acetic acids, as well as hydrogen peroxide, formic acid, propionic acid, and diacetyl, as well as a number of antifungal metabolites, e.g., cyclic dipeptides, phenyllactic acid, proteinaceous compounds, and 3-hydroxylated fatty acids (Schnürer and Magnusson, 2005).

Substantial work has been conducted to verify the effectiveness of bacteriocins against microbial spoilage. For example, treatment of raw meats and poultry meats with pediocins (especially pediocin PA-1/Ach) can inhibit or delay growth of spoilage Gram-positive bacteria (such as *B. thermosphacta*) and/or reduce *L. monocytogenes* populations. In fruit juices and drinks, bacteriocin addition (nisin, enterocins) has been suggested for inactivation of endospore-forming bacteria causing spoilage such as *Alicyclobacillus acidoterrestris* and thermophilic spore formers such as *G. stearothermophilus* (Gálvez et al., 2014). Similarly, increasing interest has been shown in the use of essential oils as natural antimicrobial agents for food preservation. For example, essential oils from *Cinnamomum cassia* and *Cinnamomum verum* and their main constituent cinnamaldehyde clearly worked well against *P. fluorescens* (Mith et al., 2014).

MAP is accomplished by enclosing the food in gas-barrier materials with a controlled gaseous environment. Through removing or replacing the original atmosphere with a new atmosphere, different foods have been observed to remain edible for long periods of time. Usually the atmosphere is modified by reducing oxygen content while increasing the presence of nitrogen and CO_2 (Fu et al., 2016). In general, aerobic microorganisms are sensitive to CO_2 and it is this, along with their requirement for O_2, that is utilized in MAP to control the spoilage of foods. Gram-negative bacteria are generally more sensitive to CO_2 than Gram-positive bacteria. In chill-stored proteinaceous foods such as meat and fish, this generally results in the inhibition of the Gram-negative *Pseudomonas*, *Enterobacteriaceae*, and *Acinetobacterl/Moraxella*, while the Gram-positive LAB and *B. thermosphacta* become the dominant organisms. As molds have an absolute requirement for O_2, in foods where mold spoilage is the major concern, e.g., bakery products or hard cheese, packaging in an anaerobic MAP can be extremely successful in delaying spoilage. If CO_2 is used to produce the MAP there is also the additional benefit of the antibacterial and mold activity of CO_2 (Davies, 1995). In general, the effect of CO_2 is explained by the inhibition of enzyme synthesis, the decrease of the intracellular pH, alteration of cell membrane functions (nutrient uptake, absorption), and changes to the physical-chemical properties of proteins. These effects lead to an increase in the length of the lag phase and a decrease of the growth rate. Generally, the effect of carbon dioxide depends on the CO_2 concentration and on the temperature, as increasing temperature levels reduce the effect of CO_2 (Kreyenschmidt and Ibald, 2012).

Antimicrobial packaging active against spoilage microorganisms is accomplished by: (1) incorporation of volatile and nonvolatile antimicrobial agents directly into polymers; (2) coating or adsorbing antimicrobials onto polymer surfaces; as well as (3) immobilization of antimicrobials to polymers by ionic or covalent linkages. Besides, sachets containing antimicrobials can be added to the package; alternatively, polymer films that are inherently antimicrobial can also be used (Siró, 2012). A whole range of antimicrobial compounds have been proposed for the development of antimicrobial packaging, including organic acids and their salts, alcohols, metals, enzymes, chelators, fungicides, plant extracts and essential oils, as well as polysaccharides (Siró, 2012). Because of their antimicrobial properties, bacteriocins have been used to develop antimicrobial packaging active against Gram-positive food spoilage bacteria. However, once again the combination with other substances improves and expands their antimicrobial properties (Mauriello, 2016).

In achieving the desired safety by only one hurdle, high severity in processing needs to be applied. This caused significant damages to the nutritional and sensory quality of foods. For this reason, it is important to have a multihurdles approach for developing safe and wholesome food products. The hurdle effect has fundamental importance for the preservation of foods, since the hurdles provide a control to microbial spoilage (Rahman, 2015).

The hurdle concept was developed in 1978 by Lothar Leistner at the Federal Center for Meat Research in Kulmach, Germany, as a semisystematic approach to improve the safety of foods. This was first applied to shelf-stable foods and then extended to other food products (Davidson and Critzer, 2012). The advantages of the hurdle concept are: (1) it can avoid the severity of one hurdle for preservation, (2) it can give synergy of combination, and (3) many of the hurdles come from past experience (i.e., tradition or culture). Currently huge numbers of products are being developed based on the multihurdles. The combinations and the levels of hurdles are determined based on the empirical experiments. However, it is a challenge to food scientists and engineers to have unified concept or approach for determining food stability considering multihurdles, such as heat treatment, water content, pH, salt, spices, preservatives, packaging, and storage temperature (Rahman, 2015).

REFERENCES

Argyri, A.A., Panagou, E.Z., Nychas, G.-J., 2014. Monitoring microbial spoilage of foods by vibrational spectroscopy (FT-IR & Raman). In: Boziaris, I.S. (Ed.), Novel Food Preservation and Microbial Assessment Techniques. Taylor & Francis, New York, NY, pp. 386–434.

Barbosa-Cánovas, G.V., Bermúdez-Aguirre, D., 2011. Introduction. In: Zhang, H.Q., Barbosa-Cánovas, G.V., Balasubramaniam, V.M., Dunne, C.P., Farkas, D.F., Yuan, J.T.C. (Eds.), Nonthermal Processing Technologies for Food. Wiley, Chichester, pp. xx–xxx.

Baron, F., Gautier, M., 2016. Microbial spoilage. In: Jeantet, R., Croguennec, T., Schuck, P., Brule, G. (Eds.), Handbook of Food Science and Technology 1: Food Alteration and Food Quality. Wiley, New York, NY, pp. 55–97.

Benner Jr., R.A., 2014. Organisms of concern but not foodborne or confirmed foodborne: spoilage microorganisms. Volume 2: hazards and diseases. In: Motarjemi, Y., Moy, J., Todd, E. (Eds.), Encyclopedia of Food Safety. Elsevier, Amsterdam, The Netherlands, pp. 245–250.

Blackburn, CdeW., 2006. Introduction. In: Blackburn, C. deW. (Ed.), Food Spoilage Microorganisms. Woodhead, Cambridge, pp. xvii–xxiii.

Bondi, M., Messi, P., Halami, P.M., Papadopoulou, C., de Niederhausern, S., 2014. Emerging microbial concerns in food safety and new control measures. Biomed. Res. Int. 2014, Article ID 251512.

Brandelli, A., 2015. Nanobiotechnology strategies for delivery of antimicrobials in agriculture and food. In: Rai, M., Ribeiro, C., Mattoso, L., Duran, N. (Eds.), Nanotechnologies in Food and Agriculture. Springer, New York, NY, pp. 119–139.

Braun, P.G., Sutherland, J.P., 2006. Modelling microbial food spoilage. In: Blackburn, C. de W. (Ed.), Food Spoilage Microorganisms. Woodhead, Cambridge, pp. 86–118.

Brown, K.L., 2000. Control of bacterial spores. Br. Med. Bull. 56, 158–171.

Campos, C.A., Gliemmo, M.F., Castro, M.P., 2015. Strategies for controlling the growth of spoilage yeasts in foods. In: Rai, V.R., Bai, J.A. (Eds.), Microbial Food Safety and Preservation Techniques. Taylor & Francis, New York, NY, pp. 497–511.

Cichello, S.A., 2015. Oxygen absorbers in food preservation: a review. J. Food Sci. Technol. 52, 1889–1895.

Davidson, P.M., Critzer, F.M., 2012. Interventions to inhibit or inactivate bacterial pathogens in foods. In: Oyarzabal, O.A., Backert, S. (Eds.), Microbial Food Safety: An Introduction. Springer, New York, NY, pp. 189–202.

Davies, A.R., 1995. Advances in modified-atmosphere packaging. In: Gould, G.W. (Ed.), New Methods of Food Preservation. Springer, New York, NY, pp. 304–320.

Di Renzo, L., Colica, C., Carraro, A., Cenci Goga, B., Marsella, L.T., Botta, R., et al., 2015. Food safety and nutritional quality for the prevention of non communicable diseases: the nutrient, hazard analysis and critical control point process (NACCP). J. Transl. Med. 13, 128.

Dousset, X., Jaffrès, E., Zagorec, M., 2016. Spoilage: bacterial spoilage. In: Caballero, B., Finglas, P.M., Toldrá, F. (Eds.), Encyclopedia of Food and Health, vol. 5. Academic Press, Oxford, pp. 106–112.

Forsythe, F.J., 2000. The microbial flora of food. In: Forsythe, F.J. (Ed.), The Microbiology of Safe Food. Blackwell, London, pp. 96–141.

Fu, L., Valentino, H.R., Wang, Y., 2016. Bacterial contamination in food production. In: Barros-Velázquez, J. (Ed.), Antimicrobial Food Packaging. Elsevier, San Diego, CA, pp. 35–43.

Galimberti, A., Sandionigi, A., Bruno, A., Bruni, I., Barbuto, M., Casiraghi, M., et al., 2016. Towards a universal molecular approach for the quality control of new foodstuffs. In: Ravishankar Rai, V. (Ed.), Advances in food biotechnology. Wiley, New York, NY, pp. 37–59.

Gálvez, A., Grande Burgos, M.J., López, R.L., Pérez Pulido, R., 2012. Food Biopreservation. Springer, New York, NY, 118 pp.

Heyndrickx, M., 2011. The importance of endospore-forming bacteria originating from soil for contamination of industrial food processing. Appl. Environ. Soil Sci. 2011, Article ID 561975.

Howell, K., 2016. Spoilage: yeast spoilage of food and beverages. In: Caballero, B., Finglas, P.M., Toldrá, F. (Eds.), Encyclopedia of Food and Health, vol. 5. Academic Press, Oxford, pp. 113–117.

Kreyenschmidt, J., Ibald, R., 2012. Modeling shelf life using microbial indicators. In: Nicoli, M.C. (Ed.), Shelf Life Assessment of Food. CRC Press, Boca Raton, FL, pp. 127–168.

Linares, D.M., del Río, B., Redruello, B., Fernández, M., Martín, M.C., Ladero, V., et al., 2014. The use of qPCR-based methods to identify and quantify food spoilage microorganisms. In: Boziaris, I.S. (Ed.), Novel Food Preservation and Microbial Assessment Techniques. Taylor & Francis, New York, NY, pp. 313–334.

Lu, S., Xia, Q., 2012. Effects of combined treatments with modified-atmosphere packaging on shelf-life improvement of food products. In: Bhat, R., Alias, A.K., Paliyath, G. (Eds.), Progress in Food Preservation. Wiley, Chichester, pp. 67–109.

Martínez-Monteagudo, S.I., Balasubramaniam, V.M., 2016. Fundamentals and applications of high- pressure processing technology. In: Balasubramaniam, V.M., Barbosa-Cánovas, G.V., Lelieveld, H.L.M. (Eds.), High Pressure Processing of Food. Springer, New York, NY, pp. 3–17.

Mauriello, G., 2016. Control of microbial activity using antimicrobial packaging. In: Barros-Velázquez, J. (Ed.), Antimicrobial Food Packaging. Elsevier, San Diego, CA, pp. 141–152.

Miescher Schwenninger, S., Meile, L., Lacroix, C., 2011. Antifungal lactic acid bacteria and propionibacteria for food biopreservation. In: Lacroix, C. (Ed.), Protective Cultures, Antimicrobial Metabolites and Bacteriophages for Food and Beverage Biopreservation. Woodhead, Cambridge, pp. 27–62.

Mith, H., Duré, R., Delcenserie, V., Zhiri, A., Daube, G., Clinquart, A., 2014. Antimicrobial activities of commercial essential oils and their components against food-borne pathogens and food spoilage bacteria. Food Sci. Nutr. 2, 403–416.

Modi, H.A., 2009. An introduction to microbial spoilage of foods. In: Modi, H.A. (Ed.), Microbial Spoilage of Foods. Aavishkar Publishers, Jaipur, India, pp. 43–66.

Nguyen Van Long, N., Joly, C., Dantigny, P., 2016. Active packaging with antifungal activities. Int. J. Food Microbiol. 220, 73–90.

Nychas, G.-J.E., Panagou, E., 2011. Microbiological spoilage of foods and beverages. In: Kilcast, D., Subramaniam, P. (Eds.), Food and Beverage Stability and Shelf Life. Woodhead, Cambridge, pp. 3–28.

Quintela-Baluja, M., Böhme, K., Fernández-No, I.C., Alnakip, M.E., Caamaño, S., Barros-Velázquez, J., et al., 2014. MALDI-TOF mass spectrometry, a rapid and reliable method for the identification of bacterial species in food-microbiology laboratories. In: Boziaris, I.S. (Ed.), Novel Food Preservation and Microbial Assessment Techniques. Taylor & Francis, New York, NY, pp. 353–385.

Rahman, M.S., 2015. Hurdle technology in food preservation. In: Siddiqui, M.W., Rahman, M.S. (Eds.), Minimally Processed Foods - Technologies for Safety, Quality, and Convenience. Springer, Cham, Switzerland, pp. 17–33.

Rawat, S., 2015. Food spoilage: microorganisms and their prevention. Asian J. Plant Sci. Res. 5, 47–56.

Remenant, B., Jaffrès, E., Dousset, X., Pilet, M.-F., Zagorec, M., 2015. Bacterial spoilers of food: behavior, fitness and functional properties. Food Microbiol. 45, 45–53.

Schnürer, J., Magnusson, J., 2005. Antifungal lactic acid bacteria as biopreservatives. Trends Food Sci. Technol. 16, 70–78.

Setlow, P., Johnson, E.A., 2013. Spores and their significance. In: Doyle, M.P., Buchanan, R.L. (Eds.), Food Microbiology: Fundamentals and Frontiers. ASM Press, Washington DC, pp. 45–79.

Siró, I., 2012. Active and intelligent packaging of food. In: Bhat, R., Alias, A.K., Paliyath, G. (Eds.), Progress in Food Preservation. Wiley, Chichester, pp. 23–48.

Sperber, W.H., 2009. Introduction to the microbiological spoilage of foods and beverages. In: Sperber, W.H., Doyle, M.P. (Eds.), Compendium of the Microbiological Spoilage of Foods and Beverages. Springer, New York, NY, pp. 1–40.

Van Derlinden, E., Mertens, L., Van Impe, J.F., 2013. Predictive microbiology. In: Doyle, M.P., Buchanan, R.L. (Eds.), Food Microbiology: Fundamentals and Frontiers. ASM Press, Washington, DC, pp. 997–1022.

Wang, Y., Li, Y., Yang, J., Ruan, J., Sun, C., 2016. Microbial volatile organic compounds and their application in microorganism identification in foodstuff. TrAC Trends Anal. Chem. 78, 1–16.

Part I

The Spoilers

Chapter 2

Pseudomonas and Related Genera

Nadia Andrea Andreani and Luca Fasolato
University of Padova, Padova, PD, Italy

2.1 INTRODUCTION TO *PSEUDOMONAS*: HISTORICAL CUES AND SIGNIFICANCE OF THE GENUS

The name *Pseudomonas* was proposed initially by Professor W.E.F.A. Migula of the Karlsruhe Institute of Germany at the end of the 19th century (Migula, 1894, 1900; Palleroni, 2010) and it was reported for the first time in the Bergey's Manual of Determinative Bacteriology in 1923. The choice of name seems to be due to its similarity in size and motility to the nanoflagellate Monas (from the Greek: "pseude" = false).

The first description of *Pseudomonas* was inaccurate: Professor Migula described *Pseudomonas* as "... cells with polar organs of motility. Formation of spores occurs in some species, but it is rare...". *Pseudomonas pyocyanea* (now *P. aeruginosa*) was proposed as the type species.

In 1926, the extreme versatility of *Pseudomonas* was highlighted by L.E. den Dooren de Jong in his thesis (den Dooren de Jong, 1926; Palleroni, 2010) which focused on soil bacteria.

By the middle of 1900, more than 800 species had been ascribed to *Pseudomonas*, creating a confusing background for researchers interested in the genus. The major cause of this erroneous classification was the trend to categorize any Gram-negative, strictly aerobic, nonsporulating, motile bacillus as a representative of the genus *Pseudomonas* (Scales et al., 2014). The turning point in this view was the development of first biomolecular approaches that, alongside classical microbiology, unraveled the difficult classification of the genus strains. Around the beginning of the third quarter of the 20th century, DNA/DNA hybridization revealed deep differences among phenotypically similar strains (Pecknold and Grogan, 1973). Subsequently, RNA/DNA hybridization

The Microbiological Quality of Food. DOI: http://dx.doi.org/10.1016/B978-0-08-100502-6.00005-4

showed the presence of 5 different rRNA groups (rRNA group I, II, III, IV, and V; Palleroni et al., 1972). *Pseudomonas* rRNA group I (called *Pseudomonas sensu stricto*) comprised *P. aeruginosa*, all the fluorescent *Pseudomonas* and some nonfluorescent *Pseudomonas* (such as *P. stutzeri*, *P. alcaligenes*, *P. pseudoalcaligenes*, and *P. mendocina*). A more in-depth analysis of genetic differences among *Pseudomonas* species was conducted with the study of 16S sequence homologies: despite the low discriminatory power of rRNA, the study allowed the identification of distinct phylogenetic groups (Laguerre et al., 1994; Anzai et al., 2000).

In 2000, great advance in the study of *Pseudomonas* was made with a siderophore study by Meyer and colleagues that provided an excellent characterization of several species (Meyer, 2000; Meyer et al., 2002).

At the present time (December 2015) the genus comprises 244 species, as reported at http://www.bacterio.net/, having different characteristics.

In this chapter, the importance of the genus *Pseudomonas* and related genera as food spoilers is described. Taxonomic organization, identification methods, spoilage mechanisms, and control plans are reported, with the goal of highlighting the extreme complexity of the spoilage potential of the genera *Pseudomonas, Xanthomonas*, and *Shewanella*.

2.2 IDENTIFICATION AND TRACKING METHODS OF FOODBORNE PSEUDOMONADS: FROM CLASSICAL MICROBIOLOGY TOWARD -OMICS

Several traditional methods have been applied to isolate and characterize *Pseudomonas* spp. strains from foodstuffs; however, recently these approaches have been replaced by more exhaustive and effective nucleic acid-based investigation methods (de Jonghe et al., 2011).

Three ISO standards are available for the enumeration of *Pseudomonas* species. ISO 16266:2006 reports enumeration methods of *P. aeruginosa* from bottled water and is based on the plating on cetrimide agar medium with nalidixic acid; a double incubation is suggested (4°C and 42°C). ISO/TS 11059:2009 and ISO 13720:2010 are official methods of isolating and counting *Pseudomonas* spp. in milk and dairy products and in meat and meat products, respectively. ISO 13720:2010 reports the use of *Pseudomonas* Agar Base, a selective agar medium containing cetrimide, fucidin, and cephalosporin (*CFC supplement*), three antibiotics selective for other non-*Pseudomonas* strains, and incubation at 25°C for 44 hours (Tryfinopoulou et al., 2001). On the other hand, ISO 11059:2009 suggests incubation in a medium containing penicillin and pimaricin at 25°C for 48 hours.

Routinely applied tests for *Pseudomonas* are based on the identification of morphological, phenotypic, and biochemical characteristics of the strains. Some commercial kits are available for strain identification, allowing a great saving of money and time. The most common kit is the API 20 NE system

(No-Enterobacteriaceae; Biomerièux, Lyon, France). The API 20 NE system was proposed by Peladan and Monteil in 1988 as an effective method to characterize *Pseudomonas*, and it was applied in further studies, e.g., the investigation of microbial diversity in milk and dairy plants conducted by Dogan and Boor (2003). As an alternative, BIOLOG GN microtiter plates can be used to characterize *Pseudomonas* strains (Johnsen et al., 1996; Arnaut-Rollier et al., 1999; Martin et al., 2011). Other analyses are based on the growth of strains on differential agarized media, on Gram staining, on strain nutritional requirements, on motility tests, and on the ability to grow at different temperatures.

The recent increasing interest in molecular approaches led to the development of several clustering and typing methods that avoid the lack of information inherent in classical methods. These new approaches allowed in some cases not only the identification of the species within the genus, but also the description of phylogenetic relationships among species belonging to specific bacterial clusters. Examples are the application of REP-PCR (Repetitive Extragenic Palindromic PCR; Johnsen et al., 1996), RFLP (Restriction Fragment Length Polymorphism; Franzetti and Scarpellini, 2007), DGGE (Denaturing Gradient Gel Electrophoresis, De Jonghe et al., 2011), and PFGE (Pulsed Field Gel Electrophoresis; Nogarol et al., 2013) to the identification of foodborne *Pseudomonas* species. Sequencing of 16S rRNA genes has been widely used to identify and categorize several species of the genus isolated from food or other environments (Laguerre et al., 1994; Moore et al., 1996; Anzai et al., 2000); however, because of the relatively high conservation of the 16S gene, not all species can be discriminated using this approach (Moore et al., 1996; Anzai et al., 2000; Yamamoto et al., 2000). For this reason, the investigation by sequencing of 16S rRNA *locus* has been replaced by the combined analysis of several sequences of defined housekeeping genes for identification, characterization, and spoiler tracking purposes. Some studies have focused on the phylogeny of the whole genus (Yamamoto et al., 2000; Hilario et al., 2004; Mulet et al., 2010), while others have concentrated on subgroups of the genus, such as the investigation of blue-pigmenting strains isolated from food products and belonging to the *Pseudomonas fluorescens* group by Andreani and collaborators (2014). Both classical microbiological and biomolecular approaches require the isolation of the strains from their original environments (culture-dependent methods). For this reason, these techniques have to take into account the problem of viable but nonculturable bacteria (VBNC). VBNC are bacteria whose metabolic activity is too low to allow reproduction. Sometimes, the induction of VBNC forms is due to stresses to which vital cells are subjected within the food chain (e.g., refrigeration of fresh products, use of sanitizers, or heating treatments).

Apart from culture-dependent biomolecular methods (based on the study of single cells isolated and grown as planktonic cultures), recently, the spread of next-generation sequencing (NGS) approaches (e.g., 454 Pyrosequencing and

Illumina sequencing) has allowed a more in-depth knowledge about food and its spoilers (Holt et al., 2008; Andrews-Polymenis et al., 2009; Gilmour et al., 2010). The ever-decreasing cost of NGS techniques has made it easier to obtain complete or draft genomes of bacterial species (Diaz-Sanchez et al., 2012; Solieri et al., 2012). Consider that, taking into account an average *P. fluorescens* genome size of 6 Mbases, in a single run of MiSeq 300 pair-end (Illumina Technology) with an estimated output of 15 Gbases (data obtained from Illumina website in December 2015), up to 10 genomes could be multiplexed to obtain a coverage that allows robust genomic and comparative investigations (Kisand and Lettieri, 2013). Recently, several draft genomes of strains belonging to the *P. fluorescens* species complex have been published, among which there are some food spoilers (Andreani et al., 2015; Lo et al., 2015). Investigation of such data might disclose the genetic information underlying the spoilage potential of *Pseudomonas* spp. strains. Similarly, NGS methods can be used to investigate the expression of genes involved in food spoilage. This kind of investigation could enhance understanding of food spoilage, building on the limited amount of information obtained by Mohareb and coworkers (2015) by applying the microarray technology to identify potential biomarkers of meat alteration in *Pseudomonas putida*. Bacterial communities can also now be investigated through NGS technologies: metagenomics studies (namely the sequencing of all genetic material extracted from different types of sample, such as food samples) are expected to be one of the future strategies for genome sequencing of interesting microbes. It is clear that the innovation of NGS approaches has not been exploited only for the purposes of food microbiology.

2.3 GENUS *PSEUDOMONAS*: CHARACTERISTICS AND ECOLOGY

The genus *Pseudomonas* is composed of a heterogeneous group of bacteria characterized by important ecological significance (de Oliveira et al., 2015). The strains ascribed to the genus *Pseudomonas* are Gram-negative, straight or slightly curved rods, 0.5 to 1.0 μm by 1.5 to 5.0 μm in length, catalase-positive, and are motile by means of one or several polar flagella. These bacteria are characterized by an obligate respiratory metabolism, a lack of gas formation from glucose and a positive result from the oxidase test. Even if they are aerobic strains, some *Pseudomonas* can grow under anoxic conditions in the presence of nitrate, fumarate or other electron acceptors (Palleroni, 1984).

As they are mesophilic bacteria, their optimal growth temperature is about 25–35°C; however, most of them are psychrotrophic microorganisms, which means they are able to grow at refrigeration temperatures and do not tolerate high temperatures. Cold adaptation of *Pseudomonas* spp. is linked to high levels of unsaturated lipid in the cell membrane and to a wide range of mechanisms that allow the tolerance of cold-induced stresses (Fonseca et al., 2011; Samaržija et al., 2012; Moreno and Rojo, 2014).

The optimum pH environment for *Pseudomonas* strains is above 5.8: lower pH values slow down bacterial growth.

Pseudomonas strains have very simple nutritional requirements and have the ability to use different organic compounds as carbon and energy sources. For these reasons, *Pseudomonas* can be defined as an ubiquitous bacterium. *Pseudomonas* are present in a wide range of different environments such as organic material under decomposition, atmospheric dust, and vegetation, and it has various animal and plant hosts (Anzai et al., 2000; Frapolli et al., 2007).

Some strains are important soil and water bacteria and are responsible for the degradation of several molecules. For this reason, they are very important in the mineralization process in nature, in sewage treatment, and in the degradation of xenobiotic substances, such as pesticides or chemicals. They are also important agents in bioremediation of the environment (Stanier et al., 1966; Wasi et al., 2013).

Some species are human pathogens, such as the well-known *P. aeruginosa*, but also *P. paucimobilis*, *P. putida*, *P. fluorescens*, and *P. acidovorans* (Tümmler et al., 2014). Other species are plant pathogens, e.g., *P. pseudoalcaligenes*, *P. savastanoi*, *P. syringae*, and others are animal pathogens, mainly against fishes and birds (*P. anguilliseptica*, *P. chlororaphis*, *P. aeruginosa*; Peix et al., 2009). Some *Pseudomonas* are well-known food spoilers, such as *P. fluorescens* which is responsible for the spoilage of dairy products, raw fish, raw meat, and eggs, because of their psychrotolerant character (Jay, 2003; Palleroni, 2010).

2.3.1 *Pseudomonas fluorescens* Group

A recent investigation based on the sequencing of four *loci* (16S rRNA, *gyrB*, *rpoB*, and *rpoD*) by Mulet and collaborators (2010) of 107 *Pseudomonas* allowed the identification of two major lineages (*P. aeruginosa* and *P. fluorescens*). The *P. fluorescens* lineage has itself been divided into six groups (*P. syringae*, *P. lutea*, *P. putida*, *P. anguilliseptica*, *P. straminea*, and *P. fluorescens*). In particular, the *P. fluorescens* group has been identified as a well-separated monophyletic group, comprising nine further subgroups and characterized by strains with high spoilage potential. Since the recent classification by Mulet and colleagues, at least 52 species are now considered as part of the *P. fluorescens* group (2010). The schematic representation of the lineages, groups, and subgroups within the genus is reported in Fig. 2.1.

In the history of microbiology, the *P. fluorescens* group has not been intensively studied (Palleroni, 2010) and only recently has this group been seriously considered, due to its high spoilage potential.

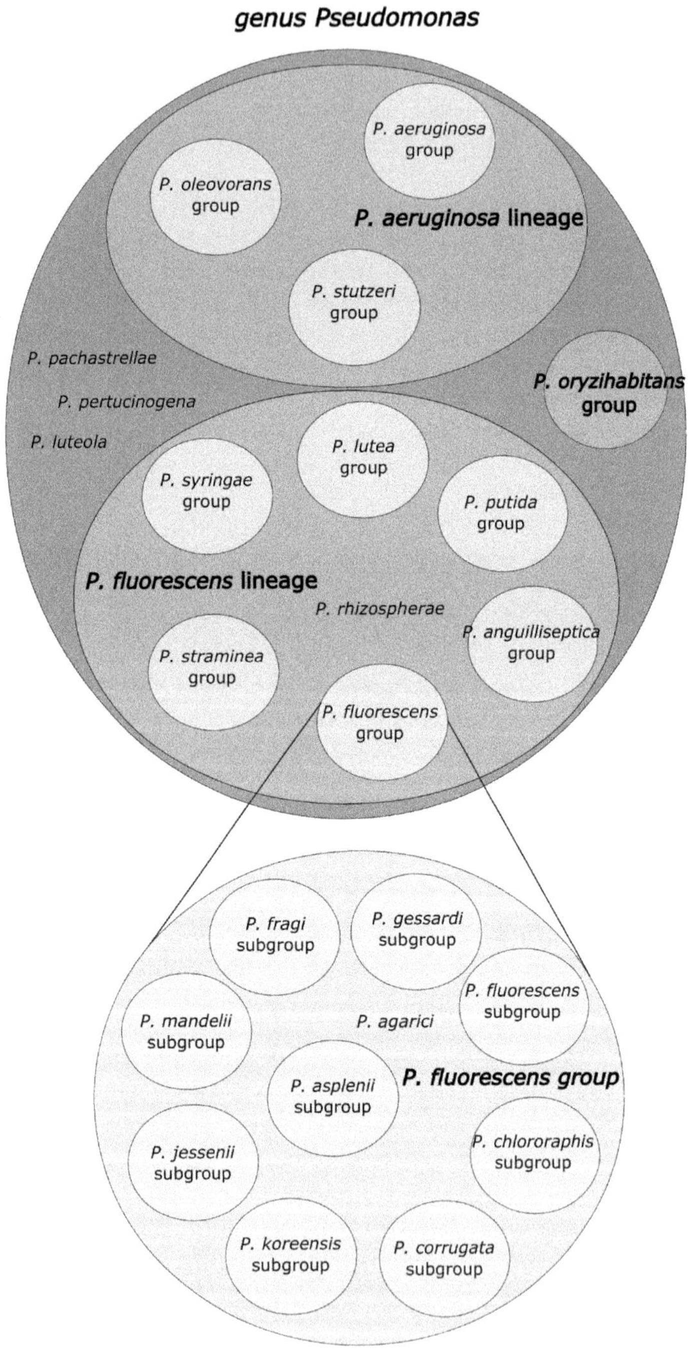

FIGURE 2.1 The genus *Pseudomonas*. Schematic representation of lineages, groups, and subgroups of the genus *Pseudomonas*, as reported by Mulet et al. (2010).

Pseudomonas fluorescens possesses a number of functional traits that allow it to grow in several environments, such as the production of several secondary metabolites or siderophores (Cornelis, 2010), the presence of several secretion systems, and the ability to form biofilms, together with the adaptability of its genome (Spiers et al., 2000; Silby et al., 2011).

The major issue concerning *P. fluorescens* group strains (and specifically *P. fluorescens* subgroup and *P. fluorescens*) is food spoilage and their importance is related to public hygiene and the food trade. In fact, even though *P. fluorescens* group strains are easily destroyed by the heat treatments usually applied to food, they are able to produce a wide range of enzymes (mainly proteases, lipases, and lecithinases), or pigments, that are in some cases thermostable. Most of the reported spoilage cases by *P. fluorescens* are attributable to postprocess contaminations.

An early investigation into the pathogenicity of *P. fluorescens* identified it as a harmless microorganism (Baader and Garre, 1887). In this study, *P. fluorescens* was defined as nonpathogenic as no reaction took place when a culture was applied subcutaneously to animals or injected into the peritoneum. Additionally, ingestion of cultures resulted in no harm to the gastro enteric tract. Actually, a recent review by Scales and coworkers (2014) reported an intriguing and complex correlation between *P. fluorescens* and human diseases, such as pulmonary and Crohn's diseases. Usually, bacteremia cases involving *Pseudomonas* are milder than those involving its cognate *P. aeruginosa* and they are usually related to nosocomial infections.

2.4 SPOILAGE POTENTIAL OF THE *PSEUDOMONAS* STRAINS: A FOCUS ON THE *PSEUDOMONAS FLUORESCENS* GROUP

Pseudomonas spp. strains are reported to be some of the most common food spoilers (Annual Report of RASFF, 2010). Strains belonging to the *P. fluorescens* species complex have been isolated from a wide range of foods, even when the spoilage was not visible in the food matrices.

Because of its mesophilic and psychrotolerant nature, *Pseudomonas* are often found in a wide range of foods. Even though its optimum temperature is substantially higher, *P. fluorescens* can replicate at the refrigeration temperatures at which fresh food is usually stored during its shelf life. This favors psychrotrophs, as competition with mesophilic bacteria that are effectively controlled by cooling is reduced. Furthermore, storage in inappropriate thermal conditions could further enhance the growth of these spoilers. The combination of increasing shelf life, to meet consumer and marketing demand, and low temperatures has been demonstrated to enhance the growth of psychrotrophs (Munsch-Alatossava and Alatossava, 2005). Additionally, *Pseudomonas* are characterized by a brief generation time (less than 4 hours) that is shortest between 0°C and 7°C; this characteristic further worsens food spoilage (Sørhaug and Stepaniak, 1997; Samaržija et al., 2012). *Pseudomonas* spp. are particularly predominant in aerobically packaged refrigerated and protein rich aliments (Gram et al., 2002).

Thanks to their favorable nutrient composition and the ability of *Pseudomonas* to degrade pectin, *Pseudomonas* strains find an optimal environment in ready-to-eat vegetables. A great variety of *Pseudomonas* species have been isolated from prepared vegetables, with a clear predominance of *P. fluorescens* and *P. fluorescens* group strains (Andreani et al., 2014; Caldera and Franzetti, 2014), but with other subgroups also represented. The main reported species in Minimally Processed Vegetables are usually *P. fluorescens*, *P. putida*, *Pseudomonas chicorii*, *P. maltophilia*, and *P. marginalis* (Franzetti and Scarpellini, 2007).

High-protein content and neutral pH foods, like tofu, are also advantageous matrices for *Pseudomonas* strains (Stoops et al., 2012).

Pseudomonas are commonly part of the microflora of aerobically stored fish (Gram and Huss, 1996) and a specific spoiler of fresh and marine water tropical fish (Gram et al., 1990; Gram, 1992) and its growth should usually be inhibited by the use of Modified Atmosphere Packages (MAPs). Actually, *Pseudomonas* and H_2S (hydrogen sulfide gas) producing bacteria were reported to dominate some products such as sea bass both in the presence of air and under MAP conditions (Parlapani et al., 2015). Some species are also fish opportunists (e.g., *P. aeruginosa* and *P. fluorescens*) or fish pathogens (such as *P. putida*, *P. fluorescens*) as reported by different studies (Altinok et al., 2006; Franzetti and Scarpellini, 2007; Zhang et al., 2014; Sun and Sun, 2015).

Meat represents another favorable environment for the growth of *Pseudomonas*. *Pseudomonas fluorescens*, *P. lundensis*, and *P. fragi* were the most represented groups in poultry, both in processed meat, on the skin of slaughtered broilers, and in processing plants (Arnaut-Rollier et al., 1999; Mellor et al., 2011; Oakley et al., 2013; Grewal et al., 2015). *Pseudomonas* spp., and especially *P. lundensis*, have also been isolated from goat and lamb slaughterhouse surfaces. Most of them were demonstrated to be resistant to at least one antimicrobial compound (Lavilla Lerma et al., 2012, 2014, 2015). *Pseudomonas* strains have also been isolated from beef. Doulgeraki and Nychas (2013) reported *P. fragi* to be the predominant species during the storage of minced beef, especially at 15°C and in the presence of oxygen; at lower temperatures (5 and 10°C), *P. fragi* prevails, while under a MAP environment at 15°C *P. fragi* and *P. fluorescens* dominate the microflora. Mastromatteo and colleagues reported *Pseudomonas* to be responsible for unacceptability of pork sausage if no MAP was applied in the packaging of the meat product (2011). Finally, a gene expression investigation conducted by Mohareb and coworkers identified specific genes involved in meat spoilage, specifically regulated by glucose concentration and storage temperature (Mohareb et al., 2015).

A great advantage for several *Pseudomonas* strains is the ability to compete with food microflora for iron (Cornelis, 2010). For example, milk is characterized by a low free iron content. Due to the production of siderophores, *Pseudomonas* strains are able to obtain iron from lactoferrin (McPhee and Griffiths, 2002). Due to their ubiquitous presence, *Pseudomonas* spp. (and primarily *P. fluorescens*) have been often chosen as model organisms to investigate spoilage mechanisms of psychrotrophs in milk and dairy products and to evaluate control measures for reducing food contamination (de Oliveira et al., 2015). Milk and dairy products are optimal matrices for the growth of *Pseudomonas* spp., thanks to their nutritional value, water content, and pH close to neutrality. *Pseudomonas* strains have been demonstrated to be the majority of bacteria in milk silos during a monitoring study in Scotland, representing 70% of all bacteria isolated (McPhee and Griffiths, 2002). In milk, *P. lundensis*, *P. fragi*, *P. fluorescens*, and *P. gessardii* are the most commonly found species (McPhee and Griffiths, 2002; Marchand et al., 2009b). The great ability of this bacterial group to spoil food is not only related to its ability to grow at refrigeration temperatures, but also its aptitude to adapt to several different environmental conditions (De Jonghe et al., 2011).

Several spoilage phenotypes of *Pseudomonas* seem to be related to quorum sensing, the complex system of communication among cells, based on the production of signaling molecules (Nychas et al., 2009). It should be remembered that spoilage activity is often the consequence of a complex population, whose interactions determine the alteration of the final products.

Pseudomonas are considered a Specific Spoilage Organism (SSO) for several types of food, mainly pork, poultry, and seafood (Gram and Dalgaard, 2002; Raab et al., 2008). The designation SSO is largely applied for the purpose of building predictive models for refrigerated food, because of these organisms' direct involvement in spoilage. Generally, it is assumed that a level of around 7 Log_{10} CFU as a threshold corresponds to the end of the shelf life (Raab et al., 2008).

The main mechanisms through which *Pseudomonas* implement spoilage activity are:

- Production of pigmented molecules
- Production of enzymes
- Production of other spoilage molecules (organoleptic spoilage)

2.4.1 Pigment Production

Pseudomonas strains are well-known producers of pigmented molecules that can be involved in several spoilage events; however, not all of the known pigments have been reported in food. Few scientific papers are available except for a few technical reports of official laboratories responsible for food control. Table 2.1 reports cases of bacterial discoloration caused by *Pseudomonas* species. It is quite evident that a more in depth investigation is necessary to clearly understand the chemical nature of the pigments involved in such unpleasant spoilage effects. The most outstanding event was represented by a discoloration reported in 2010 in mozzarella cheese. Official reports of the Rapid Alert System for Food and Feed (the system that comes into play when a quick notification of presumed dangerous food contaminations is necessary to defend human health) of 2010 identified the cause in *Pseudomonas* strains (and specifically *P. tolasii* and *P. libanensis*). In the

TABLE 2.1 Discoloration Caused by *Pseudomonas* in Food Products
Summary of recent reports of discoloration caused by Pseudomonas *strains. The table reports the identified pigment, if available, the color, the food matrix of isolation, the identified species, and the reference.*

Color	Pigment	Food	Species	References
Blue	Unknown	Rabbit meat	*P. gessardi* (*P. fluorescens* subgroup)	Sarale et al. (2011)
Blue	Indigoidine	Mozzarella cheese: fresh, low-acid cheese	*P. lemonnieri* now *P. fluorescens* biovar IV (*P. corrugata* subgroup)	Martin et al. (2011), Caputo et al. (2015)
Blue	Melanin	Pork and beef	*P. fluorescens* (*P. fluorescens* group)	Kröckel (2009)
Blue-black	Unknown (indigo-derivative)	Mozzarella cheese Pork Beef Ricotta cheese	*P. fluorescens* subgroup (*P. fluorescens* group)	Cantoni et al. (2006)
Dark orange-red	Unknown	Mozzarella cheese	*P. brassicacearum* (*P. fluorescens* group)	Cantoni et al. (2003)
Yellow-purple	Unknown	Mozzarella cheese	*P. gessardi* (*P. fluorescens* subgroup)	Cantoni et al. (2006)
Green-blue	Pyoverdine & indigoidine	Mozzarella cheese	*P. putida*, *P. fluorescens*	Cantoni et al. (2006)
Green	Pyoverdine	Mozzarella cheese	*P. fluorescens* biovar I	Cantoni et al. (2006)

preliminary study conducted by Caputo and colleagues (2015), the pigment was identified as indigoidine (as already reported by Martin et al., 2011 reporting a case of blue discoloration in a Latin-style fresh cheese) and the strains responsible for the spoilage defect were suggested to be *P. fluorescens* biovar IV (previously *P. lemonnieri*). Actually, these results differ from the investigation conducted by Andreani and collaborators that placed the blue-pigmenting strains into a well-separated branch of the *P. fluorescens* subgroup (Andreani et al., 2014, 2015) based on the construction and application of a Multilocus Sequence Typing (MLST) scheme. They also concluded that the blue pigment could not be indigoidine, but rather an indigo-derivative. In all of these studies, only a high microbiological load was able to induce the blue discoloration (about 10^7 CFU/g). Interestingly, the pigment seemed to turn blue in the presence of oxygen (meaning that it is derived from an uncolored parent compound).

Blue discoloration has also been reported in other food products, such as ricotta cheese, bacon, and beef (Andreani et al., 2014), as a result of the growth of strains belonging to the *P. fluorescens* subgroup.

2.4.2 Enzyme Production

Food alteration defects can also be induced by *Pseudomonas* strains through the production of a wide range of enzymes. As a matter of fact, not only do the resident and living microflora produce enzymes, but due to the thermoresistant nature and activity in low a_w environments of some enzymes, they can also retain their activity in finished products. Hydrolytic enzymes, namely proteases, lipases, and lecithinases, may maintain up to 70% of their activity after heat treatments (Samaržija et al., 2012), contributing to the shortening of the shelf life of food products by altering the physicochemical characteristics of the products, or by affecting food functionality and organoleptic qualities (Teh et al., 2014).

The major spoilage enzymes are proteases. One of the most outstanding examples of this type of spoilage is visible in UHT milk and dairy products, where proteases are active against caseins (the most represented proteins in this type of fresh products). Proteolysis of UHT milk during shelf life causes different changes in organoleptic characteristics, such as gelation of milk, that causes a rise in viscosity, an unpleasant gray discoloration and flavor changes that induce a bitter, unclean, fruity, yeasty, and metallic taste (Datta and Deeth, 2003; Arslan et al., 2011; Samaržija et al., 2012). *Pseudomonas* proteases showed activity against κ-, α_{s1}-, and β- caseins, with the subsequent coagulation of the newly formed complexes, forming a gel structure in milk (Datta and Deeth, 2003). It should be mentioned that this alteration might be also induced by the activity of an endogenous milk proteinase, namely plasmin (Datta and Deeth, 2001). Production of proteinases affects cheese yield by destabilizing casein, as demonstrated by several studies

(Mitchell and Marshall, 1989). In particular, plasmin and plasminogen are released from casein micelles, changing the cheese yield, as plasmin is a well-known actor in cheese manufacture, as well as affecting the flavor and texture of the final product (Samaržija et al., 2012).

Marchand and colleagues reported *P. fragi*, *P. lundensis*, and other minor species of the *P. fluorescens* group as prolific producers of proteases that are in most cases heat-stable (2009a,b).

A recent study based on the phenotypic investigation of 87 *P. fluorescens* species complex strains revealed that almost all the strains (94%) were able to induce proteolysis on Nutrient Agar Plates with 2% UHT Milk at 22°C, while about 72% were able to induce the defect at refrigeration temperatures (Andreani et al., 2014), emphasizing the high prevalence of proteolytic strains within the *P. fluorescens* group.

The most reported family of thermostable proteases within the genus *Pseudomonas* is the serralysin protease family, a highly conserved protein group whose main representatives are members of the AprX family, an alkaline zinc metalloprotease family with molecular masses between 39.2 ± 0.7 and 45.3 ± 1.3 kDa (Dufour et al., 2008; Marchand et al., 2009b; Teh et al., 2014). Recent advances in molecular biology have allowed the investigation of genetic and regulation mechanisms underlying protease production. The *aprX* gene was detected in almost all *P. fragi* and several reference strains of *Pseudomonas* species (such as *P. fluorescens, P. rhodesiae, P. tolaasii, P. chlororaphis*, and *P. gessardi*), but not in *P. lundensis*. Furthermore, a high degree of heterogeneity among amino acid sequences of AprX proteins was reported (Marchand et al., 2009b). It must be noted that the presence of the gene is not synonymous with certain expression; most protease production depends on several environmental parameters, such as temperature, oxygen concentration, quorum sensing, nutrient composition of the media (i.e., presence of iron), and phase variation (Marchand et al., 2009b; Teh et al., 2014). Interestingly, high storage temperatures seem to inhibit protease production in *P. fragi* (Marchand et al., 2009b); however, strong heat treatments, such as boiling of milk for 30 minutes, seem to induce a massive release of proteases, inducing a more palpable spoilage effect (Samaržija et al., 2012). In some strains of *Pseudomonas* spp., AprX has been identified as the sole protease involved in food spoilage (Woods et al., 2001).

Proteases are mainly produced at the end of the exponential phase, when cell density is substantially greater, highlighting the involvement of quorum sensing mechanisms in spoilage activity (Bai and Rai, 2011), although some other studies (Liu et al., 2007; Pinto et al., 2010) failed to detect any signaling molecules when investigating spoiled milk.

Lipolytic spoilage is less common than the proteolytic type. Lipases, triacylglycerol hydrolases, act on the fat constituents of food and induce the release of fatty acids and glycerol. Free short-chain fatty acids induce unpleasant flavors, usually described as rancid, while medium-chain fatty acids are responsible for unclean, soapy, or bitter flavors (Samaržija et al., 2012). No particular unpleasant effect seems to be caused by the production of long-chain fatty acids. It has been extensively reported that lipolytic activity is more remarkable at refrigeration temperatures (low-temperature regulation; Woods et al., 2001; Rajmohan et al., 2002).

In cheese, lipases are adsorbed in fat globules and persist in the cheese, inducing spoilage effects during the ripening of semi-hard and hard cheeses (Samaržija et al., 2012). The heat-stability of lipases seems to be enhanced in the presence of multiple lipases (Teh et al., 2014).

Interestingly, in *P. fluorescens* B52, genes coding for the thermostable protease (*aprX*) and the lipase (*lipA*) are located in the same operon, separated by other genes involved in secretion of the extracellular protease (Woods et al., 2001; McCarthy et al., 2004). Therefore, gene expression of both hydrolases is correlated, dependent on low temperature, and inhibited by iron (Woods et al., 2001).

Another important class of extracellular enzymes is constituted by lecithinases and other phospholipases. These enzymes disrupt the fat globules of milk and make fat content available for further lipase activity (Koka and Weimer, 2000; Samaržija et al., 2012). Their final products are diglycerides and substituted phosphoric acids. The most commonly represented enzyme family is phospholipase C. In milk, spoilage defects related to the production of phospholipases are sweet curdling, bitterness, and feathering (Koka and Weimer, 2000). As these kinds of enzymes are not usually present as endogenous enzymes in milk, phospholipase-mediated spoilage is usually uniquely related to psychrotrophs such as *Pseudomonas* spp. strains.

Some strains have been reported to be able to produce the whole range of spoilage enzymes (proteases, lipases, and lecithinases) in a plate assay (Andreani et al., 2014).

Table 2.2 reports the most common spoilage defects related to enzyme production as an effect of *Pseudomonas* growth in different food matrices.

The detection of these enzymes in the final products might help in monitoring the spoilage effect related to microbial contamination. The most common methods are the use of caseinate agar (to detect proteolytic activity), but also electrophoresis, HPLC, UV-Vis spectrometry, fluorimetric, and immunological methods to detect enzymes directly in food (Teh et al., 2014).

In fact, *Pseudomonas* should not be the only target of the screening. Teh and colleagues (2012) demonstrated that enzymatic activity of *P. fragi* BC5 was enhanced in coculture with other strains, such as *Bacillus licheniformis*.

2.4.3 Other Types of Spoilage

Besides enzymatic and pigment spoilage, the growth of *Pseudomonas* spp. on several food matrices can induce further quality modifications making the

TABLE 2.2 Enzymatic Spoilage Caused by *Pseudomonas*

Summary of the main spoilage activities in food products involving enzyme production. The table reports the enzymes responsible for the spoilage, the food involved in the defect, and the characteristics of the defect. Finally, responsible species and references are reported.

Enzyme	Food	Alteration	Species	References
Lipases	Sterilized milk	Rancidity, off-flavors, bitterness, soapy off-flavors	*Pseudomonas* spp., *P. fluorescens*	Samaržija et al. (2012)
Lipases	Cheese	Enhance of coagulation time due to high amount of free fatty acids	*Pseudomonas* spp.	Samaržija et al. (2012)
Lipases	Cheese	Undesirable lipolysis at the end of ripening	*Pseudomonas* spp.	Samaržija et al. (2012)
Lipases and proteases	Cream Butter	Rancidity, off-flavors, fruity bitterness, soapy off-flavors	*P. fragi*, *P. putrefaciens*, *Pseudomonas* spp.	Jay (2000), McPhee and Griffiths (2002), Samaržija et al. (2012)
Proteases	Pasteurized milk	Gelation due to enhanced viscosity of milk	*P. fluorescens*, *P. fragi*	Datta and Deeth (2003), Arslan et al. (2011), Samaržija et al. (2012)
Proteases	Pasteurized milk	Fouling in heat exchangers	*Pseudomonas* spp.	Samaržija et al. (2012)
Proteases	Cheese	Reduction of cheese yield due to destabilization of casein fraction Reduction of cheese yield due to a higher heat sensitivity of milk proteins	*Pseudomonas* spp.	Mitchell and Marshall (1989), Samaržija et al. (2012)
Proteases	Cheese	Reduction in coagulation time due to high amount of free amino acids	*Pseudomonas* spp.	Samaržija et al. (2012)

(*Continued*)

TABLE 2.2 (Continued)

Enzyme	Food	Alteration	Species	References
Proteases	Yogurt	Firmer texture, higher viscosity, and greater syneresis Enhanced growth of microflora due to higher free amino acid contents Bitter, rancid, unclean, and fruity off-flavors	*Pseudomonas* spp.	Gassem and Frank (1991), Sørhaug and Stepaniak (1997)
Proteases	Semi-hard and hard cheese	Atypical flavor during ripening due to the persistence of spoilage proteases	*Pseudomonas* spp.	Fox (1989), Samaržija et al. (2012)
Proteases	Fresh cheese	Gelation and slimy texture due to the activity of enzymes	*P. fluorescens*, *P. fragi*, *P. putida*	Fox (1989), Samaržija et al. (2012)
Proteases	Soft cheese	Poor texture due to the activity of enzymes	*P. fluorescens*, *P. fragi*, *P. putida*	Samaržija et al. (2012)

Source: Adapted and implemented from Samaržija, D., Zamberlin, Š., Pogačić, T., 2012. Psychrotrophic bacteria and milk and dairy products quality. Mljekarstvo 62, 77–95.

product undesirable for human consumption (organoleptic spoilage). In these cases, the spoilage involves the ability of bacterial strains to use nutrients in food and to produce molecules that induce modification of the food quality.

As an example, several studies have reported that the growth of *Pseudomonas* spp. on iced fish is related to fruity, rotten, and sulfydryl odors and flavors as a result of the synergy of newly produced acetate, butyrate, and hexanoate (Miller et al., 1973; Gram, 1992). Furthermore, strains of the genus *Pseudomonas* produce several volatile aldehydes, ketones, esters, and sulfides (Miller et al., 1973; Dainty et al., 1984). According to Gram and Huss (1996), the fruity off-odors produced by *P. fragi* originate from monoamino-monocarboxylic amino acids. Moreover, *P. fragi* develops apple-like esters inducing rancidity in butter (Samaržija et al., 2012). Ercolini and coworkers revealed through a challenge test that the volatile compounds produced after 15 days of storage of inoculated beef with a

P. fragi strain were highly heterogeneous and showed their highest diversity in ketones and higher-molecular weight alcohols, while only a few esters were detected (Ercolini et al., 2009). Recently, Bhadra and colleagues (2015) correlated volatile compounds such as trimethylamine, dimethylamine, and ammonia with the enhancement of total viable count that after 12 hours is completely represented by *Pseudomonas*. *Pseudomonas putrefaciens* is able to release a wide range of organic acids, and mainly isovaleric acid, inducing a putrid odor in butter (Jay, 2000; Samaržija et al., 2012). It seems clear that the production of volatile compounds is a strain-dependent characteristic, and a common pattern for all of the spoiler *Pseudomonas* cannot be identified, as has already been reported in single-species investigations (Ercolini et al., 2009).

Some strains are able to produce a surface film: e.g., strains of *P. fluorescens* have been detected on the surface of cottage cheese during its shelf life at 7°C (Brocklehurst and Lund, 1985).

Pseudomonas spp. also induce putrefactive odors and slime on fresh chilled meat in the presence of a population with at least 10^7 CFU/cm^2 (Sofos, 1994).

In 2010, Motoyama and coworkers reported *Pseudomonas* spp. and specifically *P. fragi* as responsible for beef reddening in raw meat due to the conversion of metmyoglobin into deoxymyoglobin, together with slime production.

2.5 RELATED GENERA: *XANTHOMONAS* AND *SHEWANELLA*

Other genera belonging to *Gammaproteobacteria* and involved in food spoilage, such as *Shewanella* and *Xanthomonas*, were previously considered as related genera to *Pseudomonas* (Liao, 2006). Nowadays, *Xanthomonas* is considered as a monophyletic lineage and a distinct genus from the other family members of *Xanthomonadaceae*. Species belonging to the genus *Xanthomonas* are reported as causative agents of disease in plants. Proteases and toxins are involved in host colonization (having mainly cellulolytic and pectolytic activities) while the presence of lipopolysaccharide or the release of extracellular polysaccharides could be involved in the pathogenicity of different pathovars and species. The maintenance of the genus in the environment seems to be related to seed transmission or persistence in plant tissues. However, some species also exhibit a saprophytic lifestyle and are related to the postharvest decay of fruits and vegetables.

On the other hand, bacteria ascribed to the genus *Shewanella* show heterogeneous phenotypic characteristics, mainly related to the origin habitat of each species. The ecological niches occupied by *Shewanella* range from aquatic sediments to animal gut, demonstrating their capability to adapt to anoxic environments (Hau and Gralnick, 2007). Despite the heterogeneity of these environments, water (marine and fresh water) can be considered as the preferred environment for *Shewanella*. These waterborne bacteria are involved in food contamination, especially of seafood, but also of poultry and other meat products (Parker and Levin, 1983; Nychas et al., 2007; Dainty et al., 1989). Moreover, Janda recently reported some species of *Shewanella* as human pathogens (i.e., *S. algae*; Janda, 2014).

Due to their ability to induce changes in the odor profile (off-odors) and in relation to the wide range of metabolites produced, in the last decade some members of this genus have been proposed as SSOs of food and have been used as shelf life predictors (e.g., in seafood products; Gram and Dalgaard, 2002).

2.5.1 *Shewanella* spp.

Shewanella is the type genus and, at the present, the sole genus of the family Shewanellaceae ascribed to the order Alteromonadales, with around 50 described species (Satomi, 2014). Taking into account the phenotypic variability of the genus, species characterization and reclassification are based on DNA/DNA hybridization and 16S rRNA gene sequencing (Hau and Gralnick, 2007). The sequence of additional genes, such as *gyrB*, confirmed the species delineation (Satomi et al., 2003; Dehaut et al., 2014). Moreover, molecular typing of strains can reduce misinterpretation and has allowed the development of intra-species studies through different approaches such as Random Amplification of Polymorphic DNA, Matrix-Assisted Laser Desorption/Ionization Time of Flight (MALDI-TOF), MLST, and Whole Genome Sequencing (Serio et al., 2014; Satomi, 2014; Deng et al., 2014). As for other bacterial taxa, NGS techniques have allowed a deeper investigation of the genomic information related to its phenotypes. To date (December 2015), 25 complete genomes and 25 draft genomes are available on GenBank. Annotations conducted by Dikow (2011) revealed that *Shewanella* possesses about 5000 genes and that its genome consists of approximately 5 Mbp.

These Gram-negative bacteria are oxidase positive, nonendospore forming bacteria characterized by rod-shaped morphology. All the members are motile (e.g., swimming motility) due to a unique polar flagellum (Wu et al., 2011; Satomi, 2014). Generally, a first phenotypic separation among psychrotrophic and mesophilic species can be delivered by testing the growth at 4°C, 37°C, and 42°C (Fonnesbech Vogel et al., 2005). Careful attention is required to the isolation temperature in order to differentiate the three major phenotypic lineages: (1) psychrotolerant nonhalophilic bacteria, (2) psychrophilic species requiring salt for growth, and (3) mesophilic species (Satomi, 2014). The use of different salt concentrations in combination with the use of different carbon sources and enzymatic activities (e.g., amino acid decarboxylation) has been widely proposed for strain characterization, taxonomic keys, and numerical taxonomy purposes (Fonnesbech Vogel et al., 2005;

Satomi et al., 2006, 2007; Dehaut et al., 2014; Serio et al., 2014). Basically, the aforementioned phenotypic traits together with the production of H_2S and nonfermentative metabolism on oxidative-fermentative medium are indicative of the genus. The work of Satomi (2014) retained the most useful phenotypic analyses for species delineation. The species are defined as aerobes/facultative anaerobes. Moreover, one of the evolutionary keys to the worldwide diffusion of *Shewanella* is related to the wide number of electron acceptors and metabolic pathways adopted under limited oxygen conditions (Hau and Gralnick, 2007). The use of several compounds such as trimethylamine-*N*-oxide (TMAO), sulfur, and other organic compounds under anaerobic conditions and their psychrotrophic habits make *Shewanella* one of the major spoilers of refrigerated food such as seafood products. The spoiler species of *Shewanella* can typically be described as H_2S producers and media containing thiosulfate (such as iron agar (IA) plates) are commonly applied for their first isolation and enumeration (Fonnesbech Vogel et al., 2005; Dehaut et al., 2014; Serio et al., 2014). On the other hand, the use of common media such as Plate Count Agar could lead to a lack of recovery of species or strains that prefer salt (e.g., some strains of *S. frigidimarina*; Broekaert et al., 2011).

2.5.2 Spoilage Potential of *Shewanella* spp.

From the first isolation on the surface of tainted butter in the 1931, the genus *Shewanella* was associated with food spoilage (Derby and Hammer, 1931). Despite their relevance as SSO in aquatic food products, some species are also involved in the deterioration of meat and poultry. The spoilage potential of *Shewanella* has been extensively reviewed in products stored in aerobic conditions, but also in low oxygen conditions such as MAP and vacuum packaging (Russell et al., 1995; Nychas et al., 2007; Vihavainen and Björkroth, 2010). In refrigerated food rich in amino acid and other nitrogen sources, such as fish or meat stored at low temperature, some members of the genus *Shewanella* can dominate the microbiota and produce several compounds according to their flexible metabolism. However, dominance in the spoilage population is not necessary in order to exert spoilage effects. The genome sequence of *Shewanella oneidensis* revealed that *Shewanella* is able to metabolize a wide range of different substrates as the sole carbon sources (Fredrickson et al., 2008). For this reason, these bacteria are among the most important microorganisms involved in the degradation of organic matter. The catabolic pathways described in *Shewanella oneidensis* revealed the ability to use substrates commonly present in food (e.g., fatty acids, amino acids, and several carbohydrates). Moreover, the study of functional properties through genomic approaches could lead to a new comprehension of the spoilage activities of *Shewanella* species. The in silico analysis of *S. putrefaciens* genomes has suggested an alternative functional pathway involved in the production of amines (Remenant et al., 2015). According to this comparative analysis, putrescine production seems related to the ability to transform L-ornithine derived from glutamate, while some strains also have the potential ability to convert putrescine to spermidine. Clearly, the production of foetid amines and diamines is strain-dependent and this approach could elucidate specific strains' food spoilage ability (Remenant et al., 2015). The decarboxylation of amino acids such as lysine and ornithine is a phenotypic marker for strains involved in the production of cadaverine and putrescine, while the use of histidine seemed to be less common, with low risk of histamine production (Serio et al., 2014). The role of *Shewanella* as an SSO of fish, especially marine fish, is related to the production of two spoilage markers involved in the classical alteration of odor: trimethylamine (TMA) and H_2S. The use of TMAO as a final electron acceptor during anaerobic metabolism would be consistent with fishy odor production due to its conversion to TMA (Fonnesbech Vogel et al., 2005), while the putrid odor caused by the H_2S formation is related to the use of sulfite and other molecules such as polysulfide and thiosulfate (Hau and Gralnick, 2007). In fresh meat of warm-blooded animals, especially in vacuum packaged meat at high pH values, the production of H_2S can induce the greening discoloration of muscle due to the formation of a pigment complex called sulfmyoglobin. This is probably due to the catabolism of sulfur-containing amino acids with the formation of sulfide (Garcia-Lopez et al., 1998; Nychas et al., 2007, 2008). Other additional enzymatic activities of *Shewanella* can induce spoilage: e.g., extracellular DNAse activity is involved in nucleotide degradation (Satomi, 2014; Serio et al., 2014). The breakdown of hypoxanthine (Hx) or the conversion of Inosine MonoPhosphate and inosine into Hx coincide with bitter flavors and the related decay of fish freshness (Haard, 2002; Serio et al., 2014). Finally, proteolytic activities could also favor deterioration and the spread of bacteria into the muscles due to the hydrolysis of collagen and other structural proteins (Gennari et al., 1999; Fonnesbech Vogel et al., 2005; Serio et al., 2014). On the basis of their metabolism, the spoilage potential of *Shewanella* is then related to the substantial production of chemical compounds with low odor thresholds. H_2S, TMA, Hx, and other volatile compounds such as alcohols and aldehydes contribute to the formation of the volatilome of food during its shelf life (Casaburi et al., 2015).

These malodorous molecules are common indicators of spoilage; however, not all the *Shewanella* species are able to produce this wide range of compounds. Ten years ago, the major species described during the deterioration of food, especially seafood, was *S. putrefaciens* (Fonnesbech Vogel et al., 2005). Fonnesbech Vogel and coworkers (2005) on the basis of phenotypic taxonomy, G-C content, DNA/DNA hybridization, and phylogenetic analysis of 16S rRNA highlighted that a dominant species among the H_2S producers is *Shewanella baltica*. Recent studies have confirmed that *S. baltica* plays a key role in the deterioration of finfish and freshwater fish, together with *S. putrefaciens* (Dehaut et al., 2014; Serio et al., 2014; Beaz-Hidalgo et al., 2015).

Several other species are H_2S producers such as *S. algae*, *S. colwelliana* (Fonnesbech Vogel et al., 2005) and other well-characterized species as reported in the works of Satomi and coworkers (Satomi et al., 2006, 2007), although only *S. putrefaciens* and *S. baltica* are widely isolated and used as SSOs, as documented by recent studies on H_2S-producing bacteria (see Table 2.3 for details).

Table 2.3 summarizes some recent works related to *Shewanella* contamination, reporting both studies cited above and some minor ones. Nearly all studies applied IA as a first isolation medium followed by different characterization methods. The use of H_2S producing bacteria as markers is also commonly applied in predictive microbiology particularly for finfish (Giuffrida et al., 2013; Carrascosa et al., 2015).

Recent findings have suggested a certain heterogeneity of the phenotypic traits of *S. baltica*. Some strains are unable to produce H_2S, and this could lead to misinterpretations when applying common media used for *Shewanella* enumeration (Dehaut et al., 2014). On the other hand, other genera with morphologies similar to *Shewanella* species, such as *Serratia*, are able to grow in IA (Serio et al., 2014). The same authors suggested the application of additional selective media for the enumeration of *Shewanella*. An increase in salt concentration could enhance the isolation and differentiation of halophilic species. At the same time, a shift from the media with thiosulfate to the media that highlight TMAO reduction or pigment synthesis (e.g., salmon pink color) could improve the detection of different *Shewanella* species (Dehaut et al., 2014; Serio et al., 2014). For instance, other additional substrates such as cysteine in substitution or in addition to thiosulfate could better elucidate the H_2S production pattern (Macé et al., 2014; Serio et al., 2014). Moreover, species identification requires the complementary use of classical phenotypic keys in combination with different genetic targets (Tryfinopoulou et al., 2007). As previously mentioned, some species, such as *S. algae*, have been also reported as putative human pathogens (Janda and Abbott, 2015). Furthermore, new insights suggested that *S. algae* was also involved in the production of tetrodotoxin during cases of food poisoning (Wang et al., 2013).

2.5.3 *Xanthomonas* spp.

Species belonging to the genus *Xanthomonas* are able to colonize a broad range of hosts (400 species of plants); however, these bacteria show a high level of specialization according to the host species and tissues (Ryan et al., 2011). Whole-genome analyses have revealed that this specialization is related to a reduction in genome size, where the species able to colonize

TABLE 2.3 *Shewanella* in Food Products

Food	Storage Condition	Media of Isolation/ Enumeration	Method of Characterization	TMAO/ H_2S	Species	References
Sea bream skin and flesh Gills	Not reported	Iron agar Lyngby	Phenotypic	+/+	H_2S producing bacteria/putative *S. putrefaciens*	Carrascosa et al. (2015)
Whole, ungutted sea bass (*Dicentrarchus labrax*)	Not reported	Iron agar Lyngby	Phenotypic	+/+	H_2S producing bacteria/putative *S. putrefaciens*	Carrascosa et al. (2015)
Spoiled salmon fillets	MAP	Iron agar with 0.04% of L-cysteine	16S rRNA	Not reported	*S. baltica*	Macé et al. (2014)
Spoiled cooked whole tropical shrimps	MAP	Iron agar	16S rRNA gene	Not reported	*S. baltica*	Macé et al. (2014)
Red tuna and swordfish steaks	Refrigerated in aerobiosis	Kligler iron agar (KIA)	16S rDNA, RAPD- and REP-PCR	+/+	*S. baltica*, *S. putrefaciens*, *S. morhuae*	Serio et al. (2014)
Atlantic horse mackerel (*Trachurus trachurus*)	MAP	Kligler iron agar Long and hammer plate	16S rRNA	Not reported	Putative *S. baltica*, *S. glacialipiscicola*, *Shewanella* spp.	Alfaro and Hernandez (2013)

(Continued)

TABLE 2.3 (Continued)

Food	Storage Condition	Media of Isolation/ Enumeration	Method of Characterization	TMAO/ H_2S	Species	References
King scallop (*Pecten maximus*)	Aerobic, vacuum packed MAP	Iron agar	16S rRNA	Not reported	*S. frigidimarina*, *S. baltica*, *S. putrefaciens*	Coton et al. (2013)
North American catfish fillets (*Ictalurus punctatus*)	Air	DNase test agar	16S rRNA	Not reported	*S. baltica*	Hickey et al. (2013)
Pacific white shrimp (*Litopenaeus vannamei*)	MAP	Plate count agar and triple sugar iron	Phenotypic	Not reported	*Shewanella* spp.	Qian et al. (2013)
Whole ungutted sea bream	On ice	Iron agar	Culture independent by 16S rRNA cloning procedure, culture-dependent biochemical identification	+/+	*S. putrefaciens*-like	Parlapani et al. (2013)
Restaurant knife, marinated beef, poisoned patients	Epidemiological survey after tetrodotoxin intoxication	WS salmonella agar, salmonella shigella agar, eosin methylene blue agar, thiosulfate citrate bile salts sucrose agar	16S rDNA	Not reported	*S. algae*, *Shewanella* spp.	Wang et al. (2013)

Broiler carcasses	Air-packaged	Not reported	PCR-DGGE	Not reported	*Shewanella* spp.	Zhang et al. (2013)
Surface of the salmon pieces	MAP	Nutrient agar long and hammer agar	16S rRNA gene	Not reported	*S. baltica*, *S. morhuae*, *Shewanella* spp.	Powell and Tamplin (2012)
Marine finfish	Not reported	Iron agar with 0.5% NaCl (L-cysteine and sodium thiosulfate)	REP-PCR and phenotypic, 16S rRNA, *gyrB* gene	Not reported	*S. frigidimarina*, *S. vesiculosa*, *S. baltica*, *S. glacialipiscicola*	Broekaert et al. (2011)
Mediterranean hake (*Merluccius merluccius*)	Not reported	Iron agar Lyngby	API 20NE	Not reported	Presumptive, *S. putrefaciens*	Baixas-Nogueras et al. (2009)
North- Atlantic cod	Air or MAP, superchilled temperatures, brined	Iron agar	16S rRNA clone analysis	Not reported	*Shewanella* spp., *S. benthica*	Reynisson et al. (2009)
Atlantic cod (*Gadus morhua*)	MAP	Iron agar Lyngby	DGGE	Not reported	*S. putrefaciens*	Hovda et al. (2007)
Sparus aurata	Air, MAP	Iron agar	SDS-PAGE, 16S rRNA, Phenotypic	+/+	*S. baltica* (phenotypic clusters: *S. baltica*, *S. oneidensis* and *S. putrefaciens*)	Tryfinopoulou et al. (2007)

Overview of some recent articles (2007–15) concerning the isolation and characterization of H_2S- producing bacteria (or putative *Shewanella* spp.) from food matrices.

seed surfaces (*X. oryzae* pv. *oryzae*) differ from species associated with xylem invasion (e.g., *X. fastidiosa*).

Xanthomonas can be a cause of postharvest diseases and was reported as one of the genera isolated from spoiled fruits and vegetables (Barth et al., 2009). As for other Gammaproteobacteria, the taxonomy of *Xanthomonas* has been progressively rebuilt during the last two decades. Particularly, *X. campestris* pathovars were split into new species taking into account their host range (Vorhölter et al., 2008). One of the main characteristics that identifies the genus is the production of yellow pigments (xanthomonadins), conferring the name to the genus: *xanthos* is yellow in ancient Greek (Rajagopal et al., 1997). Despite their diverse and adaptable pathogenicity, the genus shows heterogeneous phenotypic and physiological traits that lead to difficult phenotypic classification (Meyer and Bogdanove, 2009). The bacteria are motile due to the presence of a single polar flagellum, catalase positive, and some species are able to produce the exopolysaccharide xanthan that is largely applied in the food industry as emulsifier (Vorhölter et al., 2008; Palaniraj and Jayaraman, 2011). Several genotyping methods were proposed for a polyphasic approach classification of *Xanthomonas* together with genome sequencing (Meyer and Bogdanove, 2009). However, the study of full metabolic profiles using different carbon sources (e.g., the BIOLOG system) could be applied routinely in order to classify common species in large datasets (Stoyanova et al., 2014).

2.5.4 Spoilage Potential of *Xanthomonas* spp.

Liao and Wells (1987) provided experimental evidence of the pectolytic activity of some strains of *Xanthomonas campestris*. Some strains showed the capacity to liquefy pectate gels in in vitro tests; moreover, in fresh plant material the strains were able to macerate vegetable tissue, suggesting that *X. campestris* plays an active role in the postharvest rot of plant crops (e.g., potato tubers, carrot, cucumber, cauliflower). *Xanthomonas* species are also involved in fruit alterations that are directly associated with their pathogenic behavior such as cankers and necrotic lesions. *Xanthomonas arboricola* pv. *pruni* is the disease agent of the bacterial spot of stone fruit that was spread around world due to human migrations (Boudon et al., 2005). The lesions on fruits depend on early infections after petal fall, with different aspects ranging from large open-deep lesions into the pulp to the coalescent spotting on the fruit surface (Ritchie, 1995). In this case, as for other fruits, each specific pathovar or species was involved in the bacterial spot on both leaves and fruits (e.g., *Xanthomonas axonopodis* pv. *passiflorae* on the passion fruit); outbreaks can lead to economic loss due to direct damage to the fruit and plant (Munhoz et al., 2011).

2.6 CONCLUSION AND FUTURE PERSPECTIVES: BIOFILM FORMATION AND CONTROL OF SPOILAGE DEFECTS

Besides the taxonomic history of *Pseudomonas*, *Shewanella*, and *Xanthomonas*, that often groups the three genera altogether along with their spoilage activity, these taxa are also associated because of their biofilm formation and their psychrotrophic nature.

For these reasons, maintenance of the cold chain might not be the solution to all of the issues related to their growth and spoilage. It has been reported that a further reduction of storage temperature from 4–6°C to 2°C might be helpful to reduce the spoilage activity of psychrotrophs during longer and longer shelf life times (Kumarsan et al., 2007).

As already reported, the three genera are able to produce biofilms: for this reason, the identification of safe and uncontaminated raw materials and final products is not synonymous with nonspoiled food. Indeed, working surfaces, packaging, transport, and storage equipment might be contaminated by bacterial biofilms, which are very difficult to remove during sanitization. Particular attention must be paid to postprocess contaminations. Adequate sanitization is fundamental, taking into account the need to deal with complex structures as biofilms. Because of these structures, bacteria are able to resist a wide range of stresses, especially sanitizers applied for the decontamination of processing surfaces. Furthermore, biofilms might act as a reservoir of secreted enzymes, protecting them from the activity of detergents and sanitization methods and enhancing enzyme activity (Teh et al., 2014).

The management of packaging processes, taking into account the aerobic nature of *Pseudomonas* and *Xanthomonas*, can be helpful in controlling proliferation. MAP, with a low or null percentage of oxygen, should reduce their count; however, recently Stoops and colleagues (2012) revealed that a strain of *Pseudomonas* isolated from the microflora of tofu was not completely inhibited by packaging in 100% CO_2.

Interestingly, natural compounds have been applied to food products. For example, Quintieri and coworkers applied bovine lactoferrin B and its pepsin-digested hydrolysate to control *Pseudomonas* in governing liquid and on mozzarella cheese to reduce growth and pigment production, demonstrating the efficacy of the hydrolysate in the control of spoilage strains (Quintieri et al., 2012). Similarly, an innovative active packaging system was proposed combining lemon extract, brine and a gel solution made of sodium alginate, which demonstrated a significant enhancement of the shelf life of mozzarella cheese (Conte et al., 2007). A natural compound from the essential oil of *Satureja thymbra* is particularly active against *Pseudomonas* biofilms, being more effective than several acid-based disinfectants (Chorianopoulos et al., 2008). In a similar way, recently Zhu and colleagues reported the effective activity of green tea polyphenols against quorum sensing and biofilm formation in *Shewanella baltica*, substantially reducing its food spoilage potential (Zhu et al., 2015).

Recently, bacteriophages have emerged as new and safe options for the treatment of pathogens and spoilers in a wide range of foods. In 2010, Sillankorva and colleagues reported the high efficacy of a phage against *P. fluorescens* in mono and dual species biofilms.

The scenario described in this chapter is complex and reveals a large variety of mechanisms through which *Pseudomonas*, *Xanthomonas*, and *Shewanella* induce spoilage effects. It is evident that only a complete understanding of all of these systems can lead to the identification of targets for the effective control of all spoilage species, with the ultimate aim of reaching high microbiological quality standards in all food products.

ACKNOWLEDGMENTS

The authors are grateful to Professor Emmanuel Coton for his useful comments and advices.

REFERENCES

Alfaro, B., Hernandez, I., 2013. Evolution of the indigenous microbiota in modified atmosphere packaged Atlantic horse mackerel (*Trachurus trachurus*) identified by conventional and molecular methods. Int. J. Food Microbiol. 167, 117–123.

Altinok, I., Kayis, S., Capkin, E., 2006. *Pseudomonas putida* infection in rainbow trout. Aquaculture 261, 850–855.

Andreani, N.A., Martino, M.E., Fasolato, L., Carraro, L., Montemurro, F., Mioni, R., et al., 2014. Tracking the blue: a MLST approach to characterise the *Pseudomonas fluorescens* group. Food Microbiol. 39, 116–126.

Andreani, N.A., Carraro, L., Martino, M.E., Fondi, M., Fasolato, L., Miotto, G., et al., 2015. A genomic and transcriptomic approach to investigate the blue pigment phenotype in *Pseudomonas fluorescens*. Int. J. Food Microbiol. 213, 88–98.

Andrews-Polymenis, H.L., Santiviago, C.A., McClelland, M., 2009. Novel genetic tools for studying food borne *Salmonella*. Curr. Opin. Biotechnol. 20, 149–157.

Anzai, Y., Kim, H., Park, J.-Y., Wakabayashi, H., Oyazu, H., 2000. Phylogenetic affiliation of the pseudomonads based on 16S rRNA sequences. Int. J. Syst. Evol. Microbiol. 50, 1563–1589.

Arnaut-Rollier, I., Vauterin, L., De Vos, P., Massart, D.L., Devriese, L.A., De Zutter, L., et al., 1999. A numerical taxonomic study of the *Pseudomonas* flora isolated from poultry meat. J. Appl. Microbiol. 87, 15–28.

Arslan, S., Eyi, A., Özdemir, F., 2011. Spoilage potentials and antimicrobial resistance of *Pseudomonas* spp. isolated from cheeses. J. Dairy Sci. 94, 5851–5856.

Baader, A., Garre, C., 1887. Über Antagonisten unter den Bacterien. Corresp. Bl. Schweiz. Ärzte 13, 385–392.

Bai, A.J., Rai, V.R., 2011. Bacterial quorum sensing and food industry. Compr. Rev. Food Sci. Food Saf. 10, 183–193.

Baixas-Nogueras, S., Bover-Cid, S., Veciana-Nogués, T., Vidal-Carou, M.C.M., 2009. Effect of gutting on microbial loads, sensory properties, and volatile and biogenic amine contents of European hake (*Merluccius merluccius* var. *mediterraneus*) stored in ice. J. Food Prot. 72, 1671–1676.

Barth, M., Hankinson, T.R., Zhuang, H., Breidt, F., 2009. Microbiological spoilage of fruits and vegetables. In: Sperber, W.H., Doyle, M.P. (Eds.), Compendium of the Microbiological Spoilage of Foods and Beverages. Food Microbiology and Food Safety, Springer New York, New York, NY, pp. 135–183.

Beaz-Hidalgo, R., Agüeria, D., Latif-Eugenín, F., Yeannes, M.I., Figueras, M.J., 2015. Molecular characterization of *Shewanella* and *Aeromonas* isolates associated with spoilage of Common carp (*Cyprinus carpio*). FEMS Microbiol. Lett. 362, 1–8.

Bergey's Manual of Determinative Bacteriology, 1923. Am. J. Public Health (New York, NY: 1912) 13 (12), 1042.

Bhadra, S., Narvaez, C., Thomson, D.J., Bridges, G.E., 2015. Non-destructive detection of fish spoilage using a wireless basic volatile sensor. Talanta 134, 718–723.

Boudon, S., Manceau, C., Nottéghem, J.L., 2005. Structure and origin of *Xanthomonas arboricola* pv. *pruni* populations causing bacterial spot of stone fruit trees in Western Europe. Phytopathology 95, 1081–1088.

Brocklehurst, T.F., Lund, B.M., 1985. Microbiological changes in cottage cheese varieties during storage at +7°C. Food Microbiol. 2, 207–233.

Broekaert, K., Heyndrickx, M., Herman, L., Devlieghere, F., Vlaemynck, G., 2011. Seafood quality analysis: molecular identification of dominant microbiota after ice storage on several general growth media. Food Microbiol. 28, 1162–1169.

Caldera, L., Franzetti, L., 2014. Effect of storage temperature on the microbial composition of ready-to-use vegetables. Curr. Microbiol. 68, 133–139.

Cantoni, C., Iacumin, L., Comi, G., 2003. Alterazione giallo-arancio di mozzarella. Industrie Alimentari 42, 134–136.

Cantoni, C., Soncin, G., Milesi, S., Cocolin, L., Iacumin, L., Comi, G., 2006. Colorazioni anomale e rigonfiamento di formaggi fusi e Mozzarelle. Industrie Alimentari 45, 276–281.

Caputo, L., Quintieri, L., Bianchi, D.M., Decastelli, L., Monaci, L., Visconti, A., et al., 2015. Pepsin-digested bovine lactoferrin prevents mozzarella cheese blue discoloration caused by *Pseudomonas fluorescens*. Food Microbiol. 46, 15–24.

Carrascosa, C., Millán, R., Saavedra, P., Jaber, J.R., Raposo, A., Pérez, E., et al., 2015. Microbiological evolution of gilthead sea bream (*Sparus aurata*) in Canary Islands during ice storage. J. Food Sci. Technol. 52, 1586–1593.

Casaburi, A., Piombino, P., Nychas, G.J., Villani, F., Ercolini, D., 2015. Bacterial populations and the volatilome associated to meat spoilage. Food Microbiol. 45, 83–102.

Chorianopoulos, N.G., Giaouris, E.D., Skandamis, P.N., Haroutounian, S.A., Nychas, G.J., 2008. Disinfectant test against monoculture and mixed-culture biofilms composed of technological, spoilage and pathogenic bacteria: bactericidal effect of essential oil and hydrosol of *Satureja thymbra* and comparison with standard acid-base sanitizers. J. Appl. Microbiol. 104, 1586–1596.

Conte, A., Scrocco, C., Sinigaglia, M., Del Nobile, M.A., 2007. Innovative active packaging systems to prolong the shelf life of mozzarella cheese. J. Dairy Sci. 90, 2126–2131.

Cornelis, P., 2010. Iron uptake and metabolism in pseudomonads. Appl. Microbiol. Biotechnol. 86, 1637–1645.

Coton, M., Joffraud, J.J., Mekhtiche, L., Leroi, F., Coton, E., 2013. Biodiversity and dynamics of the bacterial community of packaged king scallop (*Pecten maximus*) meat during cold storage. Food Microbiol. 35, 99–107.

Dainty, R.H., Edwards, R.A., Hibbard, C.M., 1984. Volatile compounds associated with the aerobic growth of some *Pseudomonas* species on beef. J. Appl. Bacteriol. 57, 75–81.

Dainty, R.H., Edwards, R.A., Hibbard, C.M., Marnewick, J.J., 1989. Volatile compounds associated with microbial growth on normal and high pH beef stored at chill temperatures. J. Appl. Bacteriol. 66, 281–289.

Datta, N., Deeth, H.C., 2001. Age gelation of UHT milk. Transaction of the institute of chemical engineers C. Food Bioprod. Process. 79, 197–210.

Datta, N., Deeth, H.C., 2003. Diagnosing the cause of proteolysis in UHT milk. LWT-Food Sci. Technol. 36, 173–182.

Dehaut, A., Midelet-Bourdin, G., Brisabois, A., Duflos, G., 2014. Phenotypic and genotypic characterization of H_2 S-positive and H_2 S-negative strains of *Shewanella baltica* isolated from spoiled whiting (*Merlangius merlangus*). Lett. Appl. Microbiol. 59, 542–548.

De Jonghe, V., Coorevits, A., Van Hoorde, K., Messens, W., Van Landschoot, A., De Vos, P., et al., 2011. Influence of storage conditions on the growth of *Pseudomonas* species in refrigerated raw milk. Appl. Environ. Microbiol. 77, 460–470.

den Dooren de Jong, L.E., 1926. Bijdrage tot de kennis Van het Mineralisatieproces. Nijgh & van Ditmar Uitgevers-Mij, Rotterdam, The Netherlands.

Deng, J., Brettar, I., Luo, C., Auchtung, J., Konstantinidis, K.T., Rodrigues, J.L., et al., 2014. Stability, genotypic and phenotypic diversity of *Shewanella baltica* in the redox transition zone of the Baltic Sea. Environ. Microbiol. 16, 1854–1866.

De Oliveira, G.B., Favarin, L., Luchese, R.H., McIntosh, D., 2015. Psychrotrophic bacteria in milk: how much do we really know? Braz. J. Microbiol. 46, 313–321.

Derby, H.A., Hammer, H.A., 1931. Bacteriology of butter. IV. Bacteriological studies on surface taint butter. Iowa Agric. Exp. Stat. Res. Bull. 145, 389–416.

Diaz-Sanchez, S., Hanning, I., Pendleton, S., D'Souza, D., 2012. Next-generation sequencing: the future of molecular genetics in poultry production and food safety. Poult. Sci. 92, 562–572.

Dikow, R.B., 2011. Genome-level homology and phylogeny of *Shewanella* (Gammaproteobacteria: Alteromonadales: Shewanellaceae). BMC Genomics 12, 237.

Dogan, B., Boor, K.J., 2003. Genetic diversity and spoilage potentials among *Pseudomonas* spp. isolated from fluid milk products and dairy processing plants. Appl. Environ. Microbiol. 69, 130–138.

Doulgeraki, A.I., Nychas, G.J., 2013. Monitoring the succession of the biota grown on a selective medium for pseudomonads during storage of minced beef with molecular-based methods. Food Microbiol. 34, 62–69.

Dufour, D., Nicodème, M., Perrin, C., Driou, A., Brusseaux, E., Humbert, G., et al., 2008. Molecular typing of industrial strains of *Pseudomonas* spp. isolated from milk and genetical and biochemical characterization of an extracellular protease produced by one of them. Int. J. Food Microbiol. 125, 188–196.

Ercolini, D., Russo, F., Nasi, A., Ferranti, P., Villani, F., 2009. Mesophilic and psychrotrophic bacteria from meat and their spoilage potential in vitro and in beef. Appl. Environ. Microbiol. 75, 1990–2001.

Fonnesbech Vogel, B., Venkateswaran, K., Satomi, M., Gram, L., 2005. Identification of *Shewanella baltica* as the most important H_2S-producing species during iced storage of Danish marine fish. Appl. Environ. Microbiol. 71, 6689–6697.

Fonseca, P., Moreno, R., Rojo, F., 2011. Growth of *Pseudomonas putida* at low temperature: global transcriptomic and proteomic analyses. Environ. Microbiol. Rep. 3, 329–339.

Fox, F., 1989. Proteolysis during cheese manufacture and ripening. J. Dairy Sci. 72, 1379–1400.

Franzetti, L., Scarpellini, M., 2007. Characterisation of *Pseudomonas* spp. isolated from foods. Ann. Microbiol. 57, 39–47.

Frapolli, M., Défago, G., Moënne-Loccoz, Y., 2007. Multilocus sequence analysis of biocontrol fluorescent *Pseudomonas* spp. producing the antifungal compound 2,4-diacetylphloroglucinol. Environ. Microbiol. 9, 1939–1955.

Fredrickson, J.K., Romine, M.F., Beliaev, A.S., Auchtung, J.M., Driscoll, M.E., Gardner, T.S., et al., 2008. Towards environmental systems biology of *Shewanella*. Nat. Rev. Microbiol. 6, 592–603.

Garcia-Lopez, M.L., Prieto, M., Otero, A., 1998. The Physiological Attributes of Gram-Negative Bacteria Associated with Spoilage of Meat and Meat Products. Blackie Academic and Professional, London, UK.

Gassem, M.A., Frank, J.F., 1991. Physical properties of yogurt made from milk treated with proteolytic enzymes. J. Dairy Sci. 74, 1503–1511.

Gennari, M., Tomaselli, S., Cotrona, V., 1999. The microflora of fresh and spoiled sardines (*Sardina pilchardus*) caught in Adriatic (Mediterranean) sea and stored in ice. Food Microbiol. 16, 15–28.

Gilmour, M.W., Graham, M., Van Domselaar, G., Tyler, S., Kent, H., Trout-Yakel, K.M., et al., 2010. High-throughput genome sequencing of two *Listeria monocytogenes* clinical isolates during a large foodborne outbreak. BMC Genomics 11, 120.

Giuffrida, A., Valenti, D., Giarratana, F., Ziino, G., Panebianco, A., 2013. A new approach to modelling the shelf life of Gilthead seabream (*Sparus aurata*). Int. J. Food Sci. Technol. 48, 1235–1242.

Gram, L., 1992. Spoilage of three Senegalese fish species stored in ice at ambient temperature. In: Bligh, E.H. (Ed.), Seafood Science and Technology. Fishing News Books, Blackwell, Oxford, pp. 225–233.

Gram, L., Dalgaard, P., 2002. Fish spoilage bacteria - problems and solutions. Curr. Opin. Biotechnol. 13, 262–266.

Gram, L., Huss, H.H., 1996. Microbiological spoilage of fish and fish products. Int. J. Food Microbiol. 33, 121–137.

Gram, L., Wedell-Neergaard, C., Huss, H.H., 1990. The bacteriology of fresh and spoiling Lake Victorian Nile perch (*Lates niloticus*). Int. J. Food Microbiol. 10, 303–316.

Gram, L., Ravn, L., Rasch, M., Bruhn, J.B., Christensen, A.B., Givskov, M., 2002. Food spoilage--interactions between food spoilage bacteria. Int. J. Food Microbiol. 78, 79–97.

Grewal, M.K., Jaiswal, P., Jha, S.N., 2015. Detection of poultry meat specific bacteria using FTIR spectroscopy and chemometrics. J. Food Sci. Technol. 52, 3859–3869.

Haard, N., 2002. The role of enzymes in determining seafood color and texture. In: Bremner, A.H. (Ed.), Safety and quality issues in food processing, Woodhead Publishing Series in Food Science, Technology and Nutrition, Elsevier Science, pp. 220–253, CRC Press, Boca Raton, FL.

Hau, H.H., Gralnick, J.A., 2007. Ecology and biotechnology of the genus *Shewanella*. Annu. Rev. Microbiol. 61, 237–258.

Hickey, M.E., Besong, S.A., Kalavacharla, V., Lee, J.-L., 2013. Identification of extracellular DNase-producing bacterial populations on catfish fillets during refrigerated storage. Food Sci. Biotechnol. 22, 87–92.

Hilario, E., Buckley, T.R., Young, J.M., 2004. Improved resolution on the phylogenetic relationships among *Pseudomonas* by the combined analysis of *atpD*, *carA*, *recA*, and 16S rDNA. Antoine van Leeuwenhoek 86, 51–64.

Holt, K.E., Parkhill, J., Mazzoni, C.J., Roumagnac, P., Weill, F.X., Goodhead, I., et al., 2008. High-throughput sequencing provides insights into genome variation and evolution in *Salmonella* Typhi. Nat. Genet. 40, 987–993.

Hovda, M.B., Lunestad, B.T., Sivertsvik, M., Rosnes, J.T., 2007. Characterisation of the bacterial flora of modified atmosphere packaged farmed Atlantic cod (*Gadus morhua*) by PCR-DGGE of conserved 16S rRNA gene regions. Int. J. Food Microbiol. 117, 68–75.

Janda, J.M., 2014. *Shewanella*: a marine pathogen as an emerging cause of human disease. Clin. Microbiol. Newsl. 36, 25–29.

Janda, J.M., Abbott, S.L., 2015. The genus *Shewanella*: from the briny depths below to human pathogen (Review). Crit. Rev. Microbiol. 40, 293–312.

Jay, J.M., 2000. Modern Food Microbiology, sixth ed. Aspen Publishers, Inc., Gaithersburg, MD.

Jay, J.M., 2003. A review of recent taxonomic changes in seven genera of bacteria commonly found in foods. J. Food Prot. 66, 1304–1309.

Johnsen, K., Andersen, S., Jacobsen, C.S., 1996. Phenotypic and genotypic characterization of phenanthrene-degrading fluorescent *Pseudomonas* biovars. Appl. Environ. Microbiol. 62, 3818–3825.

Kisand, V., Lettieri, T., 2013. Genome sequencing of bacteria: sequencing, de novo assembly and rapid analysis using open source tools. BMC Genomics 14, 211.

Koka, R., Weimer, B.C., 2000. Isolation and characterization of a protease from *Pseudomonas fluorescens* RO98. J. Appl. Microbiol. 89, 280–288.

Kröckel, L., 2009. Black spots on adipose tissue of pork and beef caused by melanin producing *Pseudomonas fluorescens*. Fleischwirtschaft 89, 89–92.

Kumarsan, G., Annalvilli, R., Sivakumar, K., 2007. Psychrotrophic spoilage of raw milk at different temperatures of storage. J. Appl. Sci. Res. 3, 1383–1387.

Laguerre, G., Rigottier-Gois, L., Lemanceau, P., 1994. Fluorescent *Pseudomonas* species categorized by using polymerase chain reaction (PCR)/restriction fragment analysis of 16S rDNA. Mol. Ecol. 3, 479–487.

Lavilla Lerma, L., Benomar, N., Gálvez, A., Abriouel, H., 2012. Prevalence of bacteria resistant to antibiotics and/or biocides on meat processing plant surfaces throughout meat chain production. Int. J. Food Microbiol. 161, 97–106.

Lavilla Lerma, L., Benomar, N., Casado Muñoz Mdel, C., Gálvez, A., Abriouel, H., 2014. Antibiotic multiresistance analysis of mesophilic and psychrotrophic *Pseudomonas* spp. isolated from goat and lamb slaughterhouse surfaces throughout the meat production process. Appl. Environ. Microbiol. 80, 6792–6806.

Lavilla Lerma, L., Benomar, N., Casado Muñoz Mdel, C., Gálvez, A., Abriouel, H., 2015. Correlation between antibiotic and biocide resistance in mesophilic and psychrotrophic *Pseudomonas* spp. isolated from slaughterhouse surfaces throughout meat chain production. Food Microbiol. 51, 33–44.

Liao, C.-H., 2006. *Pseudomonas* and related genera. In: Food Spoilage Microorganisms, Cambridge, UK: CRC Press & Woodhead Publ. Ltd, 507–540.

Liao, C.H., Wells, J.M., 1987. Association of pectolytic strains of *Xanthomonas campestris* with soft rots of fruits and vegetables at retail markets. Phytopathology 77, 418–422.

Liu, M., Wang, H., Griffiths, M.W., 2007. Regulation of alkaline metalloprotease promoter by *N*-acyl homoserine lactone quorum sensing in *Pseudomonas fluorescens*. J. Appl. Microbiol. 103, 2174–2184.

Lo, R., Stanton-Cook, M.J., Beatson, S.A., Turner, M.S., Bansal, N., 2015. Draft genome sequence of *Pseudomonas fluorescens* SRM1, an isolate from spoiled raw milk. Genome Announc. 3, e00138−15.

Macé, S., Cardinal, M., Jaffrès, E., Cornet, J., Lalanne, V., Chevalier, F., et al., 2014. Evaluation of the spoilage potential of bacteria isolated from spoiled cooked whole tropical shrimp (*Penaeus vannamei*) stored under modified atmosphere packaging. Food Microbiol. 40, 9−17.

Marchand, S., Heylen, K., Messens, W., Coudijzer, K., De Vos, P., Dewettinck, K., et al., 2009a. Seasonal influence on heat-resistant proteolytic capacity of *Pseudomonas lundensis* and *Pseudomonas fragi*, predominant milk spoilers isolated from Belgian raw milk samples. Environ. Microbiol. 11, 467−482.

Marchand, S., Vandriesche, G., Coorevits, A., Coudijzer, K., De Jonghe, V., Dewettinck, K., et al., 2009b. Heterogeneity of heat-resistant proteases from milk *Pseudomonas* species. Int. J. Food Microbiol. 133, 68−77.

Martin, N.H., Murphy, S.C., Ralyea, R.D., Wiedmann, M., Boor, K.J., 2011. When cheese gets the blues: *Pseudomonas fluorescens* as the causative agent of cheese spoilage. J. Dairy Sci. 94, 3176−3183.

Mastromatteo, M., Incoronato, A.L., Conte, A., Del Nobile, M.A., 2011. Shelf life of reduced pork back-fat content sausages as affected by antimicrobial compounds and modified atmosphere packaging. Int. J. Food Microbiol. 150, 1−7.

McCarthy, C.N., Woods, R.G., Beacham, I.R., 2004. Regulation of the *aprX-lipA* operon of *Pseudomonas fluorescens* B52: differential regulation of the proximal and distal genes, encoding protease and lipase, by *ompR-envZ*. FEMS Microbiol. Lett. 241, 243−248.

McPhee, J.D., Griffiths, M.W., 2002. *Pseudomonas* spp. In: Roginski, H., Fuquay, W.J., Fox, F.P. (Eds.), Encyclopaedia of Dairy Sciences, vol. 4. Academic Press, New York, NY, USA. pp. 2340−2350.

Mellor, G.E., Bentley, J.A., Dykes, G.A., 2011. Evidence for a role of biosurfactants produced by *Pseudomonas fluorescens* in the spoilage of fresh aerobically stored chicken meat. Food Microbiol. 28, 1101−1104.

Meyer, D.F., Bogdanove, A.J., 2009. Genomics-driven advance in *Xanthomonas* biology. In: Jackson, R.W. (Ed.), Plant Pathogenic Bacteria: Genomics and Molecular Biology (Book). Caister Academic Press/School of Biological Sciences, University of Reading, Whiteknights, Reading, UK.

Meyer, J.M., 2000. Pyoverdines: pigments, siderophores and potential taxonomic marker of fluorescent *Pseudomonas* species. Arch. Microbiol. 174, 135−142.

Meyer, J.M., Geoffroy, V.A., Baida, N., Gardan, L., Izard, D., Lemanceau, P., et al., 2002. Siderophore typing, a powerful tool for the identification of fluorescent e non fluorescent pseudomonads. Appl. Environ. Microbiol. 68, 2745−2753.

Migula, W., 1894. Über ein neues system der Bakterien. Arb. Bakteriol. Inst. Karlsruhe 1, 235−328.

Migula, W., 1900. System der Bakterien, vol. 2. Gustav Fischer, Jena, Germany.

Miller, A., Scanlan, R.A., Lee, J.S., Libbey, L.M., 1973. Volatile compounds produced in sterile fish muscle (*Sebastes melanops*) by *Pseudomonas putrefaciens*, *Pseudomonas fluorescens*, and an *Achromobacter* species. Appl. Microbiol. 26, 18−21.

Mitchell, S.L., Marshall, R.T., 1989. Properties of heat-stable proteases of *Pseudomonas fluorescens*: characterization and hydrolysis of milk proteins. J. Dairy Sci. 72, 864−874.

Mohareb, F., Iriondo, M., Doulgeraki, A.I., Van Hoek, A., Aarts, H., Cauchi, M., et al., 2015. Identification of meat spoilage gene biomarkers in *Pseudomonas putida* using gene profiling. Food Control 57, 152–160.

Moore, E.R.B., Mau, M., Arnscheidt, A., Böttger, E.C., Hutson, R.A., Collins, M.D., et al., 1996. The determination and comparison of the 16S rRNA gene sequences of species of the genus *Pseudomonas* (*sensu stricto*) and estimation of the natural intrageneric relationships. Syst. Appl. Microbiol. 19, 478–492.

Moreno, R., Rojo, F., 2014. Features of pseudomonads growing at low temperatures: another facet of their versatility. Environ. Microbiol. Rep. 6, 417–426.

Motoyama, M., Kobayashi, M., Sasaki, K., Nomura, M., Mitsumoto, M., 2010. *Pseudomonas* spp. convert metmyoglobin into deoxymyoglobin. Meat Sci. 84, 202–207.

Mulet, M., Lalucat, J., Garcìa-Valdes, E., 2010. DNA sequence-based analysis of the *Pseudomonas* species. Environ. Microbiol. 12, 1513–1530.

Munhoz, C.F., Weiss, B., Hanai, L.R., Zucchi, M.I., Fungaro, M.H., Oliveira, A.L., et al., 2011. Genetic diversity and a PCR-based method for *Xanthomonas axonopodis* detection in passion fruit. Phytopathology 101, 416–424.

Munsch-Alatossava, P., Alatossava, T., 2005. Phenotypic characterization of raw milk-associated psychrotrophic bacteria. Microbiol. Res. 161, 334–346.

Nogarol, C., Acutis, P.L., Bianchi, D.M., Maurella, C., Peletto, S., Gallina, S., et al., 2013. Molecular characterization of *Pseudomonas fluorescens* isolates involved in the Italian "blue mozzarella" event. J. Food Prot. 76, 500–504.

Nychas, G.J.E., Marshall, D., Sofos, J., 2007. Food microbiology fundamentals and frontiers. In: Doyle, M.P., Beuchat, L.R., Montville, T.J. (Eds.), Meat Poultry and Seafood, third ed. ASM Press, Washington, DC, pp. 105–140.

Nychas, G.J., Skandamis, P.N., Tassou, C.C., Koutsoumanis, K.P., 2008. Meat spoilage during distribution. Meat Sci. 78, 77–89.

Nychas, G.J., Dourou, D., Skandamis, P., Koutsoumanis, K., Baranyi, J., Sofos, J., 2009. Effect of microbial cell-free meat extract on the growth of spoilage bacteria. J. Appl. Microbiol. 107, 1819–1829.

Oakley, B.B., Morales, C.A., Line, J., Berrang, M.E., Meinersmann, R.J., Tillman, G.E., et al., 2013. The poultry-associated microbiome: network analysis and farm-to-fork characterizations. PLoS ONE 8, e57190.

Palaniraj, A., Jayaraman, V., 2011. Production, recovery and applications of xanthan gum by *Xanthomonas campestris*. J. Food Eng. 106, 1–12.

Palleroni, N.J., 1984. Genus *Pseudomonas*. In: Krieg, N.R., Holt, J.G. (Eds.), Bergey's Manual of Systematic Bacteriology, vol. 1. Williams & Wilkins, Baltimore, MD, pp. 141–199.

Palleroni, N.J., 2010. The *Pseudomonas* story. Environ. Microbiol. 12, 1377–1383.

Palleroni, N.J., Ballard, R.W., Ralston, E., Doudoroff, M., 1972. Deoxyribonucleic acid homologies among some *Pseudomonas* species. J. Bacteriol. 110, 1–11.

Parker, L.L., Levin, R.E., 1983. Relative incidence of *Alteromonas putrefaciens* and *Pseudomonas putrefaciens* in ground beef. Appl. Environ. Microbiol. 45, 796–799.

Parlapani, F.F., Meziti, A., Kormas, K.A., Boziaris, I.S., 2013. Indigenous and spoilage microbiota of farmed sea bream stored in ice identified by phenotypic and 16S rRNA gene analysis. Food Microbiol. 33, 85–89.

Parlapani, F.F., Haroutounian, S.A., Nychas, G.J., Boziaris, I.S., 2015. Microbiological spoilage and volatiles production of gutted European sea bass stored under air and commercial modified atmosphere package at 2°C. Food Microbiol. 50, 44–53.

Pecknold, P.C., Grogan, R.G., 1973. Deoxy ribonucleic acid homology groups among phytopathogenic *Pseudomonas* species. Int. J. Syst. Bacteriol. 23, 111–121.

Peix, A., Ramírez-Bahena, M., Velázquez, E., 2009. Historical evolution and current status of the taxonomy of genus *Pseudomonas*. Sci. Direct 9, 1132–1147.

Peladan, F., Monteil, H., 1988. Identification of *Pseudomonas*, *Flavobacterium*, and *Alcaligenes* with the API 20 NE system. Pathol. Biol. (Paris) 36, 187–192.

Pinto, U.M., Costa, E.D., Mantovani, H.C., Vanetti, M.C.D., 2010. The proteolytic activity of *Pseudomonas fluorescens* 07A isolated from milk is not regulated by quorum sensing signals. Braz. J. Microbiol. 41, 91–96.

Powell, S.M., Tamplin, M.L., 2012. Microbial communities on Australian modified atmosphere packaged Atlantic salmon. Food Microbiol. 30, 226–232.

Qian, Y.F., Yang, S.P., Xie, J., Xiong, Q., Gao, Z.L., 2013. Impact of the O_2 concentrations on bacterial communities and quality of modified atmosphere packaged Pacific white shrimp (*Litopenaeus vannamei*). J. Food Sci. 78, M1878–M1884.

Quintieri, L., Caputo, L., Monaci, L., Deserio, D., Morea, M., Baruzzi, F., 2012. Antimicrobial efficacy of pepsin-digested bovine lactoferrin on spoilage bacteria contaminating traditional mozzarella cheese. Food Microbiol. 31, 64–71.

Raab, V., Bruckner, S., Beierle, E., Kampmann, Y., Petersen, B., Kreyenschmidt, J., 2008. Generic model for the prediction of remaining shelf life in support of cold chain management in pork and poultry supply chains. J. Chain Network Sci. 8, 59–73.

Rajagopal, L., Sundari, C.S., Balasubramanian, D., Sonti, R.V., 1997. The bacterial pigment xanthomonadin offers protection against photodamage. FEBS Lett. 415, 125–128.

Rajmohan, S., Dodd, C.E.R., Waites, W.M., 2002. Enzymes from isolates of *Pseudomonas fluorescens* involved in food spoilage. J. Appl. Microbiol. 93, 205–213.

Rapid Alert System for Food and Feed (RASFF). Annual Report, 2010, European Communities, 2011, Luxembourg: Office for Official Publications of the European Communities.

Remenant, B., Jaffres, E., Dousset, X., Pilet, M.F., Zagorec, M., 2015. Bacterial spoilers of food: behavior, fitness and functional properties. Food Microbiol. 45, 45–53.

Reynisson, E., Lauzon, H.L., Magnússon, H., Jónsdóttir, R., Olafsdóttir, G., Marteinsson, V., et al., 2009. Bacterial composition and succession during storage of North-Atlantic cod (*Gadus morhua*) at superchilled temperatures. BMC Microbiol. 9, 250.

Ritchie, D.F., 1995. Bacterial spot. In: Ogawa, J.M., Zehr, E.I., Bird, G.W., RitchieDF, Uriu, K., Uyemoto, J.K. (Eds.), Compendium of Stone Fruit Diseases. American Phytopathological Society, St Paul, MN, pp. 50–52.

Russell, S.M., Fletcher, D.L., Cox, N.A., 1995. Spoilage bacteria of fresh broiler chicken carcasses. Poult. Sci. 74, 2041–2047.

Ryan, R.P., Vorhölter, F.J., Potnis, N., Jones, J.B., Van Sluys, M.A., Bogdanove, A.J., et al., 2011. Pathogenomics of *Xanthomonas*: understanding bacterium-plant interactions. Nat. Rev. Microbiol. 9, 344–355.

Samaržija, D., Zamberlin, Š., Pogačić, T., 2012. Psychrotrophic bacteria and milk and dairy products quality. Mljekarstvo 62, 77–95.

Sarale, A., Grassi, M.A., Civera, T., 2011. *Pseudomonas gessardii* causative agent for blue spot formation on rabbit carcasses. A.I.V.I online, June 2011 1(1), 261–262.

Satomi, M., 2014. The family *Shewanellaceae*. In: Rosemberg et al. (Eds.), The prokaryotes - Gammaproteobacteria, Springer-Verlag Berlin Heidelberg, pp. 597–619.

Satomi, M., Oikawa, H., Yano, Y., 2003. *Shewanella marinintestina* sp. nov., *Shewanella schlegeliana* sp. nov. and *Shewanella sairae* sp. nov., novel eicosapentaenoic-acid-producing marine bacteria isolated from sea-animal intestines. Int. J. Syst. Evol. Microbiol. 53, 491–499.

Satomi, M., Vogel, B.F., Gram, L., Venkateswaran, K., 2006. *Shewanella hafniensis* sp. nov. and *Shewanella morhuae* sp. nov., isolated from marine fish of the Baltic Sea. Int. J. Syst. Evol. Microbiol. 56, 243–249.

Satomi, M., Vogel, B.F., Venkateswaran, K., Gram, L., 2007. Description of *Shewanella glacialipiscicola* sp. nov. and *Shewanella algidipiscicola* sp. nov., isolated from marine fish of the Danish Baltic Sea, and proposal that *Shewanella affinis* is a later heterotypic synonym of *Shewanella colwelliana*. Int. J. Syst. Evol. Microbiol. 57, 347–352.

Scales, B.S., Dickson, R.P., LiPuma, J.J., Huffnagle, G.B., 2014. Microbiology, genomics, and clinical significance of the *Pseudomonas fluorescens* species complex, an unappreciated colonizer of humans. Clin. Microbiol. Rev. 27, 927–948.

Serio, A., Fusella, G.H., Chaves López, C., Sacchetti, G., Paparella, A., 2014. A survey on bacteria isolated as hydrogen sulfide-producers from marine fish. Food Control 39, 111–118.

Silby, M.W., Winstanley, C., Godfrey, S.A.C., Levy, S.B., Jackson, R.W., 2011. *Pseudomonas* genomes: diverse and adaptable. FEMS Microbiol. Rev. 35, 652–680.

Sillankorva, S., Neubauer, P., Azeredo, J., 2010. Phage control of dual species biofilms of *Pseudomonas fluorescens* and *Staphylococcus lentus*. Biofouling 26, 567–575.

Sofos, J.N., 1994. Microbial growth and its control in meat, poultry and fish. Chapter in Quality Attributes and their Measurement in Meat, Poultry and Fish Products, Volume 9 of the series Advances in Meat Research, Springer US, pp. 359–403.

Solieri, L., Dakal, T.C., Giudici, P., 2012. Next-generation sequencing and its potential impact on food microbial genomics. Ann. Microbiol. 63, 21–37.

Sørhaug, T., Stepaniak, L., 1997. Psychrotrophs and their enzymes in milk and dairy products: quality aspects. Trends Food Sci. Technol. 8, 35–41.

Spiers, A.J., Buckling, A., Rainey, P.B., 2000. The causes of *Pseudomonas* diversity. Microbiology 146, 2345–2350.

Stanier, R.Y., Palleroni, N.J., Doudoroff, M., 1966. The aerobic pseudomonads: a taxonomic study. J. Gen. Microbiol. 43, 159–271.

Stoops, J., Maes, P., Claes, J., Van Campenhout, L., 2012. Growth of *Pseudomonas fluorescens* in modified atmosphere packaged tofu. Lett. Appl. Microbiol. 54, 195–202.

Stoyanova, M., Vancheva, T., Moncheva, P., Bogatzevska, N., 2014. Differentiation of *Xanthomonas* spp. Causing bacterial spot in bulgaria based on biolog system. Int. J. Microbiol. 2014, 495476.

Sun, Y.Y., Sun, L., 2015. *Pseudomonas fluorescens*: iron-responsive proteins and their involvement in host infection. Vet. Microbiol. 176, 309–320.

Teh, K.H., Flint, S., Palmer, J., Andrewes, P., Bremer, P., Lindsay, D., 2012. Proteolysis produced within biofilms of bacterial isolates from raw milk tankers. Int. J. Food Microbiol. 157, 28–34.

Teh, K.H., Flint, S., Palmer, J., Andrewes, P., Bremer, P., Lindsay, D., 2014. Biofilm - an unrecognised source of spoilage enzymes in dairy products? Int. Dairy J. 34, 32–40.

Tryfinopoulou, P., Drosinos, E.H., Nychas, G.J., 2001. Performance of *Pseudomonas* CFC-selective medium in the fish storage ecosystems. J. Microbiol. Methods 47, 243–247.

Tryfinopoulou, P., Tsakalidou, E., Vancanneyt, M., Hoste, B., Swings, J., Nychas, G.-J.E., 2007. Diversity of *Shewanella* population in fish *Sparus aurata* harvested in the Aegean Sea. J. Appl. Microbiol. 103, 711–721.

Tümmler, B., Wiehlmann, L., Klockgether, J., Cramer, N., 2014. Advances in understanding *Pseudomonas*. F1000Prime Rep. 6, 9.

Vihavainen, E.J., Björkroth, J., 2010. Microbial ecology and spoilage of poultry meat and poultry meat products. In: Guerrero-Legarreta (Ed.), Handbook of Poultry Science and Technology, Secondary Processing, vol. 2. Blackwell-Wiley, New York, NY, pp. 485–493.

Vorhölter, F.J., Schneiker, S., Goesmann, A., Krause, L., Bekel, T., Kaiser, O., et al., 2008. The genome of *Xanthomonas campestris* pv. *campestris* B100 and its use for the reconstruction of metabolic pathways involved in xanthan biosynthesis. J. Biotechnol. 134, 33–45.

Wang, D., Wang, Y., Huang, H., Lin, J., Xiao, D., Kan, B., 2013. Identification of tetrodotoxin-producing *Shewanella* spp. from feces of food poisoning patients and food samples. Gut Pathog. 5, 15.

Wasi, S., Tabrez, S., Ahmad, M., 2013. Use of *Pseudomonas* spp. for the bioremediation of environmental pollutants: a review. Environ. Monit. Assess. 185, 8147–8155.

Woods, R.G., Burger, M., Beven, C.A., Beacham, I.R., 2001. The *aprX-lipA* operon of *Pseudomonas fluorescens* B52: a molecular analysis of metalloprotease and lipase production. Microbiology 147, 345–354.

Wu, L., Wang, J., Tang, P., Chen, H., Gao, H., 2011. Genetic and molecular characterization of flagellar assembly in *Shewanella oneidensis*. PLoS ONE 6, e21479.

Yamamoto, S., Kasai, H., Arnold, D.L., Jackson, R.W., Vivian, A., Harayama, S., 2000. Phylogeny of the genus *Pseudomonas*: intragenic structure reconstructed from the nucleotide sequences of *gyrB* and *rpoD* genes. Microbiology 146, 2385–2394.

Zhang, H., Fu, H., Wang, J., Sun, L., Jiang, Y., Zhang, L., et al., 2013. Impacts of nitrate and nitrite on physiology of *Shewanella oneidensis*. PLoS ONE 8, e62629.

Zhang, S.R., Zhang, L., Sun, L., 2014. Identification and analysis of three virulence-associated TonB-dependent outer membrane receptors of *Pseudomonas fluorescens*. Dis. Aquat. Organ. 110, 181–191.

Zhu, J., Huang, X., Zhanmg, F., Feng, L., Li, J., 2015. Inhibition of quorum sensing, biofilm, and spoilage potential in *Shewanella baltica* by green tea polyphenols. J. Microbiol. 53, 829–836.

Chapter 3

Specific Spoilage Organisms (SSOs) in Fish

Ioannis S. Boziaris and Foteini F. Parlapani
School of Agricultural Sciences, University of Thessaly, Volos, Greece

3.1 INTRODUCTION

Fish and seafood are very perishable products. Their quality deteriorates very fast due to three mechanisms; microbial activity, chemical oxidations, and enzymatic reactions (Ashie et al., 1996; Gram and Huss, 1996). Straight after death of fish autolysis occurs. Autolysis is the result of the action of various endogenous enzymes. Initially, enzymatic degradation of adenosine-triphosphate and other related compounds takes place which causes loss of characteristic fresh odor and taste and subsequently the action of various proteolytic enzymes softens fish flesh (Ashie et al., 1996). Enzymes are also responsible for color changes. Browning of crustacean carapace is due to the action of polyphenol-oxidase on tyrosine and its derivatives such as tyramine (Martinez-Alvarez et al., 2007). Oxidation of lipids is the main spoilage mechanism of chemical nature. Oxygen limitation or exclusion retards or eliminates respectively the oxidative rancidity of fish lipids (Ashie et al., 1996). Contamination of fish flesh with microorganisms also occurs straight after death. The high content in nonprotein low-molecular weight nitrogenous compounds Nonprotein nitrogen (NPN) and the low acidity ($pH > 6$) of fish flesh favor the growth of spoilage bacteria. As the bacteria grow they produce metabolic by-products and their accumulation causes the organoleptic rejection (Gram and Huss, 1996). All these mechanisms advance almost simultaneously. However, microbial spoilage is the mechanism that affects markedly the quality deterioration of fresh and lightly preserved fish, while in processed seafood products, wherein microbial growth is retarded or inhibited, the nonmicrobial mechanisms play a more determinative role.

The initial microbiota of fish/seafood (microbiota at the beginning of product shelf life) consists of the indigenous and exogenous microbiota.

The Microbiological Quality of Food. DOI: http://dx.doi.org/10.1016/B978-0-08-100502-6.00006-6

Indigenous microbiota refers to the natural bacterial populations that can be found in skin, gills, and digestive tract and are related to the waters in which fish live (Shewan, 1962; Liston, 1980). The exogenous microbiota originates from the contamination of the product with microorganisms from terrestrial environment, food contact surfaces, workers etc., during product handling. Only a part of the initial microbiota prevails under the applied storage (temperature, atmosphere) and processing conditions (salting, acidification, thermal treatments, smoking, antimicrobials, etc.). This grows faster than the rest of microorganisms, and becomes the dominant spoilage microbiota (Gram and Huss, 1996). The fraction of the dominant microbiota that poses spoilage potential (qualitative ability to produce off-odors) and spoilage activity (quantitative ability to produce metabolites) are known as specific spoilage organisms (SSOs) (Gram and Dalgaard, 2002).

3.2 FISH MICROBIOTA

3.2.1 Initial Microbiota

Alive and healthy fish do not contain microorganisms in their flesh. Microorganisms can be found only in skin, gills, and digestive tract (Shewan, 1962; Liston, 1980). Only after fish death do the microorganisms colonize the flesh. The initial microbiota of fish flesh is characterized by its low abundance and high diversity. Numerous works have concluded that the total viable counts of fish flesh is usually in the range of 3–4 log cfu/g, while its diversity is quite high. The initial microbiota consists of various genera. *Pseudomonas*, *Shewanella*, *Psychrobacter*, *Pseudoalteromonas*, *Moraxella*, *Acinetobacter*, *Flavobacterium*, and *Vibrio*, *Photobacterium*, and *Aeromonas* are the most common Gram-negative, while lactic acid bacteria (LAB), *Micrococcus*, *Corynebacterium*, *Vagococcus*, *Bacillus*, and *Clostridium* are the most common Gram-positive genera of the indigenous microbiota (Gram et al., 1990; Gram and Huss, 1996; Gennari et al., 1999; Gram, 2009; Svanevik and Lunestad, 2011; Parlapani et al., 2015b). Enterobacteriaceae, staphylococci, *Listeria*, and other microorganisms can also be found in the initial microbial population mainly as contamination microbiota (Huss et al., 2000).

The composition of initial microbiota is markedly affected by geographic origination (temperate or tropical) and the waters (marine or fresh waters) that those aquatic organisms live in (Liston, 1980; Ashie et al., 1996; Gram and Huss, 1996; Gram, 2009). Fish from the cold temperate waters of North Atlantic or North Sea contain mainly psychrophilic and psychrotrophic Gram-negative microorganisms such as *Photobacterium*, *Shewanella*, *Psychrobacter, Pseudomonas*, while fish from the warm waters of temperate zones, such as the Mediterranean Sea, contain more or less the same genera but not psychrophilic microorganisms such as *Photobacterium*.

Photobacterium has not been detected in fish flesh from Greek waters using either classical or molecular techniques (Koutsoumanis and Nychas, 1999; Parlapani et al., 2013). Fish from tropical waters have more or less the same genera but contain a greater percentage of Gram-positives compared to fish from temperate zones (Gram, 2009). Fish from fresh waters contain mainly *Pseudomonas*, *Aeromonas*, and various Gram-positives (Gram et al., 1990; Gram, 2009).

3.2.2 Spoilage Microbiota

Although fish microbiota initially consists of a wide range of bacterial genera, the imposed processing and storage conditions, as well as other implicit factors, cause the selection of few microorganisms that while they grow they outcompete the others and finally dominate on the product (spoilage microbiota). Hence, the composition of spoilage microbiota is dependent on the diversity of the initial microbiota and the extrinsic, intrinsic, and implicit factors that affect growth of microorganisms in foods. Thus, spoilage microbiota is connected to the fish origination, processing conditions, storage temperature, and atmosphere as well as to any microbial interactions. In chilled fish stored under aerobic storage, the spoilage microbiota contains genera that are able to grow fast aerobically at low temperatures, such as *Pseudomonas* and *Shewanella*, while at reduced oxygen and elevated carbon dioxide (modified atmosphere packaging-MAP), *Photobacterium*, LAB, and *Brochothrix thermosphacta* usually dominate (Table 3.1). Processes such as salting, acidification, and smoking alter the intrinsic properties of seafood, such as pH and a_w, impose various hurdles on microbial growth (Leistner and Gorris, 1995), and inevitably affect the selection of spoilage microbiota. Processed seafood usually is dominated by LAB and yeasts and other acid-tolerant or salt-tolerant microorganisms according to the product type (Gram, 2009; Boziaris et al., 2013).

3.3 THE SSO CONCEPT

Among the microorganisms that dominate the spoilage microbiota, only those with the ability to produce metabolites (spoilage potential) in adequate quantities (spoilage activity) to cause organoleptic rejection can be considered as the main cause of spoilage, and those microorganisms are the so-called SSOs (Gram and Huss, 1996; Gram and Dalgaard, 2002). The SSOs might belong only to one microbial genus or species in contrast to the spoilage microbiota which might be more than one microbial group, genus or species (Dalgaard, 2003).

The SSOs are a small fraction of the initial microbiota which grows fast under the particular storage conditions, reaches high numbers, becomes the large fraction of the total microbial population, and produces metabolites

TABLE 3.1 SSO/Dominant Microorganisms in Various Fish and Seafood Stored Under Various Conditions and the Methodology of Their Determination

SSO/Dominants	Fish/Seafood	Origination	Storage Conditions	Method of Determination	References
P. phosphoreum	Cod fillets (*G. morhua*)	North Sea (Denmark)	0°C/ MAP (various CO_2 levels), vacuum	Plates/phenotypic analysis	Dalgaard et al. (1993)
P. phosphoreum	Atlantic salmon (*Salmon salar*)	North Sea (Norway)	2°C/MAP: CO_2/N_2 (60/40)	Plates/classical	Emborg et al. (2002)
P. phosphoreum	Coalfish fillets	North Sea	1°C & 5°C/MAP: CO_2/N_2 (60/40)	Plates & tissue/T-RFLP, cloning & sequencing	Rudi et al. (2004)
C. divergens, C. piscicola, B. thermosphacta	Salmon fillets	North Sea	1°C & 5°C/MAP: CO_2/N_2 (60/40)	Plates & tissue/T-RFLP, cloning & sequencing	Rudi et al. (2004)
H2S-producing bacteria	Atlantic salmon (*S. salar*)	North Sea (Iceland)	Ice/AIR	Plates/classical	Sveinsdottir et al. (2002)
Lactobacillus spp., *Photobacterium* spp.	Cold-smoked salmon	North Sea (Sweden)	7°C/vacuum	Plates & tissue/16S rRNA gene analysis (cloning & sequencing)	Olofsson et al. (2007)
B. thermosphacta P. phosphoreum	Halibut (*Hippoglossus hippoglossus*)	North Sea (Norway)	4°C/MAP: CO_2/N_2 & CO_2/O_2 (50/50)	Plates & tissue/DGGE	Hovda et al. (2007a)
Pseudomonas spp., *P. phosphoreum*	Atlantic cod (*G. morhua*)	North Sea (Norway)	0°C/MAP: CO_2/N_2 & CO_2/O_2 (50/50)	Plates & tissue/DGGE	Hovda et al. (2007b)

P. phosphoreum	Norway lobster (*N. norvegicus*)	Scotland, UK	0°C, 4°C, 8°C, 10°C, 12°C, 16°C/AIR	Plates/16S rRNA gene analysis	Gornik et al. (2011)
Yersinia intermedia, *Y. kristensenii*, *C. maltaromaticum*, *S. baltica*	Atlantic horse mackerel (*Trachurus trachurus*)	North Atlantic	6°C/MAP: $CO_2/N_2/O_2$ (48/50/2)	Plates/16S rRNA gene analysis	Alfaro and Hernandez (2013)
Psychrobacter *Pseudoalteromonas*	Brown shrimp (*Crangon crangon*)	North Sea	Ice & 7.5°C/AIR	Plates & tissue/DGGE	Broekaert et al. (2013)
Psychrobacter spp., *Pseudomonas* spp.	Norway lobster (*N. norvegicus*)	North Sea	Ice/AIR	Plates (bulk cells) & tissue/DGGE	Bekaert et al. (2015)
Proteolytic strains	Turbot (*Psetta maxima*)	North Atlantic Ocean (Spain)	Ice/AIR	Plates/classical	Rodriguez et al. (2003)
Pseudomonas sp., *Aeromonas* sp., *S. putrefaciens*	Sea bass (*D. labrax*)	Atlantic Ocean (Canary Islands, Spain)	Ice/AIR	Plates/phenotypic analysis	Carrascosa et al. (2014)
P. phosphoreum, *Lactococcus piscium*	Salmon (*S. salar*) steak	France	3 days at 2°C and transferred to 8°C for 7 days/vacuum, MAP: CO_2/N_2 (50/50)	Plates & tissue/TTGE	Macé et al. (2012)
P. fragi, *S. putrefaciens*	Sardines (*Sardina pilchardus*)	Italy	Ice/AIR	Plates/phenotypic analysis	Gennari et al. (1999)

(*Continued*)

TABLE 3.1 (Continued)

SSO/Dominants	Fish/Seafood	Origination	Storage Conditions	Method of Determination	References
S. putrefaciens, Pseudoalteromonas sp., *Pseudomonas* sp.	Squid (*Todaropsis eblanae*)	Spain	Ice/AIR	Plates/phenotypic analysis, G + C mol% and 16S rRNA gene analysis	Paarup et al. (2002)
Pseudomonas spp.	Mediterranean Boque (*Boops boops*)	Greece	0°C, 3°C, 7°C, 10°C/AIR	Plates/classical	Koutsoumanis and Nychas (1999)
Pseudomonas spp.	Gilt-head sea bream (*S. aurata*)	Greece	0°C, 5°C, 10°C, 15°C/AIR	Plates/classical	Koutsoumanis and Nychas (2000)
Pseudomonas spp.	Sea bass (*D. labrax*)	Greece	0°C, 5°C, 10°C, 15°C/AIR	Plates/classical	Koutsoumanis et al. (2002)
Pseudomonas spp.	Sea bass (*D. labrax*)	Greece	Ice/AIR	Plates/classical	Taliadourou et al. (2003); Papadopoulos et al. (2003)
Pseudomonas spp.	Norway lobster (*N. norvegicus*)	Greece	0°C, 5°C, 20°C/AIR	Plates/classical	Boziaris et al. (2011)
Pseudomonas spp.	Sea bass (*D. labrax*)	Greece	2°C/AIR	Plates/classical	Parlapani et al. (2015a)
Pseudomonas lundensis	Gilt-head sea bream (*S. aurata*)	Greece	0°C, 10°C, 20°C/AIR	Plates/phenotypic analysis and SDS-PAGE	Tryfinopoulou et al. (2002)

H_2S-producing bacteria	Sea bass (*D. labrax*)	Greece	Ice/AIR	Plates/classical	Kyrana and Lougovois (2002)
H_2S-producing bacteria	Red mullet (*M. surmuletus*)	Greece	4°C/AIR, MAP: $O_2/CO_2/N_2$ (M1: 10/20/70 M2:10/40/50, M3: 10/60/30)	Plates/classical	Pournis et al. (2005)
Pseudomonas spp.	Mussels (*Mytilus galloprovincialis*)	Greece	4°C/AIR, vacuum & MAP:	Plates/classical	Goulas et al. (2005)
H_2S-producing bacteria			M1: CO_2/N_2 (50/50), M2: CO_2/N_2 (80/20), M3: $CO_2/N_2/O_2$ (40/30/30)		
Lactic acid bacteria, *B. thermosphacta*	Chub mackerel (*Scomber colias japonicas*)	Greece	3°C, 6°C/AIR, vacuum, MAP: CO_2/N_2 (50/50)	Plates/classical	Stamatis and Arkoudelos (2007)
A. salmonicida, *Pseudomonas* spp.	Gilt-head sea bream (*S. aurata*)	Greece	Ice/AIR	Fish tissue/16S rRNA gene analysis (cloning & sequencing)	Parlapani et al. (2013)
P. fragi	Sea bream fillets (*S. aurata*)	Greece/Aegean Sea	0°C & 5°C/AIR	Plates/16S rRNA gene analysis	Parlapani et al. (2015a)
Pseudomonas veronii	Sea bream fillets (*S. aurata*)	Greece/Aegean Sea	0°C/MAP: $CO_2/O_2/N_2$ (60/10/30)	Plates/16S rRNA gene analysis	Parlapani et al. (2015a)
C. maltaromaticum	Sea bream fillets (*S. aurata*)	Greece/Aegean Sea	5°C/MAP: $CO_2/O_2/N_2$ (60/10/30)	Plates/16S rRNA gene analysis	Parlapani et al. (2015a)
P. fluorescens	Gutted sea bream (*S. aurata*)	Greece/Ionian Sea	2°C/AIR	Plates/16S rRNA gene analysis	Parlapani et al. (2015b)

(Continued)

TABLE 3.1 (Continued)

SSO/Dominants	Fish/Seafood	Origination	Storage Conditions	Method of Determination	References
P. fragi	Gilt-head sea bream (*S. aurata*)	Greece/ Aegean Sea	0°C & 5°C/AIR	Plates/16S rRNA gene analysis	Parlapani and Boziaris (2016)
P. fluorescens	Gilt-head sea bream (*S. aurata*)	Greece/ Aegean Sea	15°C/AIR	Plates/16S rRNA gene analysis	Parlapani and Boziaris (2016)
Pseudomonas sp., *P. lundensis*, *Carnobacterium* sp.	Gilt-head sea bream (*S. aurata*)	Greece/ Ionian Sea	0°C, 4°C, 8°C/AIR	Tissue/Illumina MiSeq	Parlapani et al. (under preparation)
Psychrobacter sp., *Psychrobacter namhaensis*	Gilt-head sea bream (*S. aurata*)	Greece/ Aegean Sea	0°C, 4°C, 8°C/AIR	Tissue/Illumina MiSeq	Parlapani et al. (under preparation)
A. salmonicida, *S. putrefaciens*	Common carp (*Cyprinus carpio*)	Argentina	Ice/AIR	Plates/phenotypic analysis and (ERIC)-PCR	Beaz-Hidalgo et al. (2014)

(chemical spoilage indices-CSIs) that cause the organoleptic rejection of the product. The time required for the SSOs to reach the minimum spoilage level, which coincides with the concentration of the metabolites able to cause sensory rejection, determines products shelf life (Fig. 3.1). For different fish and sets of processing and storage conditions, different SSOs, minimum spoilage levels, and CSIs apply.

To discern the SSOs from the rest of the dominant spoilage microbiota is not an easy or straightforward task, and it has not been elucidated and documented in the majority of the scientific publications. A comparison of metabolite production and off-flavors development, between inoculated sterile fish or model fish substrate with the SSOs under investigation and naturally contaminated fish, is necessary (Gram and Huss, 1996; Dalgaard, 2003). Thus, in most of the occasions, SSOs and dominant spoilage microbiota are considered conventionally as synonyms.

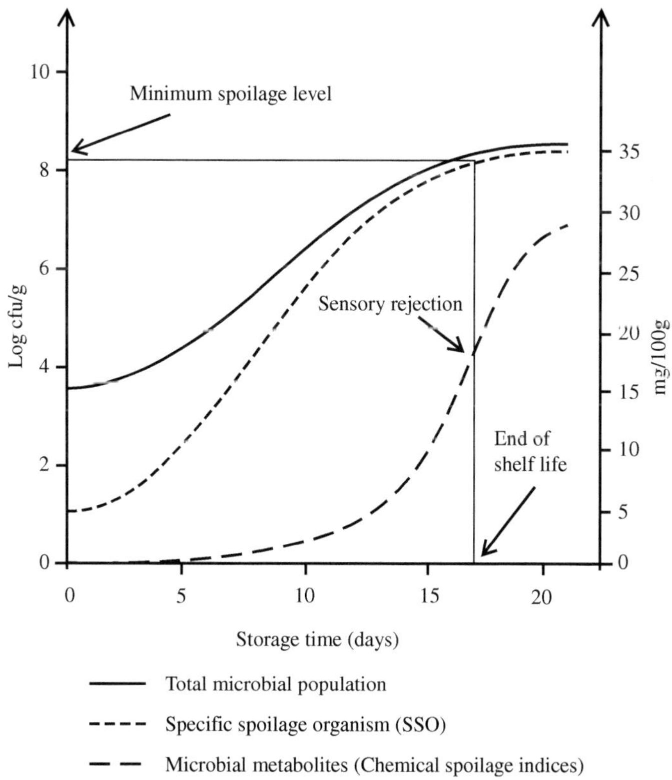

FIGURE 3.1 Microbiological spoilage and SSO concept. *Modified from Dalgaard, P., 2003. Spoilage of seafood. In: Caballero, B., Trugo, L., Finglas, P. (Eds.), Encyclopedia of Food Science and Nutrition. Academic Press, London, p. 2465.*

3.4 FACTORS AFFECTING THE SELECTION OF SSOs

The selection of the dominant spoilage microbiota and SSOs is dependent upon the composition of the initial microbiota (which in turn is dependent upon fish origination and the contamination after catch), preservation hurdles applied, and temperature and atmospheric conditions during storage and distribution. Other factors that may influence the selection are the substrate composition and microbial interactions between spoilage bacteria.

3.4.1 Fish Origination and Storage Conditions

Fish that remain for a long time at elevated temperatures, mainly due to temperature abuse during distribution, are spoiled by Enterobacteriaceae, *Vibrio,* and *Aeromonas* (Gram et al., 1987, 1990). Fish stored under chilling temperatures spoil by phychrotrophic microorganisms. The main spoilage microorganisms of chilled fish stored aerobically belong to the genera of *Pseudomonas* and *Shewanella.* Marine fish and other fishery products from cold temperate waters stored aerobically at low temperatures are spoiled by the action of primarily *Shewanella* and secondarily *Pseudomonas* species (Dalgaard, 2003; Gram, 2009). On the other hand, primarily *Pseudomonas* spp. and secondary *Shewanella putrefaciens* have been found to be the main spoilage microorganisms of fish and shellfish from warmer waters, such as the Mediterranean region, stored aerobically at low temperatures (Koutsoumanis and Nychas, 1999; Dalgaard, 2003; Tryfinopoulou et al., 2002, 2007; Boziaris et al., 2011; Parlapani et al., 2015a,b; Parlapani and Boziaris, 2016). Fish from marine tropical waters stored in ice spoil mainly due to the action of *Pseudomonas* (Dalgaard, 2003; Gram, 2009). *Psychrobacter* and *Aeromonas* may also contribute to the fish spoilage. *Psychrobacter* was found to be part of the dominant microbiota in spoiled iced angler (*Lophius piscatorius*) (Broekaert et al., 2011), *Nephrops norvegicus* (Bekaert et al., 2015), and gilt-head sea bream (*Sparus aurata*) (Parlapani et al., under preparation). Hozbor et al. (2006) and Parlapani et al. (2013), found *Aeromonas* as dominant microbiota of spoiled Atlantic sea salmon (*Pseudopercis semifasciata*) and Mediterranean gilt-head sea bream (*S. aurata*) respectively. However, the spoilage potential of *Psychrobacter* and *Aeromonas* has not been thoroughly investigated yet.

Regarding fresh water fish and shellfish from either temperate or tropical regions, mainly *Pseudomonas* (Gram et al., 1990; Chytiri et al., 2004; Lalitha and Surendram, 2006) and *Aeromonas* (Gonzalez et al., 2001; Lalitha and Surendram, 2006; Beaz-Hidalgo et al., 2014) dominate in the spoiled product. Moreover, even slight differences of geographical origination might affect the selection of dominants. Gilt-head sea bream from the Aegean Sea stored at low temperatures (FAO 37.3.1) is spoiled due to the presence of various *Psychrobacter* and *Pseudomonas* strains while from the Ionian Sea

(FAO 37.2.2) it is spoiled due to *Pseudomonas* (Parlapani et al., under preparation).

The succession of spoilage microorganisms and their final dominance or not is greatly influenced by packaging atmosphere. The elevated CO_2 and the reduced O_2 of MAP not only prolongs the shelf life of fishery products but also affects the synthesis of spoilage microbiota and the selection of SSOs (Noseda et al., 2014). Mainly *Photobacterium phosphoreum* and various LAB, including *Carnobacterium,* and *B. thermosphacta* usually predominate under MAP or vacuum package of fish from Northern seas (Dalgaard et al., 1993, 1997a; Dalgaard, 2003; Rudi et al., 2004; Hovda et al., 2007b; Macé et al., 2012). The main spoilage microorganisms of fish from the Mediterranean region stored at low temperatures under MAP belong to various *Lactobacillus*, *Carnobacterium*, and *B. thermosphacta* (Drosinos and Nychas, 1996; Koutsoumanis et al., 2000; Parlapani et al., 2015b). It has to be noted that O_2 concentration is very important for the selection of SSO in MAP. In packaging where O_2 tension remains high, *Pseudomonas* can still be dominant despite the fact that LAB and others are favored. Indeed, Kostaki et al. (2009), Pournis et al. (2005), and Parlapani et al. (2014), using gaseous composition with O_2 as high as 10%, found that *Pseudomonas* spp. and H_2S-producing bacteria (presumptive *Shewanella*) were the dominant spoilage microorganisms in organic aquacultured sea bass (*Dicentrarchus labrax*), Mediterranean mullet (*Mullus surmuletus*), and gilt-head sea bream (*S. aurata*), respectively.

3.4.2 Effect of Preservation Hurdles

Seafood processing conditions impose hurdles to microbial growth (Leistner and Gorris, 1995). Treatments such as salting, acidification, smoking, and drying alter mostly the pH and a_w of the fishery products. Those preservation hurdles favor the growth of LAB and yeasts compared to Gram-negative spoilers such as *Pseudomonas* and *Shewanella* (Boziaris et al., 2013; Gram, 2009). *Lactobacillus* were the dominant microbiota in marinated anchovies, sardines, and pacific saury (Fuselli et al., 1998; Kilinc and Cackli, 2005; Sallam, 2008), and various LAB and yeasts were dominant in marinated herring and cod (Lyhs et al., 2001; Boziaris et al., 2013). In some salted products like salted anchovies and bonito, yeasts were part of the dominant microbiota (Erkan et al., 2009; Boziaris et al., 2013), while Enterobacteriaceae and halotolerant microorganisms, such as staphylococci and halophilic pediococci, have also been reported as part of the spoilage microbiota in salted anchovies (Villar et al., 1985; Aponte et al., 2010). In cold-smoked salmon packed under vacuum the dominant spoilage microorganisms are usually *B. thermosphacta*, *Lactobacillus*, *Leuconostoc*, and *Carnobacterium* (Leroi et al., 1998; Truelstrup Hansen and Huss, 1998; Leroi et al., 2001). It has to be noted that the higher the intension of preservation hurdles the lower the microbial population in the spoiled products. This implies that in some processed fishery products the contribution of chemical oxidation and autolysis on spoilage mechanisms might be more crucial compared to microbial activity.

3.4.3 Substrate Composition and Microbial Interaction

Bacterial antagonism of nutrients plays a crucial role in the selection of the dominant bacterial consortium of spoiled fish. Bacteria compete for assimilation of nutrients (e.g., glucose, lactate, amino-acids, and other nitrogen compounds) or metals (e.g., iron). In single and coculture experiments of three of the main fish spoilage microorganisms, *Pseudomonas* sp., *S. putrefaciens*, and *B. thermosphacta*, the glucose assimilation was accelerated in coculture of *Pseudomonas* with *B. thermosphacta*, in contrast to the single cultures, which is attributed to the antagonism. In the same study, growth rate of *S. putrefaciens* was reduced in the coculture with *Pseudomonas* (Tsigarida et al., 2003). Additionally, it has been observed that *Pseudomonas* can inhibit the growth of *S. putrefaciens* due to siderophore-mediated competition for iron (Gram and Melchiorsen, 1996).

However, *S. putrefaciens* usually dominate against *Pseudomonas* in fish from Northern seas, but cannot dominate over *Pseudomonas* in fish from Greek waters. This inability is attributed to the lower content of trimethylamine oxide (TMAO), a compound used for energy production in *S. putrefaciens* (Dalgaard et al., 1993), in fish from Greek waters compared to fish from Northern Seas. Indeed, trimethylamine (TMA) development is not pronounced in fish from Mediterranean waters compared to other species from Northern Seas (Parlapani et al., 2015a; Parlapani and Boziaris, 2016). Kyrana and Lougovois (2002) determined the TMAO level in European sea bass (*D. labrax*) to the value of 22.8 mg N/100 g, which is quite significantly lower compared to 66–75 mg N/100 g found in *Gadus morhua* (Herland et al., 2009).

Finally, Gram-negative spoilage bacteria can interact by producing acylated homoserine lactones, which are communication signals. The elucidation of the role of such signaling and interaction between spoilage microorganisms may give us insight into the spoilage mechanism (Gram et al., 2002).

3.5 METABOLISM OF SPOILAGE MICROORGANISMS AND SSOs

Various metabolites with characteristic odor such as total volatile bases nitrogen (TVB-N), TMA, sulfuric compounds, alcohols, aldehydes, ketones, and esters are produced by microorganisms during fish spoilage (Gram and Huss, 1996; Olafsdottir et al., 1997; Dalgaard, 2003). Which compounds are produced is dependent on the metabolic activity of selected SSOs, which is directly affected by storage conditions, especially aeration.

Some of the precursors for the metabolic by-products are the TMAO for the formation of TMA, carbohydrates for various organic acids, the amino-acids cysteine and methionine for sulfur by-products (e.g., dimethyl sulfide), other amino acids and NPN molecules for ammonia and various carbonylic compounds (Gram and Huss, 1996). In particular, leucine and isoleucine are responsible for the formation of aldehydes 2-methylbutanal and 3-methylbutanal (Soncin et al., 2008; Casaburi et al., 2015).

TVB-N and TMA are the most common CSIs of fish (Oehlenschläger, 2014). TVB-N value includes molecules such as TMA, dimethylamine (DMA), ammonia, biogenic amines, and other volatile nitrogenous compounds (Gram and Huss, 1996). TMA is produced mainly by *P. phosphoreum* and *S. putrefaciens* through the reduction of TMAO to TMA (Dalgaard, 1995). Ammonia is produced through the deamination of amino-acids and nucleotide catabolism (Malle and Poumeyrol, 1989) by various microorganisms including *Pseudomonas* (Dainty, 1996). Biogenic amines are produced due to the decarboxylation of specific amino acids by various microorganisms (Al Bulushi et al., 2009). It has to be noted that DMA is not product of microbial metabolism but is instead the result of the action of autolytic enzymes.

In fish originated from cold waters (mainly cod), where *S. putrefaciens* and *P. phosphoreum* are the main spoilers of air-stored and vacuum/MAP fish respectively, high amounts of TVB-N and TMA have been reported (Connell and Shewan, 1980; Huss, 1988; Jørgensen et al., 1988; Dalgaard, 1995; Debevere and Boskou, 1996). Fish is considered spoiled when TVB-N and TMA exceeds 35 mg N/100 g and 15 mg N/100 g, respectively (EC 2074/2005; Connell and Shewan, 1980; Huss, 1995). In fish originated from Mediterranean Sea (mainly gilt head sea bream and European Sea bass) and stored under air, *Pseudomonas* is the main spoiler, but TVB-N values usually do not reach amounts higher than 30 mg N/100 g at the time of organoleptic rejection (Koutsoumanis and Nychas, 2000; Kyrana and Lougovois, 2002; Taliadourou et al., 2003; Castro et al., 2006; Parlapani et al., 2014, 2015a). Similar TVB-N values were also reported by other researchers for Atlantic salmon caught in Chile and Norway (Calanche et al., 2013; Aguilera-Barraza et al., 2015), silver pomfret in Kuwait (Anderson, 2008), and sardine in Spain (Calanche et al., 2013). However, large amounts of TVB-N are produced in crustaceans such as white shrimp (Huang et al., 2016), pink shrimp (Mendes et al., 2005), tropical brackish water shrimp (Dabadé et al., 2015), rock lobster muscle (Sidhu et al., 1974), and Norway lobster (Losada et al., 2006; Aubourg et al., 2007; Boziaris et al., 2011). Raw crustaceans contain higher content of NPN compared to fish (Mendes et al., 2005), offering an explanation of the faster production and higher values of TVB-N in crustaceans compared to teleost fish.

Low amounts of TMA have been reported in fish from Mediterranean waters stored at chill temperatures aerobically (Kyrana et al., 1997; Koutsoumanis and Nychas, 2000; Erkan and Özden, 2006, 2008; Erkan, 2007; Parlapani et al., 2015a) and under MAP (Drosinos and Nychas, 1996, 1997; Drosinos et al., 1997). This is attributed to the lower content of TMAO in fish from Greek waters (Koutsoumanis and Nychas, 1999; Kyrana and Lougovois, 2002; Parlapani et al., 2015a). Low amounts of TMA have been also reported for chill-stored fish from other warm waters (Reddy et al., 1994; Pastoriza et al., 1996; Gonçalves et al., 2007; Tejada et al., 2007). In freshwaters fish (farmed catfish, rainbow trout, mangrove snapper, freshwater grouper) in Europe and Asia, TMA values were found lower than those reported for marine fish (Manthey et al., 1988; Rodríguez et al., 1999; Chan et al., 2006), presumably due to lower amounts of TMAO in fresh water fish flesh (Chung and Chan, 2009).

The exploration of other volatile organic compounds (VOCs) which are metabolic products of SSOs is the new trend to evaluate the course of microbiological spoilage of fish. The molecules 3-methyl-1-butanol, 2-methyl-1-butanol ethanol, 3-methylbutanal, 2-methylbutanal, 3-hydroxy-2-butanone, dimethyl disulfide, ethyl acetate, and acetic acid are the most common among a large number of VOCs detected in various fish (Table 3.2).

It has been reported that *Pseudoalteromonas*, *Pseudomonas*, and *Shewanella* are associated with the formation of methyl mercaptan, dimethylsulfide, dimethyl disulfide, and dimethyl trisulfide (Table 3.2). Sulfur compounds are produced through the metabolism of sulfur-containing amino acids, e.g., cysteine and methionine (Varlet and Fernandez, 2010).

Alcohols, mainly ethanol, 3-methyl-1-butanol, and 2-methyl-propanol, are produced by *Pseudomonas*. However, *Pseudoalteromonas*, *Psychrobacter*, and LAB are also associated with the formation of various alcohols (Table 3.2). *Pseudoalteromonas* was responsible for the formation of 2-propanol and 2-methyl-propanol and *Psychrobacter* for 2-methyl-2-propanol in shrimps (Broekaert et al., 2013), while *Pseudomonas* and LAB for 3-methyl-1-butanol in inoculated sea bream model systems under air and MAP at chill temperatures (Parlapani et al., under preparation). Aldehydes are mainly formed by the catabolism of amino acids by various microorganisms. Then they can be reduced to alcohols or oxidized to carboxylic acids. For example, 3-methylbutanal is reduced to 3-methyl-1-butanol by alcohol dehydrogenases or oxidized to isovaleric acid by an aldehyde dehydrogenase (Filipiak et al., 2012).

Aldehydes, 3-methylbutanal, and 2-methylbutanal have been found in various seafoods (Table 3.2). The aldehyde 3-methylbutanal was reported to be associated with action of *Pseudoalteromonas* spp. in ice-stored shrimps (Broekaert et al., 2013), while Joffraud et al. (2001) and Laursen et al. (2006) attributed the production of 3-methylbutanal and 2-methylbutanal to *Carnobacterium* species. Parlapani et al. (under preparation) revealed that

TABLE 3.2 Overview of Studies Regarding Volatile Organic Compounds (VOCs) Produced by Spoilage Bacteria in Seafood

SSO/ Microorganism	Volatile Compounds	Matrix/Product	Storage Conditions	References
P. perolens	Dimethyl disulfide, dimethyl trisulfide, 3-methyl-1-butanol, 1-penten-3-ol, butanone, trimethylamine, dimethylamine, dimethyl disulfide	Inoculated sterile black rockfish (*Sebastes melanops*)	5°C, 15°C, 25°C	Miller et al. (1973a)
P. fragi	Acetaldehyde, methyl mercaptan, dimethyl disulfide, dimethyl sulfide, ethanol, butanone, acetone, heptanone, ethyl acetate, ethyl butyrate, ethyl hexanoate	Inoculated sterile black rockfish (*S. melanops*)	5°C, 15°C, 25°C	Miller et al. (1973b)
P. fragi	Methanethiol, 1-propanethiol, 2-propanethiol, 2-methyl-1-propanethiol, 2-(methylthio) propane, dimethyl sulfide, dimethyl trisulfide, bis(1-methylethyl) disulfide, ethyl acetate, acetic acid methyl ester, propanoic acid ethyl ester, 2-methyl-propionic acid ethyl ester, 3-methyl butanoic acid methyl ester, 3-methyl butanoic acid ethyl ester	Inoculated sterile prawn extract	5°C	Chinivasagam et al. (1998)
P. fluorescens	Methyl mercaptan, dimethyl disulfide	Inoculated sterile black rockfish (*S. melanops*)	0°C	Miller et al. (1973c)
Pseudoalteromonas spp.	Methanethiol, trimethylamine, ethylene oxide, dimethyl disulfide, 1,2-butanediol, 2-methyl-propanol, 3-methylbutanal, 1,2-	Inoculated sterile shrimps	Ice	Broekaert et al. (2013)

(Continued)

TABLE 3.2 (Continued)

SSO/ Microorganism	Volatile Compounds	Matrix/Product	Storage Conditions	References
	butanone, 2-butanone, 2-pentanone, acetone			
	Ammonia, methyl mercaptan, dimethyl disulfide, sulfur hydride, 2-propanol, 1,2-butanediol, butanone, 2-pentanone, acetone, ethyl acetate, acetic acid	Inoculated sterile shrimps	4°C	Broekaert et al. (2013)
Psychrobacter spp.	Trimethylamine, ethylene oxide, 2-butanone, acetone, acetaldehyde, 2-methyl-2-propanol	Inoculated sterile shrimps	Ice	Broekaert et al. (2013)
	Trimethylamine	Inoculated sterile shrimps	4°C	Broekaert et al. (2013)
P. putrefaciens	Trimethylamine, methyl mercaptan dimethyl disulfide dimethyl trisulfide, 3-methyl-1-butanol	Inoculated sterile black rockfish (*S. melanops*)	1–2°C	Miller et al. (1973c)
S. putrefaciens	Methanethial, methanethiol, 2-propanethiol, dimethyl sulfide, dimethyl trisulfide, bis (1-methylethyl) disulfide, 2-methyl-propionic acid ethyl ester, 3-methyl butanoic acid methyl ester	Inoculated sterile prawn extract	5°C	Chinivasagam et al. (1998)
P. phosphoreum	Isobutyraldehyde, acetic acid, ethyl acetate, 2-methylbutanal, 3-methylbutanal; 2 methyl-1-propanol, 3-methyl-2-butanol,	Inoculated sterile raw salmon model system	4°C for 3 days then 8°C for 7	Macé et al. (2013a)

	benzaldehyde, benzene acetaldehyde, benzene ethanol, 3-hydroxybutanone		days after a break of 2 h at 20°C	
Achromobacter	Trimethylamine, methyl mercaptan, dimethyl disulfide, 3-methyl-1-butanol	Inoculated sterile black rockfish (*S. melanops*)	1–2°C	Miller et al. (1973c)
C. divergens, *C. maltaromaticum*	2-ethyl-1-hexanol, 3-heptanol, 1-octen-3-ol, 1-pentanol, 2,3-butanedione (diacetyl), 3-heptanone, 2-heptanone, 2-hexanone, 4-methyl-2-pentanone,2,4-pentanedione, 2-pentanone, 3-pentanone, 4-methyl-3-penten-2-one, 2,4,6-trimethylpyridine, acetic acid, total volatile nitrogen/ammonia, 3-methyl-1-butanal, and 3-methyl-1-butanol	Inoculated cooked and peeled MAP shrimp	5°C	Laursen et al. (2006)
Carnobacterium spp.	2,3-butanedione, 2,3-pentanedione	Inoculated sterile blocks of cold-smoked salmon	6°C (vacuum)	Joffraud et al. (2001)
Group: Aeromonas spp. Enterobacteriaceae *S. putrefaciens*	TMA, disulfide dimethyl, 2,3-butanediol, 2-pentanol	Inoculated sterile blocks of cold-smoked salmon	6°C (vacuum)	Joffraud et al. (2001)
Lactobacillus spp.	Acetic acid, ethyl acetate, *n*-propyl acetate	Inoculated sterile blocks of cold-smoked salmon	6°C (vacuum)	Joffraud et al. (2001)

(*Continued*)

TABLE 3.2 (Continued)

SSO/ Microorganism	Volatile Compounds	Matrix/Product	Storage Conditions	References
B. thermosphacta	2-heptanone, 2-hexanone	Inoculated sterile blocks of cold-smoked salmon	6°C (vacuum)	Joffraud et al. (2001)
B. thermosphacta	2,3-butanedione (diacetyl), 3-methyl-1-butanal, 2-octene, 3-methyl-1-butanol, dimethylbenzene, butanediol	Inoculated cooked and peeled MAP shrimp	5°C	Laursen et al. (2006)
Microbial activity (general)	Trimethylamine, dimethyl disulfide, 2-methyl-propanol, 3-methyl-1-butanol, 2,3-butanediol, 3-methylbutanal, 2,3-butanedione, 3-hydroxy-2-butanone, 3-pentanone, 2-heptanone, ethyl acetate	Whiting (*Merlangius merlangus*), cod (*G. morhua*), mackerel (*Scomber scombrus*)	4°C	Duflos et al. (2006)
	Methyl mercaptan	Cod (*G. morhua*)	Ice	Herbert et al. (1975)
	Dimethyl sulfide	Mussels (*Perna canaliculus*)	6.5°C	Tuckey et al. (2013)
	Dimethyl sulfide	Cod (*G. morhua*)	Ice	Herbert et al. (1975)
	Ethanol, 3-methyl-1-butanol	Yellowfin tuna (*Thunnus albacares*)	Ice and 30°C	Edirisinghe et al. (2007)
	Ethanol, 3-methyl-1-butanol, 2-methyl-1-butanol, 2-ethyl-1-hexanol, 3-methylbutanal, 2-methylbutanal, 3-hydroxy-2-butanone, ethyl acetate, ethyl propionate, ethyl isobutyrate, ethyl-2-methylbutyrate, ethyl isovalerate	Gutted sea bass (*D. labrax*), sea bream (*S. aurata*)	2°C	Parlapani et al. (2015a,c)

	Ethyl hexanoate, butanoic acid 2-methylbutyl ester, acetic acid	Gutted sea bass (*D. labrax*)	2°C	Parlapani et al. (2015a)
	Ethyl butyrate, isoamyl butyrate	Gutted Sea bream	2°C	Parlapani et al. (2015c)
	Ethanol, 3-methyl-1-butanol, 2-ethyl-1-hexanol, 3-methylbutanal, 2-methylbutanal, 2,3-pentanedione, ethyl-2-methylbutyrate, ethyl isovalerate, ethyl hexanoate	Sea bream (*S. aurata*) fillets	0°C, 5°C, 15°C (air and/or MAP)	Parlapani et al. (2014)
	2-Methyl-1-butanol, 2-heptanone	Sea bream (*S. aurata*) fillets	0°C	Parlapani et al. (2014)
	3-Hydroxy-2-butanone, acetic acid	Sea bream (*S. aurata*) fillets	0°C and/or 5°C (MAP)	Parlapani et al. (2014)
	Ethyl acetate, ethyl propionate, ethyl isobutyrate, ethyl butyrate, acetic acid	Sea bream (*S. aurata*) fillets	5°C and/or 15°C	Parlapani et al. (2014)
	1-Propanol	Mackerel (*S. scombrus*)	4°C	Duflos et al. (2006)
	Acetic acid	Cod (*G. morhua*)	4°C	Duflos et al. (2006)
	3-Methyl-1-butanol, 3-methylbutanal, 2-methylbutanal, 2,3-pentanedione, 3-hydroxy-2-butanone	Sea bream (*S. aurata*)	Ice	Soncin et al. (2008)
	Ethanol, 2-butanone, 2-pentanone, 3-methylbutanal, 3-hydroxy-2-butanone, 3-methyl-1-butanol, acetic acid	Cold-smoked salmon	5°C and 10°C	Jonsdottir et al. (2008)

(*Continued*)

TABLE 3.2 (Continued)

SSO/ Microorganism	**Volatile Compounds**	**Matrix/Product**	**Storage Conditions**	**References**
	Ethanol, 1-propanol, 1-butanol, 2-butanol, 2-methyl-1-propanol, 2-ethyl-1-hexanol, 2-methyl-1-butanol, 3-methyl-1-butanol, (E)-2-penten-1-ol 2-butanone, 3-methylbutanal, ethyl acetate, trimethylamine	Cold-smoked salmon	5°C (vacuum)	Jørgensen et al. (2001)
	3-Methyl butanoic acid, acetic acid, 3-hydroxy-2-butanone	King salmon (*Oncorhynchus tshawytscha*)	Ice (air, MAP)	Wierda et al. (2006)
	1-Penten-3-ol and 3-methyl-1-butanol dimethyl disulfide and dimethyl trisulfide, acetic acid, trimethylamine	Sea bream (*S. aurata*)	Ice	Alasalvar et al. (2005)
	Ethanol, 2-methyl-propanol, 3-methyl-1-butanol, 3-hydroxy-2-butanone, ethyl acetate, butanoic acid ethyl ester	Haddock (*Melanogrammus aeglefinus*) fillets	Ice	Olafsdottir (2005)
	Ethanol, 2-methyl-propanol, 3-methyl-1-butanol, butanoic acid ethyl ester	Cod (*G. morhua*) fillets	0.5°C	Olafsdottir et al. (2005)
	Ethyl acetate, ethyl butanoate	Cod (*G. morhua*) fillets	0.5°C	Olafsdottir et al. (2005)
	Trimethylamine, 3-hydroxy-2-butanone	Cod (*G. morhua*) fillets	0.5°C	Olafsdottir et al. (2005)

both compounds were produced only by LAB (coculture of *Lactobacillus* spp., *Carnobacterium maltaromaticum*, and *Carnobacterium divergens*) during storage of inoculated sterile sea bream model systems under air and MAP at chill temperatures. Acetaldehyde has been found to be produced by *Psychrobacter* in ice-stored shrimps (Broekaert et al., 2013), and *Pseudomonas fragi* in sterile black rockfish (Miller et al., 1973b).

The presence of ketones in seafood is mainly associated with the action of *Pseudomonas*, *Pseudoalteromonas*, and *Psychrobacter* (Table 3.2). The ketone 3-hydroxy-2-butanone is produced by LAB under air and MAP in sterile sea bream model systems (Parlapani et al., under preparation). The formation of 3-hydroxy-2-butanone comes from the glucose catabolism of *Carnobacterium* spp. and *Lactobacillus* spp. (Casaburi et al., 2015). Acetone has been found in seafood as well and its production has been associated with *Pseudomonas, Psychrobacter*, and *Pseudoalteromonas* (Miller et al., 1973b; Broekaert et al., 2013).

Esters are among the most dominant compounds presented in seafood (Table 3.2). The formation of esters is linked with esterification of alcohols and carboxylic acids, as well by microbial esterase activity (Casaburi et al., 2015). Ethyl acetate is the most frequently detected ester in fish (Table 3.2). *Pseudoalteromonas* spp. and *Pseudomonas* are able to produce this compound in many substrates (Table 3.2). *Pseudomonas* is also responsible for the production of many other esters, e.g., ethyl butanoate, ethyl octanoate, ethyl butyrate, ethyl isovalerate, 3-methyl butanoic acid ethyl ester (Table 3.2). However, the ester 3-methylbutyl acetate was found to be produced only by LAB in inoculated sterile sea bream model systems stored at chill temperatures under air and MAP (Parlapani et al., under preparation). On the other hand, in the same study ethyl esters were attributed only to the metabolism of *Pseudomonas*. Acetic acid also appears in spoiled fish (Duflos et al., 2006; Parlapani et al., 2014, 2015a). *Pseudoalteromonas elyakovii* was found to produce a high concentration of this compound in sterile shrimps (Broekaert et al., 2013). Additionally, acetic acid was found to be produced by *Pseudomonas* and *Shewanella* in sterile sea bream model systems stored under air at chill temperatures (Parlapani et al., under preparation).

3.6 DETERMINATION OF SPOILAGE MICROBIOTA AND SSOs

Isolation, identification, and enumeration of fish spoilage microorganisms are of great importance for research and seafood quality assurance. Isolation and identification of spoilage microorganisms are important when the aim of the study is the elucidation of fish spoilage mechanisms. Enumeration of spoilage microorganisms is essential in the evaluation of fish freshness and potential shelf life.

Microbiological determination can be based on culture-dependent and culture-independent techniques. The culture-dependent techniques based on culture media are still considered as standard methods. Bacterial population is cultured using agar plates for isolation and subsequent identification using phenotypic tests or for enumeration. However, culture-dependent methods are laborious and time-consuming, the phenotypic identifications based on culture media lack the discriminatory power of polymerase chain reaction (PCR)-based methods (Nisiotou et al., 2014). Furthermore, a recent study has concluded that many potential fish spoilage microorganisms are overlooked, while others are unable to grow on the media used (Broekaert et al., 2011). Additionally, sublethally injured cells are not able to recover and grow on selective media, while other populations are inhibited by microorganisms presented in higher numbers (Hugenholtz et al., 1998). Hence, the use of culture-independent molecular techniques is continuously increasing for both identification and enumeration of SSOs.

3.6.1 Identification of Spoilage Microorganisms

3.6.1.1 Culture-Based Methodologies

Classical microbiological approaches to the identification of spoilage microorganisms in fish were first carried out in the 1960s (Shewan et al., 1960a,b). Microorganisms such as *Pseudomonas*, *Vibrio*, *Aeromonas*, *Achromobacter*, *Flavobacterium*, and Coryneforms were identified using to phenotypic tests such as Gram-reaction, oxidase test, Hugh and Leifson reaction, production of acid and/or gas from glucose, motility, and sensitivity to various compounds, Gram et al. (1987) isolated "black" colonies on Iron Agar (IA) with the ability to reduce TMAO to TMA and produce H_2S, from spoiled fish at 0°C and identified as *Alteromonas putrefaciens*. Subsequently, Jørgensen and Huss (1989) studied and identified H_2S-producing organisms isolated on IA from spoiling cod were as *S. putrefaciens*, as also described by Gram et al. (1987). Dalgaard et al. (1993) and Dalgaard (1995) isolated *P. phosphoreum* and *S. putrefaciens* from spoiled packed cod using IA. Nowadays, mesophilic, halotolerant strains are recognized as *Shewanella algae* and the psychrotrophic strains as *S. putrefaciens* and *S. baltica* (Gram and Dalgaard, 2002). Newer efforts on identification of spoilage microbiota of fish were also later conducted using both classical and molecular approaches.

3.6.1.2 Culture-Independent Methodologies

The use of phylogenetic 16S rRNA gene is the most common approach for microbial identification and microbiota analysis. Colonies of microorganisms that grow on agar plates can be easily used for 16S rRNA sequence analysis in order to identify them. However, various spoilage microorganisms in fish do not grow on various growth media (Broekaert et al., 2011). The selectivity of microorganisms growing on media usually requires undefined growth factors that are not supported by laboratory media (Mayo et al., 2014). Hence, an alternative approach to exploring spoilage microbiota is the direct extraction of DNA from fish tissue. In this case the microorganisms that escape standard plating approaches can be revealed. The first challenging task is the extraction of DNA from the samples, followed by PCR amplification of the nucleotide sequence of 16S rRNA gene. The final step is the analysis of 16S rRNA gene fragments from the mixed amplification products. This can be carried out using various approaches. The microbial composition can be mainly evaluated through fingerprinting (pattern analysis) of the PCR product by separation of amplified 16S rRNA gene products. Fingerprinting is possible through utilizing techniques such as terminal restriction fragment length polymorphism and denaturing gradient gel electrophoresis or thermal gradient gel electrophoresis. To identify differential bands obtained from the above gel-based approaches, many researchers used the separation products for sequencing to characterize the dominant microbiota (Rudi et al., 2004; Hovda et al., 2007a,b; Jaffrès et al., 2009; Svanevik and Lunestad, 2011; Macé et al., 2012). Another approach is to separate the amplified 16S rRNA gene fragments using cloning (Rudi et al., 2004; Olofsson et al., 2007; Jaffrès et al., 2009; Parlapani et al., 2013).

The introduction of new generation sequencing technology platforms such as 454-pyrosequencing and Illumina next-generation sequencing (NGS) changed the way in the exploration of microbial communities. Microbial diversity can now be studied by these high-throughput sequencing approaches (after direct extraction of DNA of samples, e.g., fish tissue). The number of nucleic acid sequences analyzed by NGS techniques is extremely higher than those analyzed by aforementioned culture-independent methods. The 454-pyrosequencing analysis of various regions of bacterial 16S rRNA gene have been mainly used to understand the microbial diversity during fish fermentation (Roh et al., 2010; Koyanagi et al., 2011, 2013; Kiyohara et al., 2012; Jung et al., 2013; Kim et al., 2014; Lee et al., 2015). Microbial communities associated with spoilage of smoked salmon, cooked peeled shrimp, salmon fillet, and cod fillet have been also described by pyrosequencing analysis (Chaillou et al., 2014). Moreover, the microbiota of fresh and spoiled fish, crustaceans, and cephalopods from Mediterranean Sea waters has been explored by Illumina analysis (Parlapani et al., under preparation).

3.6.2 Enumeration of Spoilage Microorganisms

3.6.2.1 Culture Techniques

Classical enumeration of spoilage microbiota/SSOs relies on the use of various selective media. *Pseudomonas* spp. are usually enumerated using cephaloridine-fucidin-cetrimide agar (CFC agar) or other newly designed chromogenic agar media, after 24–48 hours incubation at 20–25°C. Glutamate Starch Phenol Red (GSP) Agar (Pseudomonas–Aeromonas Selective Agar) acc. to Kielwein (1969, 1971) can be used for simultaneous determination of *Pseudomonas* and *Aeromonas*. LAB in seafood are usually enumerated on Mann–Rogosa–Sharp agar (De Man et al., 1960) or nitrite actidione polymyxin agar (Davidson and Cronin, 1973) or Elliker agar (Elliker et al., 1956) after 72–96 hours incubation at 20–25°C. *Brochothrix. thermosphacta* population is enumerated on Streptomycin Sulfate Thallous Acetate Agar spread plates after incubation at 20°C for 3–4 days. Yeast and molds, mostly found in processed seafood, can be enumerated using Rose Bengal Chloroamphenicol and Tryptose-Glucose-Yeast Extract containing Chloroamphenicol, or oxytetracycline-glucose-yeast extract agar, after 72–96 hours incubation at 25°C.

Although various media are available for enumeration of various SSOs, there are not selective media to be used for differentiation of *Shewanella* and *Photobacterium*. H_2S-producing microorganisms (presumptive *Shewanella*) are enumerated as black colonies on IA, but other H_2S-producing bacteria that belong to other genera such as Enterobacteriaceae, *Aeromonas,* and *Vibrio* can also grow on IA (Tryfinopoulou et al., 2007). *Photobacterium phosphoreum* can grow on IA or Long and Hammer Agar according to Van Spreekens (1974) with 1% w/v added NaCl after incubation at 5°C for 14 days, or 15°C for 5 days (Dalgaard et al., 1993, 1996; Dalgaard, 1995; López-Caballero et al., 2002). *Photobacterium phosphoreum* is a psychrophilic bacterium very sensitive to elevated temperatures. To get the maximum population the analysis has to be performed with diluents, agar plates, pipet tips etc., that remain chilled (below 10°C) throughout the analytical procedure (Dalgaard et al., 1993). Biochemical and/or other tests have to be performed to confirm *P. phosphoreum* and *Shewanella* counts.

3.6.2.2 PCR-Based Techniques

Real-time PCR has been tested for quantification of bacteria presented on fish tissue. This method offers the advantage of giving results in an extremely short time period (usually 3–4 hours) compared to traditional plate counts (48–72 hours). Reynisson et al. (2008) quantified *Pseudomonas* spp. in cod in 5 hours using the carbamoyl phosphate synthase gene (*car*A) as target. The quantification of *P. phosphoreum* in raw salmon using the gyrase subunit B (gyrB) gene DNA sequence from *P. phosphoreum* has been also carried out in 6 hours (Macé et al., 2013b). The total copies of genomic target presented a high degree of correlation with plate counts (Reynisson et al., 2008; Macé et al., 2013b). Duflos et al. (2010) used Real-time PCR quantified the bacteria that are able to reduce TMAO to TMA. The number of copies of the torA gene coding for TMA *N*-oxide reductase increased during fish storage. It was found that the population of bacteria which contain the enzyme TMA *N*-oxide reductase (e.g., *Shewanella* and *Photobacterium*) and which are associated with whiting and plaice spoilage, increased during storage. However, it is known that Real-time PCR has the disadvantage of determination of the DNA of alive and dead cells. Lee and Levin (2007) quantified total viable bacteria on cod fillets using ethidium bromide monoazide (EMA) for the differentiation of viable and dead cells, while Mamlouk et al. (2012) quantified *B. thermosphacta* population in cooked shrimp and salmon using propidium monoazide. A highly conserved region of 16S rRNA gene sequence was used as target for determination of total viable bacteria (370 bp) and *B. thermosphacta* (70 bp). Lee and Levin (2007) observed that the findings by Real-time PCR, either with or without EMA, correlated well to the plate counts. A high correlation between the *B. thermosphacta* counts obtained by Real-time PCR and plate counts on shrimp and salmon was also observed by Mamlouk et al. (2012). Although Real-time PCR is recognized as a power tool for quantification of SSOs, the literature currently contains only few studies for SSOs quantification in fish and seafood.

3.7 RELATION OF SSOs TO FISH FRESHNESS AND SHELF LIFE ASSESSMENT

Microbial spoilage mechanism is the main cause of fresh and lightly preserved fishery products spoilage and apparently is responsible for shelf life duration. Growth of SSOs to relatively high numbers and the concomitant accumulation of various metabolites, alters the sensory characteristics of fishery products, resulting in rejection. Minimum spoilage level varies usually from 7 to 9 logs cfu/g and depends on spoilage potential and activity of the SSOs. Dalgaard (1995) concluded that *P. phosphoreum* produces 30 times more TMA compared to *S. putrefaciens*, which means that for a fish product to spoil and become rejected it requires a population of about 10^7 cfu/g *P. phosphoreum* or 10^{8-9} cfu/g *S. putrefaciens*. Fish that spoils by *Pseudomonas*, which produce mostly volatiles nitrogenous compounds, requires a minimum spoilage level of about 10^8 cfu/g. However, in crustaceans, minimum spoilage levels in the range of 10^{5-6} cfu/g have been recorded (Robson et al., 2007; Boziaris et al., 2011) due to the higher amount of metabolic products TVB-N that are produced compared to fish (Losada et al., 2006; Boziaris et al., 2011; Dabadé et al., 2015; Huang et al., 2016).

Growth inhibition of SSOs in fresh or minimally processed fish is achieved by chill storage, and packaging with elevated CO_2 concentrations MAP. Shelf life is mainly dependent upon (1) fish and shellfish species and product type (fresh, processed, etc.), (2) storage temperature, (3) synthesis of gaseous atmosphere, and (4) population level of initial microbiota. The most common way to retard spoilage microorganisms from fast growing is the application of ice. Shelf life of iced stored fish usually varies from approx. 10 to 20 days (Gram, 2009). Other ways of chilling is the refrigeration at 2°C or 4°C, especially for gutted or filleted packed fish (Jessen et al., 2014). Fillets spoil earlier compared to the whole fish stored in the same conditions, mainly due to the increased population level of its initial microbiota which is the results of the increased contamination due to filleting and other handling. Another mean of refrigeration is super-chilling, which is the storage of fish in temperatures just below 0°C (about −1°C to −4°C), where only a percentage of fish tissue water is frozen (Kaale et al., 2011). With super-chilling shelf life is significantly extended. Application of MAP extends shelf life of fish due to the bacteriostatic activity of CO_2 (Sivertsvik et al., 2002). However, MAP not only causes the shelf life to be prolonged but also affects the selection of SSOs and their metabolic products (as was referred to in previous sections) resulting to a different type of spoilage, milder in comparison to aerobic storage (e.g., from putrid and ammonia-like to sour) (Noseda et al., 2014).

Shelf life and freshness status of fish are assessed by sensory evaluation, SSOs enumeration, and quantification of metabolic products (Olafsdottir et al., 1997). Sensory methods require trained assessors and are expensive, difficult to standardize and unsuitable for utilization as routine procedure by the seafood industry (Dainty, 1996; Dalgaard, 2003). Additionally, sensory analysis is not reliable for fish products such as fillets, due to the diminished number of assessment attributes (Duflos et al., 2006). The microbial growth is the main cause of fresh seafood quality deterioration but the currently used microbiological methods are retrospective, expensive, and time-consuming (Dainty, 1996; Dalgaard, 2003). To overcome the above-mentioned limitations, freshness status and shelf life evaluation can be determined rapidly by (1) microbial growth models in conjunction with prediction software or time-temperature indicators, (2) determination of volatile metabolic products in conjunction with electronic nose, and (3) the use of modern qPCR methods for rapid quantification of spoilage microbiota and/or SSOs.

Kinetic growth models to predict growth of SSOs have been developed (Dalgaard et al., 1997b; Koutsoumanis, 2001). Additionally, software that includes both relative spoilage rate empirical models and kinetic models of SSO is available to predict product shelf life under constant and fluctuating storage conditions (Dalgaard et al., 2002). Time−temperature integration tags, which rely on color changes in response to temperature history, have been employed to estimate product freshness and remaining shelf life (Giannakourou et al., 2005; Tsironi et al., 2008).

Chemical determination of various metabolic products of SSOs is another option for accessing freshness and shelf life. Traditional chemical methods for monitoring microbial activity in fish include the determination of TVB-N and TMA. However, TVB-N and TMA increase in fish occurs only at the late stages of storage, hence those two parameters cannot be used as spoilage markers to access fish quality and remaining shelf life, but only as rejection criterion (Oehlenschläger, 2014). VOCs have been studied as potential spoilage indicators for evaluation, considering that they usually vary significantly between the initial and rejection day of fish. A suitable compound for the assessment of spoilage should: (1) be a microbial metabolite produced by the dominant spoilage microorganisms, (2) be initially absent or at low levels in food, (3) increase during storage, and (4) show good correlation with microbial growth, sensory score, and remaining shelf life (Jay, 1986). Recent research has shown that various compounds, such as ethanol, acetic acid, 2-methylbutanal, 3-methylbutanal, 1-propanol, 2-methyl-1-butanol, 3-methyl-1-butanol, 3-hydroxy-2-butanone, 2-butanone, and ethyl esters, in various fish have the potential to be used as spoilage markers (Jørgensen et al., 2001; Alasalvar et al., 2005; Olafsdottir et al., 2005; Duflos et al., 2006; Wierda et al., 2006; Edirisinghe et al., 2007; Jonsdottir et al., 2008; Soncin et al., 2008; Moreira et al., 2013; Parlapani et al., 2014, 2015a,c). However, due to the numerous factors that govern the selection and metabolism of SSOs and the dynamic nature of spoilage, sums or rations of various metabolites might be necessary as spoilage markers (Dalgaard, 2003). Efforts to employ electronic devices (electronic nose) for rapid detection of volatile compounds in order to evaluate the level of fish freshness/spoilage have given encouraging results (Olafsdottir et al., 2005; Haugen, 2006; Di Natale, 2001).

Determination of SSOs using qPCR methodology is another promising alternative for rapid evaluation of fish quality/freshness and estimation of fish shelf life. Apart from the attempts to quantify the total microbial population (Lee and Levin, 2007), efforts to quantify SSO microorganisms, such as *Pseudomonas*, *Brochothrix*, *Photobacterium*, and bacteria that produces TMA, have been carried out by extracting prokaryotic DNA directly from fish flesh (Reynisson et al., 2008; Duflos et al., 2010; Mamlouk et al., 2012; Macé et al., 2013b). In these studies a relatively good level of correlation between PCR cycle threshold values (C_T) and SSO counts has been achieved; however, optimization of DNA extraction from fish flesh is a challenge that needs to be addressed.

3.8 EPHEMERAL SPOILAGE ORGANISMS

Spoilage of fresh or minimally processed fish is caused by microbial growth. The growth determinants which are the intrinsic (substrate attributes), extrinsic (environmental limitations), processing, and implicit factors influence the establishment of the dominant microbial fraction which cause the spoilage. It has been proposed that spoilage depends on an even smaller fraction of SSOs, called Ephemeral Spoilage Organisms (ESOs), the prevalence of which is dependent on even small differences in factors imposed during processing, storage, and distribution (Nychas et al., 2008).

The introduction of molecular biology techniques in Food Microbiology has contributed immensely to the exploration of bacterial communities on foods (Cocolin et al., 2007; Nisiotou et al., 2014). The advantage that using molecular techniques provides is the ability to determine microorganisms at species or even strains level, which provides a more detailed and in depth view of the dominant spoilage microbiota. Indeed, during the last decades, various studies have reported that *Pseudomonas* spp. are the SSO of fish from the Mediterranean Sea stored aerobically at chill temperatures. However, various *Pseudomonas* species/strains have been found to prevail under different processing and/or storage conditions using molecular techniques (Parlapani et al., 2013, 2015b,c). For example, sea bream spoilage occurred due to the domination of *P. fragi* in whole and filleted sea bream (Parlapani et al., 2013, 2015b) but *P. fluorescens* in gutted sea bream (Parlapani et al., 2015c). *Pseudomonas fragi* was found to dominate in whole chill-stored sea bream, while *P. fluorescens* dominated at a temperature of 15°C (Parlapani and Boziaris, 2016). Therefore, it seems that various species/strains of *Pseudomonas* could be characterized as ESOs of aerobically stored gilt-head sea bream, which occasionally predominate under particular processing and/or storage conditions.

3.9 FUTURE TRENDS

Spoilage of fish is a complex phenomenon which involves substrate parameters, microbial ecology, and microbial competition, as it is affected by storage and processing conditions, as well as the mechanisms of spoilage bacteria metabolism. Profound understanding of fish spoilage has to combine the information of "which are the SSOs/ESOs" with the "which is the metabolism that causes the spoilage." It is obvious that both structural and functional diversity have to be elucidated. Use of new tools based on omics technologies have to be employed to fully understand the mechanisms of spoilage and to link the genotype of SSOs with phenotype. The knowledge from using this approach will be able to increase our understanding on the dynamics of spoilage microbiota (genomics), where specific metabolites are originated from (metabolomics), how these are regulated at the cell level (genomics–transcriptomics-proteomics), and what is the effect on fish quality (sensory attributes). The profound knowledge of spoilage mechanisms and the quantification of these phenomena are of great and continuing interest for the implementation of appropriate preservation strategies and new methodologies for rapid and accurate estimation of fish freshness and shelf life.

REFERENCES

Aguilera-Barraza, F.A., Quevedo Leon, R.A., Lopez-Alvarez, P.X., 2015. Kinetics of protein and textural changes in Atlantic salmon under frozen storage. Food Chem. 182, 120–127.

Al Bulushi, I., Poole, S., Deeth, H.C., Dykes, G.A., 2009. Biogenic amines in fish: roles in toxication, spoilage, and nitrosamine formation: a review. Crit. Rev. Food Sci. Nutr. 49, 369–377.

Alasalvar, C., Taylor, K.D.A., Shahidi, F., 2005. Comparison of volatiles of cultured and wild sea bream (*Sparus aurata*) during storage in ice by dynamic headspace analysis/gas chromatography-mass spectrometry. J. Agric. Food Chem. 53, 2616–2622.

Alfaro, B., Hernandez, I., 2013. Evolution of the indigenous microbiota in modified atmosphere packaged Atlantic horse mackerel (*Trachurus trachurus*) identified by conventional and molecular methods. Int. J. Food Microbiol. 167, 117–123.

Anderson, A.K., 2008. Biogenic and volatile amine-related qualities of three popular fish species sold at Kuwait fish markets. Food Chem. 107, 761–767.

Aponte, M., Blaiotta, G., Francesca, N., Moschetti, G., 2010. Could halophilic archaea improve the traditional salted anchovies (*Engraulis encrasicholus* L.) safety and quality?. Lett. Appl. Microbiol. 51, 697–703.

Ashie, I.N.A., Smith, J.P., Simpson, B.K., 1996. Spoilage and shelf-life extension of fresh fish and shellfish. Crit. Rev. Food Sci. Nutr. 36, 87–121.

Aubourg, S.P., Losada, V., Prado, M., Miranda, J.M., Barros-Velazquez, J., 2007. Improvement of the commercial quality of chilled Norway lobster (*Nephrops norvegicus*) stored in slurry ice: effects of a preliminary treatment with an antimelanosic agent on enzymatic browning. Food Chem. 103, 741–748.

Beaz-Hidalgo, R., Agueria, D., Latif-Eugenin, F., Yeannes, M.I., Figueras, M.J., 2014. Molecular characterization of Shewanella and Aeromonas isolates associated with spoilage of Common carp (*Cyprinus carpio*). FEMS Microbiol. Ecol. 362, 1–8.

Bekaert, K., Devriese, L., Maes, S., Robbens, J., 2015. Characterization of the dominant bacterial communities during storage of Norway lobster and Norway lobster tails (*Nephrops norvegicus*) based on 16S rDNA analysis by PCR-DGGE. Food Microbiol. 46, 132–138.

Boziaris, I.S., Kordila, A., Neofitou, C., 2011. Microbial spoilage analysis and its effect on chemical changes and shelf-life of Norway lobster (*Nephrops norvegicus*) stored in air at various temperatures. Int. J. Food Sci. Technol. 46, 887–895.

Boziaris, I.S., Stamatiou, A.P., Nychas, G.-J.E., 2013. Microbiological aspects and shelf life of processed seafood products. J. Sci. Food Agric. 93, 1184–1190.

Broekaert, K., Heyndrickx, M., Herman, L., Devlieghere, F., Vlaemynck, G., 2011. Seafood quality analysis: molecular identification of dominant microbiota after ice storage on several general growth media. Food Microbiol. 28, 1162–1169.

Broekaert, K., Noseda, B., Heyndrickx, M., Vlaemynck, G., Devlieghere, F., 2013. Volatile compounds associated with *Psychrobacter* spp. and *Pseudoalteromonas* spp., the dominant microbiota of brown shrimp (*Crangon crangon*) during aerobic storage. Int. J. Food Microbiol. 166, 487–493.

Calanche, J., Samayoa, S., Alonso, V., Provincial, L., Roncales, P., Beltran, J.A., 2013. Assessing the effectiveness of a cold chain for fresh fish salmon (*Salmo salar*) and sardine (*Sardina pilchardus*) in a food processing plant. Food Control 33, 126–135.

Carrascosa, C., Millán, R., Saavedra, P., Jaber, R.J., Montenegro, T., Raposo, A., et al., 2014. Predictive models for bacterial growth in sea bass (*Dicentrarchus labrax*) stored in ice. Int. J. Food Sci. Technol. 49, 354–363.

Casaburi, A., Piombino, P., Nychas, G.-J., Villani, F., Ercolini, D., 2015. Bacterial populations and the volatilome associated to meat spoilage. Food Microbiol. 45, 83–102.

Castro, P., Padron, J.C.P., Cansino, M.J.C., Velazquez, E.S., De Larriva, R.M., 2006. Total volatile base nitrogen and its use to assess freshness in European sea bass stored in ice. Food Control. 17, 245–248.

Chaillou, S., Chaulot-Talmon, A., Caekebeke, H., Cardinal, M., Christieans, S., Denis, C., et al., 2014. Origin and ecological selection of core and food-specific bacterial communities associated with meat and seafood spoilage. ISME J.1–14.

Chan, S.T., Yao, M.W.Y., Wong, Y.C., Wong, T., Mok, C.S., Sin, D.W.M., 2006. Evaluation of chemical indicators for monitoring freshness of food and determination of volatile amines in fish by headspace solid-phase microextraction and gas chromatography-mass spectrometry. Eur. Food Res. Technol. 224, 67–74.

Chinivasagam, H.N., Bremner, H.A., Wood, A.F., Nottingham, S.M., 1998. Volatile components associated with bacterial spoilage of tropical prawns. Int. J. Food Microbiol. 42, 45–55.

Chung, S.W.C., Chan, B.T.P., 2009. Trimethylamine oxide, dimethylamine, trimethylamine and formaldehyde levels in main traded fish species in Hong Kong. Food Addit. Contam. Part B 2, 44–51.

Chytiri, S., Chouliara, I., Savvaidis, I.N., Kontominas, M.G., 2004. Microbiological, chemical and sensory assessment of iced whole and filleted aquacultures rainbow trout. Food Microbiol. 21, 157–165.

Cocolin, L., Diez, A., Urso, R., Rantsiou, K., Comi, G., Bergmaier, I., et al., 2007. Optimization of conditions for profilingbacterial populations in food by culture-independent methods. Int. J. Food Microbiol. 120, 100–109.

Commission Regulation (EC) No 2074/2005 of 5 December 2005, laying down implementing measures for certain products. Official Journal of the European Union L 338/27, 22.12.2005.

Connell, J.J., Shewan, J.M., 1980. Sensory and non-sensory assessment of fish. In: Connell, J. (Ed.), Advances in Fish Science and Technology. Fishing News Books, Farnham, p. 55556.

Dabadé, D.S., den Besten, H.M.W., Azokpota, P., Nout, M.J.R., Hounhouigan, D.J., Zwietering, M.H., 2015. Spoilage evaluation, shelf-life prediction, and potential spoilage organisms of tropical brackish water shrimp (*Penaeus notialis*) at different storage temperatures. Food Microbiol. 48, 8–16.

Dainty, R.H., 1996. Chemical/biochemical detection of spoilage. Int. J. Food Microbiol. 33, 19–33.

Dalgaard, P., 1995. Qualitative and quantitative characte;rization of spoilage bacteria from packed fish. Int. J. Food Microbiol. 26, 319–333.

Dalgaard, P., 2003. Spoilage of seafood. In: Caballero, B., Trugo, L., Finglas, P. (Eds.), Encyclopedia of Food Science and Nutrition. Academic Press, London, pp. 2462–2472.

Dalgaard, P., Gram, L., Huss, H.H., 1993. Spoilage and shelf life of cod fillets packed in vacuum or modified atmospheres. Int. J. Food Microbiol. 19, 283–294.

Dalgaard, P., Mejlholm, O., Huss, H.H., 1996. Conductance method for quantitative determination of *Photobacterium phosphoreum* in fish products. J. Appl. Bacteriol. 81, 57–64.

Dalgaard, P., Mejlholm, O., Christiansen, T.J., Huss, H.H., 1997a. Importance of *Photobacterium phosphoreum* in relation to spoilage of modified atmosphere packed fish products. Lett. Appl. Microbiol. 24, 373–378.

Dalgaard, P., Mejlholm, O., Huss, H.H., 1997b. Application of an iterative approach for development of a microbial model predicting the shelf-life of packed fish. Int. J. Food Microbiol. 38, 169–179.

Dalgaard, P., Buch, P., Silberg, S., 2002. Seafood spoilage predictor –development and distribution of a product-specific application software. Int. J. Food Microbiol. 73, 343–349.

Davidson, C.M., Cronin, F., 1973. Medium for the selective enumeration of lactic acid bacteria from foods. Appl. Microbiol. 26, 439–440.

De Man, J.C., Rogosa, M., Sharpe, M.E., 1960. A medium for the cultivation of lactobacilli. J. Appl. Bacteriol. 23, 130–135.

Debevere, J., Boskou, G., 1996. Effect of modified atmosphere packaging on the TVB/TMA-producing microflora of cod fillets. Int. J. Food Microbiol. 31, 221–229.

Di Natale, C., Olafsdottir, G., Einarsson, S., Martinelli, E., Paolesse, R., D'Amico, A., 2001. Comparison and integration of different electronic noses for freshness evaluation of cod-fish fillets. Sensors and Actuators, B: Chemical 77, 572–578.

Drosinos, E.H., Nychas, G.-J.E., 1996. *Brochothrix thermosphacta*, a dominant microorganism in Mediterranean fresh fish (*Sparus aurata*) stored under modified atmosphere. Italian J. Food Sci. 4, 323–330.

Drosinos, E.H., Nychas, G.-J.E., 1997. Production of acetate and lactate in relation to glucose content during modified atmosphere storage of gilt-head seabream (*Sparus aurata*) at 0.1°C. Food Res. Int. 30, 711–717.

Drosinos, E.H., Lambropoulou, K., Mitre, E., Nychas, G.-J.E., 1997. Attributes of fresh gilt Seabream (*Sparus aurata*) fillets treated with potassium sorbate, sodium gluconate and stored under a modified atmosphere at 0 ± 1°C. J. Appl. Microbiol. 83, 569–575.

Duflos, G., Coin, V.M., Cornu, M., Antinelli, J.F., Malle, P., 2006. Determination of volatile compounds to characterize fish spoilage using headspace/mass spectrometry and solid-phase microextraction/gas chromatography/mass spectrometry. J. Sci. Food Agric. 86, 600–611.

Duflos, G., Theraulaz, L., Giordano, G., Mejean, V., Malle, P., 2010. Quantitative PCR method for evaluating freshness of whiting (*Merlangius merlangus*) and plaice (*Pleuronectes platessa*). J. Food Prot. 73, 1344–1347.

Edirisinghe, R.K.B., Graham, A.J., Taylor, S.J., 2007. Characterisation of the volatiles of yellowfin tuna (*Thunnus albacares*) during storage by solid phase microextraction and GC–MS and their relationship to fish quality parameters. Int. J. Food Sci. Technol. 42, 1139–1147.

Elliker, P.R.A., Anderson, W., Hannesson, G., 1956. An agar culture medium for lactic acid streptococci and lactobacilli. J. Dairy Sci. 39, 1611.

Emborg, J., Laursen, B.G., Rathjen, T., Dalgaard, P., 2002. Microbial spoilage and formation of biogenic amines in fresh and thawed modified atmosphere-packed salmon (*Salmo salar*) at 2°C. J. Appl. Microbiol. 92, 790–799.

Erkan, N., 2007. Sensory, chemical, and microbiological attributes of sea bream (*Sparus aurata*): effect of washing and ice storage. Int. J. Food Prop. 10, 421–434.

Erkan, N., Özden, Ö., 2006. Gutted and un-gutted sea bass (*Dicentrarchus Labrax*) stored in ice: influence on fish quality and shelf-life. Int. J. Food Prop. 9, 331–345.

Erkan, N., Özden, Ö., 2008. Quality assessment of whole and gutted sardines (*Sardina pilchardus*) stored in ice. Int. J. Food Sci. Technol. 43, 1549–1559.

Erkan, N., Tosun, Y.Ş., Alakavuk, Ü.D., Ulusoy, Ş., 2009. Keeping quality of different packaged salted atlantic bonito 'lakerda'. J. Food Biochem. 33, 728–744.

Filipiak, W., Sponring, A., Baur, M.M., Filipiak, A., Ager, C., Wiesenhofer, H., et al., 2012. Molecular analysis of volatile metabolites released specifically by *Staphylococcus aureus* and pseudomonas aeruginosa. BMC Microbiol. 12, 113.

Fuselli, S.R., Casales, M.R., Fritz, R., Yeannes, M.I., 1998. Isolation and characterization of microorganisms associated with marinated anchovy (*Engraulis anchoita*). J. Aquat. Food Prod. Technol. 7, 29–38.

Gennari, M., Tomaselli, S., Cotrona, V., 1999. The microflora of fresh and spoiled sardines (*Sardina pilchardus*) caught in Adriatic (Mediterranean) sea and stored in ice. Food Microbiol. 16, 15–28.

Giannakourou, M.C., Koutsoumanis, K., Nychas, G.J.E., Taoukis, P.S., 2005. Field evaluation of the application of time temperature integrators for monitoring fish quality in the chill chain. Int. J. Food Microbiol. 102, 323–336.

Gonçalves, A.C., Antas, S.E., Nunes, M.L., 2007. Freshness and quality criteria of iced farmed Senegalese sole (*Solea senegalensis*). J. Agric. Food Chem. 55, 345–346.

Gonzalez, C.J., Santos, J.S.A., Garcia-Lopez, M.-L., Gonzalez, N., Otero, A., 2001. Mesophilic aeromonads in wild and aquacultured freshwater fish. J. Food Prot. 64, 687–691.

Gornik, S.G., Albalat, A., Macpherson, H., Birkbeck, H., Neil, D.M., 2011. The effect of temperature on the bacterial load and microbial composition in Norway lobster (*Nephrops norvegicus*) tail meat during storage. J. Appl. Microbiol. 111, 582–592.

Goulas, A.E., Chouliara, I., Nessi, E., Kontominas, M.G., Savvaidis, I.N., 2005. Microbiological, biochemical and sensory assessment of mussels (Mytilus galloprovincialis) stored under modified atmosphere packaging. J. Appl. Microbiol. 98, 752–760.

Gram, L., 2009. Microbiological spoilage of fish and seafood products. In: Spenser, W.H., Doyle, M.P. (Eds.), Compendium of the Microbiological Spoilage of Food and Beverages. Springer Science, New York, NY, pp. 87–119.

Gram, L., Dalgaard, P., 2002. Fish spoilage bacteria – problems and solutions. Curr. Opin. Biotechnol. 13, 262–266.

Gram, L., Huss, H.H., 1996. Microbiological spoilage of fish and fish products. Int. J. Food Microbiol. 33, 121–137.

Gram, L., Melchiorsen, J., 1996. Interaction between fish spoilage bacteria *Pseudomonas* sp. and *Shewanella putrefaciens* in fish extracts and on fish tissue. J. Appl. Bacteriol. 80, 589–595.

Gram, L., Trolle, G., Huss, H.H., 1987. Detection of specific spoilage bacteria on fish stored at high (20°C) and low (0°C) temperatures. Int. J. Food Microbiol. 4, 65–72.

Gram, L., Wedell-Neergaard, C., Huss, H.H., 1990. The bacteriology of fresh and spoiling Lake Victorian Nile perch (*Lates niloticus*). Int. J. Food Microbiol. 10, 303–316.

Gram, L., Ravn, L., Rasch, M., Bartholin Bruhn, J., Christensen, A.B., Givskov, M., 2002. Food spoilage—interactions between food spoilage bacteria. Int. J. Food Microbiol. 78, 79–97.

Haugen, J.E., 2006. The use of chemical sensor array technology, the electronic nose, for detection of boar taint. Acta Vet. Scand. 48 (Suppl. 1), S15.

Herbert, R.A., Ellis, J.R., Shewan, J.M., 1975. Isolation and identifiaction of the volatile sulphides produced during chillstorage of North Sea cod (*Gadus morhua*). J. Sci. Food Agric. 26, 1187–1194.

Herland, H., Esaiassen, M., Cooper, M., Olsen, R.L., 2009. Changes in trimethylamine oxide and trimethylamine in muscle of wild and farmed cod (*Gadus morhua*) during iced storage. Aquac. Res. 41, 95–102.

Hovda, M.B., Lunestad, B.T., Sivertsvik, M., Rosnes, J.T., 2007a. Characterisation of the bacterial flora of modified atmosphere packaged farmed Atlantic cod (*Gadus morhua*) by PCR-DGGE of conserved 16S rRNA gene regions. Int. J. Food Microbiol. 117, 68–75.

Hovda, M.B., Sivertsvik, M., Lunestad, B.T., Lorentzen, G., Rosnes, J.T., 2007b. Characterisation of the dominant bacterial population in modified atmosphere packaged farmed halibut (*Hippoglossus hippoglossus*) based on 16S rDNA-DGGE. Food Microbiol. 24, 362–371.

Hozbor, M.C., Saiz, A.I., Yeannes, M.I., Fritz, R., 2006. Microbiological changes and its correlation with quality indices during aerobic iced storage of sea salmon (*Pseudopercis semifasciata*). LWT Food Sci. Technol. 39, 99–104.

Huang, Y.-R., Guevara Zelaya, M.F., Shiau, C.-Y., 2016. Changes in biochemical compositions and quality of white shrimp (*Litopenaeus vannamei*) during storage. J. Aquat. Food Prod. Technol. 25, 35–45.

Hugenholtz, P., Goebel, B.M., Pace, N.R., 1998. Impact of culture independent studies on the emerging phylogenetic view of bacterial diversity. J. Bacteriol. 180, 4765–4774.

Huss, H.H., 1988. Fresh Fish – Quality and Quality Changes. FAO, Danish International Development Agency, Rome, Sagraf.

Huss, H.H., 1995. Quality and quality changes in fresh fish. FAO Fisheries Technical Paper, No. 348. FAO, Rome.

Huss, H.H., Reilly, A., Karim Ben Embarek, P., 2000. Prevention and control of hazards in seafood. Food Control 11, 149–156.

Jaffrès, E., Sohier, D., Leroi, F., Pilet, M.F., Prévost, H., Joffraud, J.-J., et al., 2009. Study of the bacterial ecosystem in tropical cooked and peeled shrimps using a polyphasic approach. Int. J. Food Microbiol. 131, 20–29.

Jay, J.M., 1986. Microbial spoilage indicators and metabolites. In: Pierson, M.D., Sterm, N.J. (Eds.), Food-Borne Microorganisms and their Toxins: Developing Methodology. Marcel Dekker, Inc., New York, NY, pp. 219–240.

Jessen, F., Nielsen, J., Larsen, E., 2014. Chilling and freezing of fish. In: Boziaris, I.S. (Ed.), Seafood Processing. Technology. Quality & Safety Wiley- Blackwell, IFST Advances in Food Science Series, West Sussex, UK, pp. 33–59.

Joffraud, J.J., Leroi, F., Roy, C., Berdague, J.L., 2001. Characterisation of volatile compounds produced by bacteria isolated from the spoilage flora of cold-smoked salmon. Int. J. Food Microbiol. 66, 175–181.

Jonsdottir, R., Olafsdottir, G., Chanie, E., Haugen, J.E., 2008. Volatile compounds suitable for rapid detection as quality indicators of cold smoked salmon (*Salmo salar*). Food Chem. 109, 184–195.

Jørgensen, B.R., Huss, H.H., 1989. Growth and activity of *Shewanella putrefaciens* isolated from spoiling fish. Int. J. Food Microbiol. 9, 51–62.

Jørgensen, B.R., Gibson, D.M., Huss, H.H., 1988. Microbiological quality and shelf life prediction of chilled fish. Int. J. Food Microbiol. 6, 295–307.

Jørgensen, L.V., Huss, H.H., Dalgaard, P., 2001. Significance of volatile compounds produced by spoilage bacteria in vacuum-packed cold-smoked salmon (*Salmo salar*) analysed by GC-MS and multivariate regression. J. Agric. Food Chem. 49, 2376–2381.

Jung, J.Y., Lee, S.H., Lee, H.J., Jeon, C.O., 2013. Microbial succession and metabolite changes during fermentation of saeu-jeot: traditional Korean salted seafood. Food Microbiol. 34, 360–368.

Kaale, L.D., Magne, E.T., Rustad, T., Kolsaker, K., 2011. Superchilling of food: a review. J. Food Eng. 107, 141–146.

Kielwein, G., 1969. Ein Nährboden zur selektiven Züchtung von Pseudomonaden und Aeromonaden. Arch. Lebensmittelhyg. 20, 131–133.

Kielwein, G., 1971. Pseudomonaden und Aromonaden in Trinkmilch: Ihr Nachweis und ihre Bewertung. Arch. Lebensmittelhyg. 22, 15–19.

Kilinc, B., Cackli, S., 2005. The determination of the shelf life of pasteurized and non-pasteurized sardine (*Sardina pilchardus*) marinades stored at 4°C. Int. J. Food Microbiol. 40, 265–271 (2005).

Kim, H.J., Kim, M.-J., Lee Turner, T., Lee, M.-K., 2014. Pyrosequencing-based analysis of the bacterial community during fermentation of Alaska pollock sikhae: traditional Korean seafood. J. Gen. Appl. Microbiol. 60, 227–233.

Kiyohara, M., Koyanagi, T., Matsui, H., Yamamoto, K., Take, H., Katsuyama, Y., et al., 2012. Changes in microbiota population during fermentation of narezushi as revealed by pyrosequencing analysis. Biosci. Biotechnol. Biochem. 76, 48–52.

Kostaki, M., Giatrakou, V., Savvaidis, I.N., Kontominas, M.G., 2009. Combined effect of MAP and thyme essential oil on the microbiological, chemical and sensory attributes of organically aquacultured sea bass (*Dicentrarchus labrax*) fillets. Food Microbiol. 26, 475–482.

Koutsoumanis, K., 2001. Predictive modeling of the shelf life of fish under nonisothermal conditions. Appl. Environ. Microbiol. 67, 1821–1829.

Koutsoumanis, K., Nychas, G.-J.E., 1999. Chemical and sensory changes associated with microbial flora of Mediterranean Boque (*Boops boops*) stored aerobically at 0, 3, 7 and 10°C. Appl. Environ. Microbiol. 65, 698–706.

Koutsoumanis, K., Nychas, G.-J.E., 2000. Application of a systematic experimental procedure to develop a microbial model for rapid fish shelf-life predictions. Int. J. Food Microbiol. 60, 171–184.

Koutsoumanis, K.P., Taoukis, P.S., Drosinos, E.H., Nychas, G.-J.E., 2000. Applicability of an arrhenius model for the combined effect of temperature and CO_2 packaging on the spoilage microflora of fish. Appl. Environ. Microbiol. 66, 3528–3534.

Koutsoumanis, K., Giannakourou, M.C., Taoukis, P.S., Nychas, G.-J.E., 2002. Application of shelf life decision system (SLDS) to marine cultured fish quality. Int. J. Food Microbiol. 73, 375–382.

Koyanagi, T., Kiyohara, M., Matsui, H., Yamamoto, K., Kondo, T., Katayama, T., et al., 2011. Pyrosequencing survey of the microbial diversity of 'narezushi', an archetype of modern Japanese sushi. Lett. Appl. Microbiol. 53, 635–640.

Koyanagi, T., Nakagawa, A., Kiyohara, M., Matsui, H., Yamamoto, K., Barla, F., Kumagai, H., 2013. Pyrosequencing analysis of microbiota in kaburazushi, a traditional medieval sushi in japan. Biosci. Biotechnol. Biochem. 77, 2125–2130.

Kyrana, V.R., Lougovois, V.P., 2002. Sensory, chemical and microbiological assessment of farm-raised European sea bass (*Dicentrarchus labrax*) stored in melting ice. Int. J. Food Sci. Technol. 37, 319–328.

Kyrana, V.R., Lougovois, V.P., Valsamis, D.S., 1997. Assessment of shelf-life of maricultured gilthead sea bream (*Sparus aurata*) stored in ice. Int. J. Food Sci. Technol. 32, 339–347.

Lalitha, K.V., Surendram, P.K., 2006. Microbiological changes in farm reared freshwater prawn (*Macrobrachium rosenbergii* de Man) in ice. Food Control 17, 802–807.

Laursen, B.G., Leisner, J.J., Dalgaard, P., 2006. Carnobacterium species: effect of metabolic activity and interaction with *Brochothrix thermosphacta* on sensory characteristics of modified atmosphere packed shrimp. J. Agric. Food Chem. 54, 3604–3611.

Lee, J.-L., Levin, R.E., 2007. Quantification of total viable bacteria on fish fillets by using ethidium bromide monoazide real-time polymerase chain reaction. Int. J. Food Microbiol. 118, 312–317.

Lee, S.H., Jung, J.Y., Jeon, C.O., 2015. Bacterial community dynamics and metabolite changes in myeolchi-aekjeot, a Korean traditional fermented fish sauce, during fermentation. Int. J. Food Microbiol. 203, 15–22.

Leistner, L., Gorris, L.G.M., 1995. Food preservation by hurdle technology. Trends Food Sci. Technol. 6, 41–46.

Leroi, F., Joffraud, J. -J., Chevalier, F., Cardinal, M., 1998. Study of microbial ecology of cold-smoked salmon during storage at 8°C. Int. J. Food Microbiol. 39, 111–121.

Leroi, F., Joffraud, J.J., Chevalier, F., Cardinal, M., 2001. Research of quality indices for cold-smoked salmon using a stepwise multiple regression of microbiological counts and physico-chemical parameters. J. Appl. Microbiol. 90, 578–587.

Liston, J., 1980. Microbiology in fishery science. In: Connell, J.J., Staff of Torry Research Station (Eds.), Advances in Fish Science and Technology. Fishing News Books, Farnham, Surrey, pp. 138–157.

López-Caballero, M.E., Huidobro, A., Pastor, A., Tejada, M., 2002. Microflora of gilthead seabream (*Sparus aurata*) stored in ice. Effect of washing. Eur. Food Res. Technol. 215, 396–400.

Losada, V., Rodriguez, O., Miranda, J.M., Barros-Velazquez, J., Aubourg, S.P., 2006. Development of different damage pathways in Norway lobster (*Nephrops norvegicus*) stored under different chilling systems. J. Sci. Food Agric. 86, 1552–1558.

Lyhs, U., Korkeala, H., Vandamme, P., Bjorkroth, J., 2001. *Lactobacillus allimentarius*: a specific spoilage organism in marinated herring. Int. J. Food Microbiol. 64, 355–360 (2001).

Macé, S., Cornet, J., Chevalier, F., Cardinal, M., Pilet, M.-F., Dousset, X., et al., 2012. Characterisation of the spoilage microbiota in raw salmon (*Salmo salar*) steaks stored under vacuum or modified atmosphere packaging combining conventional methods and PCR-TTGE. Food Microbiol. 30, 164–172.

Macé, S., Joffraud, J.-J., Cardinal, M., Malcheva, M., Cornet, J., Lalanne, V., et al., 2013a. Evaluation of the spoilage potential of bacteria isolated from spoiled raw salmon (*Salmo salar*) fillets stored under modified atmosphere packaging. Int. J. Food Microbiol. 160, 227–238.

Macé, S., Mamlouk, K., Chipchakova, S., Prévost, H., Joffraud, J.-J., Dalgaard, P., et al., 2013b. Development of a rapid real-time PCR method as a tool to quantify viable *Photobacterium phosphoreum* bacteria in salmon (*Salmo salar*) steaks. Appl. Environ. Microbiol. 79, 2612–2619.

Malle, P., Poumeyrol, M., 1989. A new chemical criterium for the quality control of fish: trimethylamine/total volatile basic nitrogen (%). J. Food Prot. 52, 419–423.

Mamlouk, K., Macé, S., Guilbaud, M., Jaffrès, E., Ferchichi, M., Prévost, H., et al., 2012. Quantification of viable *Brochothrix thermosphacta* in cooked shrimp and salmon by real-time PCR. Food Microbiol. 30, 173–179.

Manthey, M., Karnop, G., Rehbein, H., 1988. Quality changes of european catfish from warm-water aquaculture during storage ice. Int. J. Food Sci. Technol. 23, 1–9.

Martinez-Alvarez, O., Lopez-Caballero, M.E., Montero, P., Gomez-Guillen, M.C., 2007. Spraying of 4-hexylresorcinol based formulations to prevent enzymatic browning in Norway lobsters (*Nephrops norvegicus*) during chilled storage. Food Chem. 100, 147–155.

Mayo, B., Rachid, C.T.C.C., Alegría, Á., Leite, A.M.O., Peixoto, R.S., Delgado, S., 2014. Impact of next generation sequencing techniques in food microbiology. Curr. Genomics 15, 293–309.

Mendes, R., Goncalves, A., Pestana, J., Pestana, C., 2005. Indole production and deepwater pink shrimp (*Parapenaeus longirostris*) decomposition. Eur. Food Res. Technol. 221, 320–328.

Miller III, A., Scanlan, R.A., Lee, J.S., Libbey, L.M., Morgan, M.E., 1973a. Volatile compounds produced in sterile fish muscle (*Sebastes melanops*) by *Pseudomonas perolens*. Appl. Microbiol. 25, 257–261.

Miller III, A., Scanlan, R.A., Lee, J.S., Libbey, L.M., 1973b. Identification of the volatile compounds produced in sterile fish muscle (*Sebastes melanops*) by *Pseudomonas fragi*. Appl. Microbiol. 25, 952–955.

Miller III, A., Scanlan, R.A., Lee, J.S., Libbey, L.M., 1973c. Volatile compounds produced in sterile fish muscle (*Sebastes melanops*) by *Psueodomonas putrefaciens*, *Pseudomonas fluorescens*, and an *Achromobacter* species. Appl. Microbiol. 26, 18–21.

Moreira, N., Valente, L.M.P., Castro-Cunha, M., Cunha, L.M., de Pinho, P.G., 2013. Effect of storage time and heat processing on the volatile profile of Senegalese sole (*Solea senegalensis* Kaup, 1858) muscle. Food Chem. 138, 2365–2373.

Nisiotou, A., Parlapani, F.F., Kormas, K., Boziaris, I.S., 2014. Old Targets, New Weapons: Food Microbial Communities Revealed With Molecular Tools. In: Boziaris, I.S. (Ed.), Novel Food Preservation and Microbial Assessment Techniques. CRC Press, Taylor & Francis, pp. 277–312.

Noseda, B., Vermeulen, A., Ragaert, P., Devlieghere, F., 2014. Packaging of fish and fishery products. In: Boziaris, I.S. (Ed.), Seafood Processing. Technology. Quality & Safety Wiley-Blackwell, IFST Advances in Food Science Series, West Sussex, UK, pp. 237–261.

Nychas, G.E., Skandamis, P.N., Tassou, C.C., Koutsoumanis, K.P., 2008. Meat spoilage during distribution. Meat Sci. 78, 77–89.

Oehlenschläger, J., 2014. Seafood quality assessment. In: Boziaris, I.S. (Ed.), Seafood Processing. Technology, Quality & Safety Wiley- Blackwell, IFST Advances in Food Science Series, West Sussex, UK, pp. 361–386.

Olafsdottir, G., 2005. Volatile compounds as quality indicators in chilled fish: evaluation of microbial metabolites by an electronic nose. PhD Thesis. Faculty of Science. University of Iceland and Icelandic Fisheries Laboratories, Reykjavík.

Olafsdottir, G., Martinsdbttir, E., Oehlenschbger, J., Dalgaard, P., Jensen, B., Undeland, I., et al., 1997. Methods to evaluate fish freshness in research and industry. A review. Trends Food Sci. Technol. 8, 258–265.

Olafsdottir, G., Jonsdottir, R., Lauzon, H.L., Luten, J., Kristbergsson, K., 2005. Characterization of volatile compounds in chilled cod (*Gadus morhua*) fillets by gas chromatography and detection of quality indicators by an electronic nose. J. Agric. Food Chem. 53, 10140–10147.

Olofsson, T.C., Ahrne, S., Molin, G., 2007. The bacterial flora of vacuum-packed cold-smoked salmon stored at 7°C, identified by direct 16S rRNA gene analysis and pure culture technique. J. Appl. Microbiol. 103, 1364–5072.

Paarup, T., Sanchez, J.A., Moral, A., Christensen, H., Bisgaard, M., Gram, L., 2002. Sensory, chemical and bacteriological changes during storage of iced squid (*Todaropsis eblanae*). J. Appl. Microbiol. 92, 941–950.

Papadopoulos, V., Chouliara, I., Badeka, A., Savvaidis, I.N., Kontominas, M.G., 2003. Effect of gutting on microbiological, chemical and sensory properties of aquacultured sea bass (*Dicentrarchus labrax*) stored in ice. Food Microbiol. 20, 411–420.

Parlapani, F.F., Boziaris, I.S., 2016. Monitoring of spoilage and determination of microbial communities based on 16S rRNA gene sequence analysis of whole sea bream stored at various temperatures. LWT-Food Sci. Technol. 66, 553–559.

Parlapani, F.F., Malouchos, A., Haroutounian, S.A., Boziaris, I.S., 2014. Microbiological spoilage and investigation of volatile profile during storage of sea bream fillets under various conditions. Int. J. Food Microbiol. 189, 153–163.

Parlapani, F.F., Meziti, A., Kormas, K.A., Boziaris, I.S., 2013. Indigenous and spoilage microbiota of farmed sea bream stored in ice identified by phenotypic and 16S rRNA gene analysis. Food Microbiol. 33, 85–89.

Parlapani, F.F., Haroutounian, S.A., Nychas, G.-J.E., Boziaris, I.S., 2015a. Microbiological spoilage and volatiles production of gutted European sea bass stored under air and commercial modified atmosphere package at 2°C. Food Microbiol. 50, 44–53.

Parlapani, F.F., Kormas, K.A., Boziaris, I.S., 2015b. Microbiological changes, shelf-life and identification of initial and spoilage microbiota of sea bream fillets stored under various conditions using 16S rRNA gene analysis. J. Sci. Food Agric. 95, 2386–2394.

Parlapani, F.F., Verdos, G.I., Haroutounian, S.A., Boziaris, I.S., 2015c. The dynamics of *Pseudomonas* and volatilome during the spoilage of gutted sea bream stored at 2°C. Food Control. 55, 257–265.

Pastoriza, L., Sampedro, G., Herrera, J.R., Cabo, M.L., 1996. Effect of modified atmosphere packaging on shelf-life of iced fresh hake slices. J. Sci. Food Agric. 71, 541–547.

Pournis, N., Papavergou, A., Badeka, A., Kontominas, M.G., Savvaidis, I.N., 2005. Shelf-life extension of refrigerated mediterranean mullet (*Mullus surmuletus*) using modified atmosphere packaging. J. Food Protect. 68, 2201–2207.

Reddy, N.R., Schreiber, C.L., Buzard, K.S., Skinner, G.E., Armstrong, D.J., 1994. Shelf life of fresh tilapia fillets packaged in high barrier film with modified atmospheres. J. Food Sci. 59, 260–264.

Reynisson, E., Lauzon, H.L., Magnusson, H., Hreggvidsson, G.O., Marteinsson, V.T., 2008. Rapid quantitative monitoring method for the fish spoilage bacteria *Pseudomonas*. J. Environ. Monit. 10, 1357–1362.

Robson, A.A., Kelly, M.S., Latchford, J.W., 2007. Effect of temperature on the spoilage rate of whole, unprocessed crabs: *Carcinus maenas*, *Necora puber* and *Cancer pagurus*. Food Microbiol. 24, 419–424.

Rodríguez, C.J., Besteiro, I., Pascual, C., 1999. Biochemical changes in freshwater rainbow trout (*Oncorhynchus mykiss*) during chilled storage. J. Sci. Food Agric. 79, 1473–1480.

Rodriguez, O., Barros-Velazquez, J., Ojea, A., Pineiro, C., Aubourg, S.P., 2003. Evaluation of sensory and microbiological changes and identification of proteolytic bacteria during the iced storage of farmed turbot (*Psetta maxima*). J. Food Sci. 68, 2764–2771.

Roh, S.W., Kim, K.-H., Nam, Y.-D., Chang, H.-W., Park, E.-J., Bae, J.-W., 2010. Investigation of archaeal and bacterial diversity in fermented seafood using barcoded pyrosequencing. Multidiscipl. J. Microb. Ecol. 4, 1–16.

Rudi, K., Maugesten, T., Hannevik, S.E., Nissen, H., 2004. Explorative multivariate analyses of 16S rRNA gene data from microbial communities in modified-atmosphere-packed salmon and coalfish. Appl. Environ. Microbiol. 70, 5010–5018.

Sallam, K.I., 2008. Effect of marination process on the microbiological quality of Pacific saury (*Cololabis saira*) during vacuum-packaged storage at 4°C. Int. J. Food Sci. Technol. 43, 220–228.

Shewan, J., Hobbs, G., Hodgkiss, W., 1960a. A determinative scheme for the identification of certain genera of Gram negative bacteria, with special reference to the Pseudomonadaceae. J. Appl. Bacteriol. 23, 379–390.

Shewan, J., Hobbs, G., Hodgkiss, W., 1960b. The *Pseudomonas* and *Achromobacter* groups of bacteria in the spoilage of marine white fish. J. Appl. Bacteriol. 23, 463–468.

Shewan, J.M., 1962. The bacteriology of fresh and spoiling fish and some related chemical changes. Recent Adv. Food Sci. 1, 167–193.

Sidhu, G.S., Montgomery, W.A., Brown, M.A., 1974. Post mortem changes and spoilage in rock lobster muscle. II. Role of amino acids in bacterial spoilage and production of volatile bases in the muscle of *Jasus novae-hollandiae*. J. Food Technol. 9, 371–380.

Sivertsvik, M., Jeksrud, W.K., Rosnes, J.T., 2002. Review. A review of modified atmosphere packaging of fish and fishery products – significance of microbial growth, activities and safety. Int. J. Food Sci. Technol. 37, 107–127.

Soncin, S., Chiesa, M.L., Panseri, S., Biondi, P., Cantoni, C., 2008. Determination of volatile compounds of precooked prawn (*Penaeus vannamei*) and cultured gilthead sea bream (*Sparus aurata*) stored in ice as possible spoilage markers using solid phasemicroextraction and gas chromatography/mass spectrometry. J. Sci. Food Agric. 89, 436–442.

Stamatis, N., Arkoudelos, J., 2007. Quality assessment of *Scomber colias japonicus* under modified atmosphere and vacuum packaging. Food Control 18, 292–300.

Svanevik, C.M., Lunestad, B.T., 2011. Characterisation of the microbiota of Atlantic mackerel (*Scomber scombrus*). Int. J. Food Microbiol. 151, 164–170.

Sveinsdottir, K., Martinsdottir, E., Hyldig, G., Jorgensen, B., Kristbergsson, K., 2002. Application of quality index method (QIM) scheme in shelf-life study of farmed Atlantic salmon (*Salmo salar*). J. Food Sci. 67, 1570–1579.

Taliadourou, D., Papadopoulos, V., Domvridou, E., Savvaidis, I.N., Kontominas, M.G., 2003. Microbiological, chemical and sensory changes of whole and filleted Mediterranean aquacultured sea bass (*Dicentrarchus labrax*) stored in ice. J. Sci. Food Agric. 83, 1373–1379.

Tejada, M., Heras, C.D., Kent, M., 2007. Changes in the quality indices during ice storage of farmed Senegalese sole (*Solea senegalensis*). Eur. Food Res. Technol. 225, 225–232.

Truelstrup Hansen, L., Huss, H.H., 1998. Comparison of the microflora isolated from spoiled cold-smoked salmon from three smokehouses. Food Res. Int. 31, 703–711.

Tryfinopoulou, P., Tsakalidou, E., Nychas, G.-J.E., 2002. Characterization of *Pseudomonas* spp. associated with spoilage of gilt-head sea bream stored under various conditions. Appl. Environ. Microbiol. 68, 65–72.

Tryfinopoulou, P., Tsakalidou, E., Vancanneyt, M., Hoste, B., Swings, J., Nychas, G.-J.E., 2007. Diversity of *Shewanella* population in fish *Sparus aurata* harvested in the Aegean Sea. J. Appl. Microbiol. 103, 711–721.

Tsigarida, E., Boziaris, I.S., Nychas, G.-J.E., 2003. Bacterial synergism or antagonism in a gel cassette system. Appl. Environ. Microbiol. 69, 7204–7209.

Tsironi, T., Gogou, E., Velliou, E., Taoukis, P.S., 2008. Application and validation of the TTI based chill chain management system SMAS (safety monitoring and assurance system) on shelf life optimization of vacuum packed chilled tuna. Int. J. Food Microbiol. 128, 108–115.

Tuckey, N.P.L., Day, J.R., Miller, M.R., 2013. Determination of volatile compounds in New Zealand Greenshell™ mussels (*Perna canaliculus*) during chilled storage using solid phase microextraction gas chromatography–mass spectrometry. Food Chem. 136, 218–223.

Van Spreekens, K.J.A., 1974. The suitability of a modification of long and Hammer's medium for the enumeration of more fastidious bacteria from fresh fishery products. Arch. für Lebensmittelhygiene 25, 213–219.

Varlet, V., Fernandez, X., 2010. Sulfur-containing volatile compounds in seafood: occurrence, odorant properties and mechanisms of formation. Food Sci. Technol. Int. 16, 463–503.

Villar, M., de Ruiz Holgado, A.P., Sanchez, J.J., Trucco, R.E., Oliver, G., 1985. Isolation and characterization of *Pediococcus halophilus* from salted anchovies (*Engraulis anchoita*). Appl. Environ. Microbiol. 49, 664–666.

Wierda, R.L., Fletcher, G., Xu, L., Dufour, J.-P., 2006. Analysis of volatile compounds as spoilage indicators in fresh king salmon (*Oncorhynchus tshawytscha*) during storage using SPME-GC-MS. J. Agric. Food Chem. 54, 8480–8490.

Chapter 4

Spore-Forming Bacteria

Fabienne Remize
Université de La Réunion, ESIROI, UMR QualiSud, Saint-Denis, Réunion

4.1 INTRODUCTION AND CURRENT TAXONOMY

As a result of their ubiquitous nature, bacterial spore-formers are widely detected in all kinds of foods. Pathogenic species focused research work from the beginning of the 20th century, but rapidly, spoilage spore-forming bacteria were also studied, because of the high economic losses resulting from food spoilage. Bacterial spore-formers spoil food from the metabolic activity of their vegetative forms. No food-related ecological niche can be preserved from spoilage caused by these bacteria, due to their remarkable diversity.

Mesophilic spore-forming bacteria belong essentially to two taxonomic groups: the *Bacillales* order and the *Clostridium* genus, depending on whether they are aerobic or not. *Bacillales* gathers aerobic, facultative aerobic, or facultative anaerobic bacteria. The *Bacillaceae* family covers 19 genera including *Bacillus* and *Anoxybacillus*, which comprise several thermophilic species (Vos et al., 2009). *Bacillus subtilis*, easy to cultivate and to sporulate under laboratory conditions, became a model, in particular for the study of sporulation and germination mechanisms, and many molecular tools were developed. *Bacillus cereus* is the only species with food pathogenicity, and some of its members are able to grow at low temperatures. Other *Bacillales* families include genera involved in food spoilage, such as *Alicyclobacillus*, particularly acidophilic, and *Paenibacillus*, which means "almost *Bacillus*." The genus *Clostridium* gathers all mesophilic anaerobic spore-formers encountered in food. Several pathogens are described, among those the species *Clostridium botulinum* which focused attention due to its production of potent neurotoxins and *Clostridium perfringens*, the other member of the genus tracked for food poisoning. The anaerobic status of these bacteria is a strong driver for niche differentiation compared to *Bacillales* representatives. The genus *Clostridium* contains a very high number of species, with 168 species validly described (Vos et al., 2009). Some of them are moderate psychrophilic, thus adapted to food products stored at low temperature.

The Microbiological Quality of Food. DOI: http://dx.doi.org/10.1016/B978-0-08-100502-6.00007-8

The taxonomy of thermophilic spore-formers has evolved considerably since the 1990s. All aerobic thermophilic spore-forming bacteria are included in the *Bacillales* order. The genus *Geobacillus*, previously attached to *Bacillus* spp., was described in 2001 and emended in 2012 (Coorevits et al., 2012; Nazina et al., 2001). Used as a microbiological control for the thermal processing of food, *Geobacillus stearothermophilus* causes defects in dairy products and "flat sour" of canned food, recognized for decades as one of the main causes of economic loss in the canned food sector. As analyzed by Zeigler (2014), this bacterium has been found everywhere people have thought to search. *Anoxybacillus*, composed of 10 species, is a moderate thermophilic aerobic spore-former and *Anoxybacillus flavithermus* is a hygiene indicator for milk products. Lastly, several strains of *Bacillus* species, such as *B. smithii*, *B. licheniformis,* or *B. coagulans* are moderately thermophilic. The taxonomy of thermophilic anaerobic bacteria is much more complex (for an extensive review, see Wagner and Wiegel, 2008). It evolved from the isolation of anaerobic bacteria from environmental samples and from taxonomical classification changes within the genus *Clostridium.* Thermophilic anaerobic spore-forming bacteria found in food are gathered in the *Thermoanaerobacetales* order, but belong to many different genera, the most frequently described as food spoiler being *Moorella* spp., *Thermoanaerobacter* spp., *Thermoanaerobacterium* spp., *Desulfotomaculum* spp. Thermophilic anaerobic spore-forming bacteria are generally gas-producers, causing swelling of food packages and generating off-odors, also for acid foodstuffs.

A simplified tree, showing the main spore-forming bacteria of interest in food, and built up from 16S rRNA coding sequences, is shown (Fig. 4.1). From above taxonomical considerations, it is clear that foodborne spore-forming bacterial spoilers cannot be considered as a whole from the taxonomical point of view. This statement suggests that the development of molecular methods for the detection of these spoilers would necessarily be applicable to a sub-assembly. Hence, molecular methods were developed toward a category of food or toward a taxonomic group.

4.2 DIVERSITY OF SPORE-FORMERS

4.2.1 Spore-Formation, Resistance, and Recovery

Formation, resistance, and germination of spores have been widely studied, essentially for *Bacillus*. Sporulation occurs in response to a high cellular density, nutritional limitation or cellular communication (Piggot and Hilbert, 2004; Sonenshein, 2000). A phosphorelay is induced, involving the master regulator Spo0A, which activates several metabolic pathways and initiates an asymmetric cellular division. For *Clostridium* spp., the mechanisms are described in far less depth. The master Spo0A regulator is present, activates downstream pathways, but is not activated through a phosphorelay (Talukdar et al., 2015).

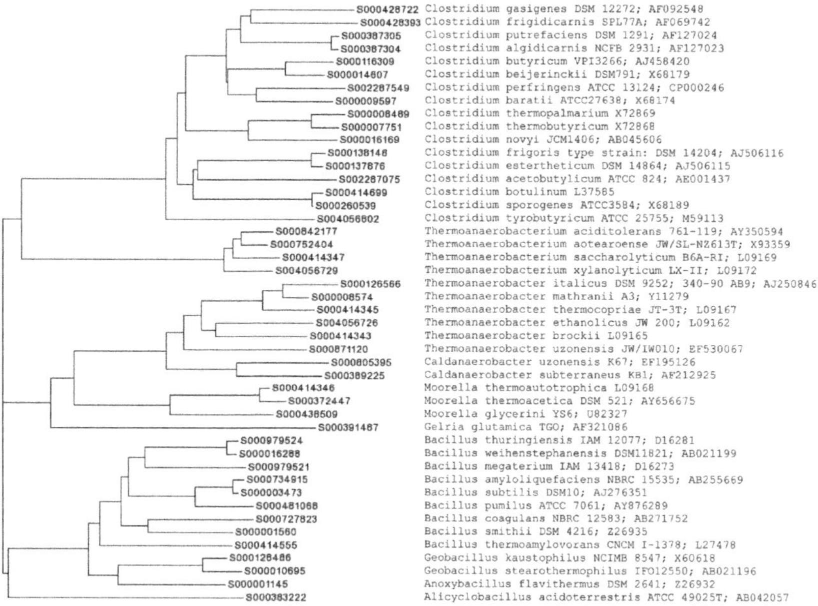

FIGURE 4.1 Neighbor-joining tree, according to Saitou and Nei (1987), from 46 spore-forming bacterium (type strains) sequences of 16S rRNA coding region.

Spore resistance is acquired during sporulation. Spore resistance originates in several spore structures, including a strongly dehydrated, calcium DPA (dipicolinic acid)-rich and SASP (small acid-soluble proteins)-rich protoplasm. Spore envelopes are multilayered, and usually are distinguished from spore core to external medium. They are composed of an internal membrane, a cortex, an external membrane, a coat, and possibly an exosporium. Each layer presents specific structural, biochemical, and permeability properties. Two very different examples of spore structures are shown in Fig. 4.2: multiple layers can be seen from the core to external medium, and the structure of these two highly heat-resistant spores of thermophiles clearly appear differently organized.

Besides being species specific, sporulation conditions largely influence the formation of spore structures. The consequences on sporulation rate and spore resistance of sporulation temperature and medium composition have been studied in a number of important species for food (Baril et al., 2011; Durand et al., 2015b; Garcia et al., 2010; Gounina-Allouane et al., 2008; Leguérinel et al., 2007; Lenz and Vogel, 2015; Molva and Baysal, 2014; Nguyen Thi Minh et al., 2008; Peck et al., 1995). The major influence of sporulation temperature is proven and a body of evidence suggests that sporulation performed at optimal vegetative growth temperature results in the highest sporulation efficiency and the highest spore resistance. Sporulation mechanisms of thermophilic species are not yet described, but

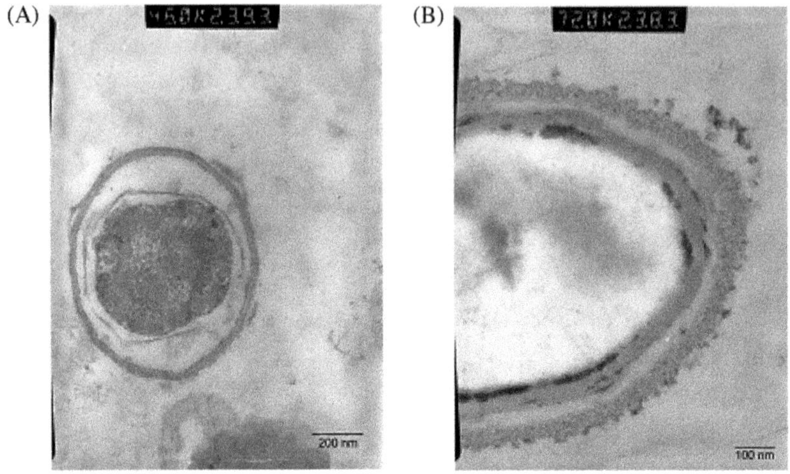

FIGURE 4.2 Electronic microscopy of spores, (A) *G. stearothermophilus*, (B) *Moorella thermoacetica. Photo: CTCPA, France.*

recent data on spore formation and resistance underline the relationship between optimal growth conditions and spore heat resistance (André et al., 2013; Mtimet et al., 2016).

The inner membrane of a spore harbors germination receptors, which differ in number and function between *Bacillus* and *Clostridium*. Germination involves DPA release and further cortex lytic enzymes activity (Keijser et al., 2007; Moir, 2006; Olguín-Araneda et al., 2015; Setlow, 2003; Xiao et al., 2011). Prediction of spore behavior in food, especially after a stabilization treatment, is challenging for shelf life determination and spoilage risk evaluation. In food, recovery is the most relevant parameter to consider, compared to survival per se. Recovery comprises germination, restoration of metabolic activity and formation of the first vegetative cell able to divide. The influence of spore sublethal treatments and of environmental conditions during treatment and recovery steps has focused attention because of consequences on food preservation. Recent attempts in modeling recovery of bacterial spores show an apparent decrease of decimal reduction rate (Bassi et al., 2016; Esteban et al., 2015; Mafart and Leguérinel, 1997; Peck et al., 1992; Ter Beek et al., 2011).

4.2.2 Phenotypic Traits of Species

In accordance with their taxonomic classification in different groups, spore-forming bacteria exhibit very diverse phenotypic traits. Due to its composition, food is not limited in carbon, nitrogen, and energy. Table 4.1 shows atmosphere, temperature, and pH conditions for growth of several pathogens and food spoilage spore-forming bacteria.

Spore-forming bacteria phenotypes cover the range of food storage temperatures. Generally, temperature is one of the first parameters used to control microbial development, and the shelf life of fresh foods is ensured by refrigeration. This is the case for meat, fish, fruit, and vegetables, as minimally processed foods. For these products, the use of modified atmosphere, e.g., low oxygen levels, impairs the growth of aerobic bacteria and molds but favors anaerobic psychrotolerant microflora. The use of modified atmosphere regarding the control of spoilage by spore-forming bacteria is not relevant and could be regarded as a risky practice since most of these bacteria

TABLE 4.1 Atmosphere, Temperature, and pH Growth Conditions of Spore-Forming Bacteria

Species	Oxygen Requirement	Temperature Range (°C)[a]	pH Range[a]
A. acidoterrestris[b]	Aerobic (microaerobiosis possible)	42–53 (25; 70)	4 (2.2; 6.5)
A. flavithermus	Facultative anaerobic	60 (aerobiosis) (30; 72)	6–9
B. subtilis	Aerobic	28–30 (5–20; 45–55)	5.5–8.5
B. cereus	Facultative anaerobic	37 (6–20; 40–45)	
B. coagulans	Facultative aerobic	40–57 (30; 57–61)	7 (4; 10.5–11)
C. estertheticum	Anaerobic	6–8 (–; 13)	6.5–7.2 (5.5; 7.8)
C. perfringens	Anaerobic	37–45 (6; 50)	5.5–8
C. tyrobutyricum	Anaerobic	30–37 (25; 45)	
G. stearothermophilus[c]	Aerobic	40–60 (30–45; 70–75)	6–8 (–; 9)
M. thermoacetica	Anaerobic	56–60 (47; 65)	6.6–6.8 (5.7; 7.65)
Thermoanaerobacterium saccharolyticum	Anaerobic	60 (45; 70)	6 (5; 7.5)

[a]*Optimum (minimum; maximum).*
[b]*Additional information from Bevilacqua et al. (2015).*
[c]*Additional information from Coorevits et al. (2012).*
Source: From Vos, P., Garrity, G., Jones, D., Krieg, N.R., Ludwig, W., Rainey, F.A., et al. (Eds.), 2009. Bergey's Manual of Systematic Bacteriology Volume 3: The Firmicutes. http://dx.doi.org/10.1007/978-0-387-68489-5.

are aero-anaerobic or strictly anaerobic. As expected, vacuum-packed food or canned food are not protected from spoilage by spore-formers by their low oxygen availability. For instance, the so-called blown-pack defect of vacuum-packed food is often caused by psychrophilic *Clostridium* (Broda et al., 1996b; Ho, 2009) and the bacteria the most frequently isolated from spoiled canned food after elevated temperature incubation are *G. stearothermophilus* and *M. thermoacetica* (André et al., 2013).

Naturally acid and acidified foods are naturally preserved from spoilage and a low pH is used to increase preservation time or to decrease thermal treatments for the same shelf life. In effect, pathogenic spore-formers do not grow below pH 4.5 and the species encountered in spoilage of acidic foods are not the same as for nonacidic foods. Low pH foods are mainly spoiled by yeasts, molds, and lactic acid bacteria. However, *Alicyclobacillus acidoterrestris* is a spore-forming bacteria found as a spoilage agent for fruit juices, fruit-based pasteurized foods, carbonated drinks, lemonade, and iced teas.

In accordance with the heat resistance of spore, thermal treatments favor spoilage by spore-forming bacteria through the inactivation of vegetative forms in refrigerated processed foods of extended durability (REPFEDs) or of the lowest resistant spores in canned food. This is also the case for dairy products such as milk powder, which involves mild heat treatment. Spoilage of such a food requires three successive events: resistance of spore to the applied treatment, germination of spore, and outgrowth in the food. Many parameters influence spore resistance. Spore-formation conditions are generally more restrictive than vegetative growth conditions and when conditions are closest to the optima, the highest resistance of spore is observed (Baril, 2011; Baril et al., 2012; Mtimet et al., 2016). Moreover, at the species level, a relationship between growth temperature optimum and spore resistance has been reported (Table 4.2).

4.2.3 Intra-Species and Other Sources of Variability

Because of its high potential impact in risk management of food spoilage, the variability of spore-forming bacteria behavior has been extensively studied. This variability relies on two different causes: variability on environmental conditions and intra-species variability. Both causes result in variability of growth parameters, of spore resistance, and of spore germination.

The effects of environmental conditions at the individual spore level are observed by single-cell dispersion of behavior from a homogeneous bacterial population. The first observations of subpopulation behavior were done from survival curves which do not fit the linear Weibullian model (Albert and Mafart, 2005; Geeraerd et al., 2000). The observation of tails and shoulders suggests that subpopulations are more or less sensitive to the inactivation stress applied. In particular, from a practical point of view, the survival of bacteria observed from tailed shaped curves would lead to increased thermal

TABLE 4.2 Optimal Growth Temperature and Spore Resistance Parameters, for Strains Isolated from Spoiled Canned Food

Species	Growth Temperature (°C)	Calculated Temperature (°C) for $D = 10$ min	$Z_{loglinear}$ Values (°C)
M. thermoacetica	55	125.3–125.9	6.1–8.2
G. stearothermophilus	55	113.6–115.6	7.4–9.4
Caladanaerobius zeae/ figensis/ polysaccharolyticus	55	118.1	6.9
Thermoanaerobacterium aotearoense	37–55	116.3–116.5	7.1–7.4
Thermoanaerobacterium thermosaccharolyticum	55	116.4	5.8
Thermoanaerobacter pseudothanolicus	55	109.1	6.9
B. smithii	37–55	105.0–113.1	6.1–7.5
C. thermopalmarium/ thermobutyricum	55	108.3	6.4
Paenibacillus sp.	37	101.4–108.0	6.2–8.0
B. subtilis	37	103.2–108.0	8.9–10.1
B. coagulans	37–55	98.6–106.8	6.2–8.5
B. licheniformis	37–55	94.9–102.4	6.7–8.7
C. haemolyticum	37	100.6	8.2
C. sporogenes	37	98.3–100.1	6.6–7.8
P. macerans	37	99.2–99.4	7.6–7.8
P. polymyxa	37	89.0–90.3	7.8–7.9
C. novii	37	84.7	7.9

Source: Adapted from André, S., Zuber, F., Remize, F., 2013. Thermophilic spore-forming bacteria isolated from spoiled canned food and their heat resistance. Results of a French ten-year survey. Int. J. Food Microbiol. 165, 134–143. http://dx.doi.org/10.1016/j.ijfoodmicro.2013.04.019.

treatments to succeed in complete spore inactivation. It became clear that the nonlinear inactivation shapes were in close relationship to environmental conditions during the heat treatment. The effect of pH or weak organic acids, but more generally of food composition, was showed on growth, heat resistance, and germination (Eijlander et al., 2011; Hornstra et al., 2009; Jagannath et al., 2005; Mafart et al., 2010; Mtimet et al., 2016).

For instance, heat treatment exerts a double effect on spore germination time, on the one hand on the extent of time variability within a population and on the other hand on the shape of distribution curves for time until outgrowth (Stringer et al., 2009, 2011). The deterministic first-order kinetics model was shown to be valid only for large populations (Corradini et al., 2010). For small populations, stochastic models, which might involve Bayesian inference to provide uncertainty distributions for the parameters (Barker et al., 2005), seem adequate.

For intra-species variability, the first studies were dedicated to pathogen species: *C. botulinum* (Esty and Meyer, 1922) and *B. cereus*. The determination of phenotypic parameters of strains which belong to different pathogenicity groups was performed, though with different approaches depending on the age of the study (Alberto et al., 2003; Carlin et al., 2013). Recently, a low diversity was shown for growth temperature, pH, and salt level of *G. stearothermophilus* strains isolated from spoiled canned food, which is promising for the development of spoilage control strategies (Durand et al., 2015b).

Several recent models developed to propose a quantitative assessment of spoilage risk take into account the two causes of variability, subpopulation variability within a homogeneous population and strain effect for a single species (Rigaux et al., 2013, 2014). On the opposite, other models do not use as input any species behavior and consider a global behavior whatever the species present (Membré and van Zuijlen, 2011). Lastly, a recent model that uses deterministic process settings, deterministic values not related to management of the process, and probabilistic inputs, was developed (Pujol et al., 2015). As stated by Augustin in 2011 (Augustin, 2011), it seems that risk assessment of spoilage by spore-forming bacteria is a never-ending story.

4.3 ECOLOGY OF SPORE-FORMING BACTERIA

4.3.1 Origin in Food

The soil is considered as the primary habitat for endospore-formers. For a single species, spore population can be as high as 10^5-10^6 spore/g of soil. Soil carriage of spores results in widespread contamination.

A primary contamination pathway is due to the presence of spores on the plant surface. These plants can be directly used as raw material or ingredients for food production. This was observed from contamination levels of vegetables, green peas, and carrots, used for canned food, and from genetic profiles of isolates found on raw materials and further on vegetables being processed (Durand et al., 2015a; Sevenier et al., 2012). This was also observed from genetic fingerprint of *B. cereus* isolates obtained from soils, processed and stored zucchini purée (Guinebretiere et al., 2003) and from *Alicyclobacillus* in fruit juices (Smit et al., 2011). Another pathway for food contamination involves feed, as silage or not, and further released animal feces. In addition, insect guts carry spore-forming bacteria, as do cow teats. As a consequence, milk and dairy products, but also meat, are contaminated by spore-forming bacteria (Carlin, 2011; Doyle et al., 2015). Air exposure did not appear to be an effective contamination flow.

As a consequence of the contamination routes, spore-forming bacteria are detected in raw materials, ingredients, and processed foods, often with high prevalence, very diverse population levels, and a high diversity of species (Burgess et al., 2010; Carlin, 2011; Ceuppens et al., 2013; Lücking et al., 2013; Postollec et al., 2012; Samapundo et al., 2014). Postollec et al. (2012) examined 90 food matrices, comprising raw materials, ingredients such as dehydrated, egg-based, and dairy ingredients, and processed foods, for the presence of several genera and species of spore-forming bacteria. They observed 100% prevalence of spore-forming bacteria from raw materials and more than 73% of foods, with or without spoilage, carried spore-formers. In most of samples, both *Bacillus* and *Clostridium* were present and the counts showed up to 10^2-10^3 cfu/g of spore-formers. In Samapundo et al. (2014), 38% of food samples, including meat containing meals, sauces, and cheeses, contained psychrotolerant spore-formers, at low level. These different studies underlined the high diversity of species detected from raw materials, ingredients, food, and the food processing environment.

4.3.2 Persistence on Processing Lines

Food processing facilities are contaminated with spore-forming bacteria by food flows. Sampling and analysis of surfaces of a processing line of soups showed a high prevalence of *Bacillus*, both as vegetative cells and as spores (Oomes et al., 2007). Similar observations were further made from a processing line dedicated to green peas and carrots canning (Durand et al., 2015a). The high prevalence, combined with elevated temperature and long residence time, suggests that sporulation occurs at zones of food debris accumulation during the working day.

In another study, it was shown that spore level was very low after cleaning (Guinebretiere et al., 2003). Effective cleaning is thus a critical point for spore spoilage management. Sporulation of mono-species or mixed *Bacillus*, organized in biofilm on a stainless steel surface, was shown (Faille et al., 2014; Ryu and Beuchat, 2005). Sporulation dynamics, but also spore heat resistance, are affected by a biofilm organization (Hayrapetyan et al., 2016). Biocide efficiency against spores, in relation to their adhesion properties and their ability to form biofilms, deserves to be investigated. In particular, very few studies investigate the biocide effect under realistic conditions during food processing lines cleaning. It was shown that certain species such as *M. thermoacetica*, adhered to stainless steel surfaces, were difficult to inactivate with commercial products application conditions (André et al., 2012).

The importance of good farming and manufacturing practices is highlighted in several studies (Arias et al., 2013; Doyle et al., 2015; Durand et al., 2015a; Miller et al., 2015a). Spore contamination levels depend on many factors, like ecological niche and diversity, animal feeding, irrigation practices, harvest methods, weather, temperature of food during processing and residence time, cross-contamination zones, such as blanching systems and pipes, microbial inactivation steps over processing which favors the highest resistant bacteria, and effective cleaning.

4.3.3 Inactivation

Several processing steps are designed to inactivate spores. Thermal treatments are still the most frequent, but irradiation and pressure-assisted treatments are increasingly used, respectively for dehydrated products and for ready-to-eat foods. The F_0 value, which is the equivalent exposure time at 121.1°C applied in the process, in reference to steam heat treatments, is crucial. F_0 is determined experimentally over retort processing by temperature record at the cold point of the food package and application of a z-value of 10°C.

To set up or to optimize a food thermal treatment, an estimate of F_0 is calculated from the Weibull model, the most frequently used model at the industrial level. The decimal reduction time, D, and the z-value or temperature coefficient, corresponding to the number of degrees necessary to change D by 10, are estimated from bibliography data, whereas initial spore load in the food is experimentally determined over time. The choice of a reference species, relevant to the processed food, is often difficult.

The reference spoilage species should have the following properties:

- Being naturally present in the food before thermal treatment
- Causing spoilage of the food, thus being able to recover in the food
- Forming spores inactivated by the process, partially or totally, and more resistant than the reference pathogen.

For instance, a reference species for low-acid canned food could be *G. stearothermophilus* and for acid fruit-based products *A. acidoterrestris*.

4.4 FOOD SPOILAGE

4.4.1 Refrigerated Food

Fresh meat can be preserved for more than 60–70 days by vacuum packaging and a strict control of storage temperature. The storage conditions are defined to ensure prevention of toxin production by *C. botulinum*. However, spoilage was observed as meat softening, formation of odors described as dairy or sulfurous and production of exudates, caused by clostridial species (Broda et al., 1996a; Cavill et al., 2011; Lawson et al., 1994; Mills et al., 2014).

This defect, called "blown pack," can occur sporadically even if no temperature abuse has occurred. Blown pack results of carbon dioxide production by cold-tolerant species, such as *Clostridium estertheticum* and *Clostridium gasigenes*. Other species, *Clostridium algidicarnis* and *Clostridium frigidicarnis*, produce off-flavors from meat, but without gas formation. For fresh or chilled meats however, spoilage by lactic acid bacteria is more frequently reported.

Surimi, which is a fish-paste obtained from cooking fish pieces with salt, sugar, and starch, is stored for long time at low temperatures. The presence of psychrotolerant spore-formers, exhibiting proteolytic and amylolytic activities, has been described (Coton et al., 2011; Tsuda et al., 2015). The described species were most frequently *Bacillus simplex*, *B. subtilis*, and *Sporosancina aquimarina*.

REPFEDs are exposed to spoilage by spore formers as cooking inactivates vegetative microorganisms. The presence of *B. cereus* group bacteria, *Bacillus pumilus* and other related species was shown. These bacteria are not gas-producers but exhibit several enzymatic activities that could result in odors and texture changes (Samapundo et al., 2014).

4.4.2 Canned Food

Heat processing of low-acid canned food was initially designed to inactivate the most resistant spores of *C. botulinum*. However, F_0 values increased over time to inactivate even more resistant spores of species that cause spoilage. Nowadays, provided that packaging is well-sealed and processing controlled, industrially produced canned food remains stable at ambient temperature for years. Observation of spoilage only occurs when canned food is stored at elevated temperature for several days. This could be the case if food is transported overseas by boat or is exposed to elevated temperatures in warehouses. Two main defects are described: flat sour and hard swell. The former results mainly from *G. stearothermophilus* recovery and outgrowth whereas the latter is the consequence of *M. thermoacetica* development. These two bacteria are the causative agents of 71% of observed spoilage cases (André et al., 2013). Other defects are due to *Thermoanaerobacterium* spp. and moderate thermophilic *Bacillus* spp. Ultra high temperature (UHT) dairy products can be contaminated by highly resistant spores of aerobic bacteria, and *G. stearothermophilus*, *B. subtilis*, and *B. coagulans* were reported as causative agents of flat sour or slimy milk or cream (Burgess et al., 2010; Pujol et al., 2015).

Acid or acidified pasteurized foods are most frequently spoiled by heat-resistant molds such as *Byssochlamys fulva* or *Byssochlamys nivea*, or by lactic acid bacteria (Aneja et al., 2014). Spoilage by endospore-forming bacteria, in particular by *Alicyclobacillus* spp., might be difficult to detect as they do not produce gas, but rather produce a smoky, medicinal, and antiseptic off-odor resulting in volatile phenols formation

(Danyluk et al., 2011; Smit et al., 2011). Several moderate acidophilic *Bacillus* spp., like *B. coagulans*, are involved in spoilage of tomato concentrates or apple juice (Thompson, 1981).

4.4.3 Cheeses

Late blowing of semi-hard and hard cheeses occurs from dihydrogen production in cheese paste, causing cracks, and slits. This defect is due to lactate, residual sugars, and citrate consumption by clostridial species. These bacteria produce butyric acid, and to a lesser extent acetic and propionic acids. *Clostridium tyrobutyricum*, *C. butyricum*, *C. beijerinckii*, and *C. sporogenes* are the species the most frequently associated with this spoilage (Brändle et al., 2016; Gómez-Torres et al., 2015; Le Bourhis et al., 2007).

Ricotta is a soft cheese made from heat coagulation of whey proteins. Ricotta pink discoloration was observed and related to *B. cereus* group, *Paenibacillus* spp. and clostridial species detection (Sattin et al., 2016).

4.4.4 Bread

Bread ropiness has been described for many years and corresponds to unpleasant odors resembling overripe pineapples, bitter taste, and later on, crumb discoloration and sticky and greasy bread (Thompson et al., 1993). It occurs within 12 hours of cooking and is more frequent in undercooked breads, bran breads, and organic breads. The defect results from amylase activity and from extracellular polysaccharides production. Several *Bacillus* species are involved in this defect, *B. licheniformis*, *B. subtilis*, and *B. pumilus* being initially described as the most frequent ones (Pepe et al., 2003; Rosenkvist, 1995). A recent study pointed out the high frequency of *B. amyloliquefaciens* and *B. cereus* group from spoiled bread of Italian origin (Valerio et al., 2012).

4.4.5 Dry Ingredients

Due to their low water activity, dry ingredients are protected from microbial spoilage. However, a lot of attention was placed on these products as they are ingredients for canned foods and REPFEDs. Dry milk, milk proteins, whey powder, flour, starch, and spices exhibit a high prevalence of spore-formers (Burgess et al., 2010; Guinebretiere et al., 2003; Iurlina et al., 2006; Miller et al., 2015b; Postollec et al., 2012; Rückert et al., 2004; Sadiq et al., 2016; Watterson et al., 2014). The most frequent species, found in dairy powders, with differing proportions depending on the study, are *G. stearothermophilus*, *B. licheniformis*, and *A. flavithermus*.

4.5 METHODS TO TRACK SPORE-FORMING BACTERIA

4.5.1 Cultural Methods

The selective effect of classical cultural methods used to enumerate or isolate spore-forming bacteria is based on incubation conditions rather than on medium composition. In certain cases, such as for *Alicyclobacillus*, media composition was optimized to increase selectivity (Smit et al., 2011). Incubation temperature and atmosphere are combined to promote the growth of groups of spore-formers. Generally, the cultivation of anaerobes takes longer than aerobe cultivation and the media used are enriched with peptone, yeast extract or meat-liver. The lack of specificity of cultural methods requires further steps for endospore-forming bacteria identification. For this purpose, phenotypical tests are still in use, but most of the time a molecular identification is performed.

A heat selection step preceding plating is used to select heat-resistant or highly heat-resistant spores: the selective effect depends on time and temperature chosen for this step. In France, a method was established (Afnor, NF V08-602) in 2011 for enumeration of bacterial spores in foods before thermal treatment of canned foods. This method involves sample preparation in buffered peptone water, followed by a heat selection step at 100°C for 10 min and further plating on two media (Dextrose Tryptone Agar and Meat Liver Glucose Agar with 2 g L^{-1} yeast extract) which are incubated at 37°C or at 55°C, under aerobiosis (Dextrose Tryptone Agar) or anaerobiosis (Meat Liver Glucose Agar with 2 g L^{-1} yeast extract).

4.5.2 Qualitative Methods

Qualitative methods used to detect spore-forming bacteria trigger vegetative forms. They often imply polymerase chain reaction (PCR), either to specifically detect a species or group of species directly from foods or processing environment samples or to identify isolates.

These methods, when applied directly to food samples, rely on two steps which have to be carefully set up: DNA extraction and PCR amplification. DNA extraction has to ensure a constant detection threshold. As food matrixes composition is very diverse, DNA extraction methods have to cope with specific DNA binding molecules and PCR inhibitors. That is the main reason for the development of methods which are food-specific (Brightwell and Clemens, 2012; Juvonen and Haikara, 2009; Reindl et al., 2014; Tilsala-Timisjarvi and Alatossava, 2004). In that view, filtration of liquids, such as milk, is an advantage in isolating bacteria from the food matrix. To avoid inhibitors, detection after an enrichment step or after isolation is often preferred.

Depending on the attempts, a single species, a single genus or larger groups of species are triggered. For instance, single species detection methods were developed for *Bacillus sporothermodurans* in dairy products

(Scheldeman et al., 2002) and for *C. tyrobutyricum* in milk (Herman et al., 1995). However, in most cases, a group of species are detected, such as psychrotolerant bacteria in vacuum-packed chilled meat (Broda et al., 2003), thermophilic bacilli in milk powders (Rueckert et al., 2005), *Thermoanaerobacterium* and *Thermoanaerobacter* in canned food (Dotzauer et al., 2002), and *Alicyclobacillus* spp. in fruit juice (Connor et al., 2005). In a perspective of detection methods automatization, simultaneous or multiplex detection of several species has been developed (Morandi et al., 2015; Pennacchia et al., 2014; Postollec et al., 2010; Prevost et al., 2010).

With these methods, low detection levels of vegetative forms, a few cells per g, were achieved after an enrichment step. The detection level increases dramatically when detection is performed directly from food, or when spores are triggered. As a consequence, these methods are very useful in identifying the bacteria which spoilt a food sample, but not applicable unless spore recovery is performed by enrichment to raw materials, ingredients or processing environment samples.

4.5.3 Quantitative Methods

Spore quantification is of crucial importance in monitoring contamination before heat treatments. Recently, a quantitative PCR method was set up, based on the Spo0A master regulator gene, to enumerate spores (Bueche et al., 2013; Ratphitagsanti et al., 2012). However, the actual detection levels, above 10^4 spores per gram, are not low enough for a direct application of the method to quantify spores from raw materials, ingredients or processing environment samples.

The most recent studies focused on alternatives to PCR methods. A method combining flow cytometry with fluorescent-labeled specific polyclonal antibodies was successfully used to detect 10 spores per mL of milk (Lavilla et al., 2010). In another study, assays for spore permeabilization in order to use fluorescent in situ hybridization techniques were applied to three *Bacillus* species (Filion et al., 2009). These approaches seem full of promise but require optimization at this stage.

4.6 FUTURE TRENDS

Monitoring of spore-forming bacteria, from farm to processed food, in relation to spoilage risk, relies essentially on enumeration or on isolation plus identification. The studies that consider the global ecosystem with respect to foods spoilt by spore formers are very few, although the use of microbiota analysis tools such as Next-Generation Sequencing is widely applied to fermentation process and fermented foods. A PCR-DGGE method adapted for *Clostridium* spp. detection in hard cheese was developed more than 10 years ago (Cocolin et al., 2004). Only recently, two studies (Bassi et al., 2015;

Sattin et al., 2016) focused their interest on the global ecosystem in cheeses and during soft cheese shelf life. As demonstrated for fermented products, the microbiota analysis is a potent tool for pointing out bacterial interactions in food communities and the possible influence of these interactions on the final quality of food. Despite a necessary improvement of methods, which have to be adapted toward a more efficient nucleic acid extraction from spores, global ecosystem analysis exhibits a high potential for finding new ways in the management of food spoilage risk by spore-forming bacteria.

REFERENCES

Albert, I., Mafart, P., 2005. A modified Weibull model for bacterial inactivation. Int. J. Food Microbiol. 100, 197–211. Available from: http://dx.doi.org/10.1016/j.ijfoodmicro.2004.10.016.

Alberto, F., Broussolle, V., Mason, D.R., Carlin, F., Peck, M.W., 2003. Variability in spore germination response by strains of proteolytic *Clostridium botulinum* types A, B and F. Lett. Appl. Microbiol. 36, 41–45.

André, S., Hédin, S., Remize, F., Zuber, F., 2012. Evaluation of peracetic acid sanitizers efficiency against spores isolated from spoiled cans in suspension and on stainless steel surfaces. J. Food Prot. 75, 371–375. Available from: http://dx.doi.org/10.4315/0362-028X.JFP-11-329.

André, S., Zuber, F., Remize, F., 2013. Thermophilic spore-forming bacteria isolated from spoiled canned food and their heat resistance. Results of a French ten-year survey. Int. J. Food Microbiol. 165, 134–143. Available from: http://dx.doi.org/10.1016/j.ijfoodmicro.2013.04.019.

Aneja, K.R., Dhiman, R., Aggarwal, N.K., Aneja, A., 2014. Emerging preservation techniques for controlling spoilage and pathogenic microorganisms in fruit juices. Int. J. Microbiol. 2014, 758942. Available from: http://dx.doi.org/10.1155/2014/758942.

Arias, C., Oliete, B., Seseña, S., Jimenez, L., Pérez-Guzmán, M.D., Arias, R., 2013. Importance of on-farm management practices on lactate-fermenting *Clostridium* spp. spore contamination of Manchega ewe milk: determination of risk factors and characterization of *Clostridium* population. Small Rumin. Res. 111, 120–128. Available from: http://dx.doi.org/10.1016/j.smallrumres.2012.11.030.

Augustin, J., 2011. Challenges in risk assessment and predictive microbiology of foodborne spore-forming bacteria. Food Control 28, 211–213. Available from: http://dx.doi.org/10.1016/j.fm.2010.05.003.

Baril, E., 2011. Quantification de l'influence de l'environnement sur la formation et la thermorésistance des spores bactériennes. Universite de Bretagne Occidentale, Brest, France.

Baril, E., Coroller, L., Postollec, F., Leguerinel, I., Boulais, C., Carlin, F., et al., 2011. The wet-heat resistance of *Bacillus weihenstephanensis* KBAB4 spores produced in a two-step sporulation process depends on sporulation temperature but not on previous cell history. Int. J. Food Microbiol. 146, 57–62. Available from: http://dx.doi.org/10.1016/j.ijfoodmicro.2011.01.042.

Baril, E., Coroller, L., Couvert, O., El Jabri, M., Leguerinel, I., Postollec, F., et al., 2012. Sporulation boundaries and spore formation kinetics of *Bacillus* spp. as a function of temperature, pH and a(w). Food Microbiol. 32, 79–86. Available from: http://dx.doi.org/10.1016/j.fm.2012.04.011.

Barker, G.C., Malakar, P.K., Peck, M.W., 2005. Germination and growth from spores: variability and uncertainty in the assessment of food borne hazards. Int. J. Food Microbiol. 100, 67–76. Available from: http://dx.doi.org/10.1016/j.ijfoodmicro.2004.10.020.

Bassi, D., Puglisi, E., Cocconcelli, P.S., 2015. Understanding the bacterial communities of hard cheese with blowing defect. Food Microbiol. 52, 106–118. Available from: http://dx.doi.org/10.1016/j.fm.2015.07.004.

Bassi, D., Colla, F., Gazzola, S., Puglisi, E., Delledonne, M., Cocconcelli, P.S., 2016. Transcriptome analysis of *Bacillus thuringiensis* spore life, germination and cell outgrowth in a vegetable-based food model. Food Microbiol. 55, 73–85. Available from: http://dx.doi.org/10.1016/j.fm.2015.11.006.

Bevilacqua, A., Mischitelli, M., Pietropaolo, V., Ciuffreda, E., Sinigaglia, M., Corbo, M.R., 2015. Genotypic and phenotypic heterogeneity in *Alicyclobacillus acidoterrestris*: a contribution to species characterization. PLoS One 10, e0141228. Available from: http://dx.doi.org/10.1371/journal.pone.0141228.

Brändle, J., Domig, K.J., Kneifel, W., 2016. Relevance and analysis of butyric acid producing clostridia in milk and cheese. Food Control 67, 96–113. Available from: http://dx.doi.org/10.1016/j.foodcont.2016.02.038.

Brightwell, G., Clemens, R., 2012. Development and validation of a real-time PCR assay specific for *Clostridium estertheticum* and *C. estertheticum*-like psychrotolerant bacteria. Meat Sci. 92, 697–703. Available from: http://dx.doi.org/10.1016/j.meatsci.2012.06.025.

Broda, D.M., DeLacy, K.M., Bell, R.G., Braggins, T.J., Cook, R.L., 1996a. Psychrotrophic *Clostridium* spp. associated with "blown pack" spoilage of chilled vacuum-packed red meats and dog rolls in gas-impermeable plastic casings. Int. J. Food Microbiol. 29, 335–352.

Broda, D.M., DeLacy, K.M., Bell, R.G., Penney, N., 1996b. Association of psychrotrophic *Clostridium* spp. with deep tissue spoilage of chilled vacuum-packed lamb. Int. J. Food Microbiol. 29, 371–378.

Broda, D.M., Boerema, J.A., Bell, R.G., 2003. PCR detection of psychrophilic *Clostridium* spp. causing "blown pack" spoilage of vacuum-packed chilled meats. J. Appl. Microbiol. 94, 515–522.

Bueche, M., Wunderlin, T., Roussel-Delif, L., Junier, T., Sauvain, L., Jeanneret, N., et al., 2013. Quantification of endospore-forming firmicutes by quantitative PCR with the functional gene spo0A. Appl. Environ. Microbiol. 79, 5302–5312. Available from: http://dx.doi.org/10.1128/AEM.01376-13.

Burgess, S.A., Lindsay, D., Flint, S.H., 2010. Thermophilic bacilli and their importance in dairy processing. Int. J. Food Microbiol. 144, 215–225. Available from: http://dx.doi.org/10.1016/j.ijfoodmicro.2010.09.027.

Carlin, F., 2011. Origin of bacterial spores contaminating foods. Food Microbiol. 28, 177–182. Available from: http://dx.doi.org/10.1016/j.fm.2010.07.008.

Carlin, F., Albagnac, C., Rida, A., Guinebretière, M.-H., Couvert, O., Nguyen-The, C., 2013. Variation of cardinal growth parameters and growth limits according to phylogenetic affiliation in the *Bacillus cereus* group. Consequences for risk assessment. Food Microbiol. 33, 69–76. Available from: http://dx.doi.org/10.1016/j.fm.2012.08.014.

Cavill, L., Renteria-Monterrubio, A.L., Helps, C.R., Corry, J.E.L., 2011. Detection of cold-tolerant clostridia other than *Clostridium estertheticum* in raw vacuum-packed chill-stored meat. Food Microbiol. 28, 957–963. Available from: http://dx.doi.org/10.1016/j.fm.2011.01.003.

Ceuppens, S., Boon, N., Uyttendaele, M., 2013. Diversity of *Bacillus cereus* group strains is reflected in their broad range of pathogenicity and diverse ecological lifestyles. FEMS Microbiol. Ecol. 84, 433–450. Available from: http://dx.doi.org/10.1111/1574-6941.12110.

Cocolin, L., Innocente, N., Biasutti, M., Comi, G., 2004. The late blowing in cheese: a new molecular approach based on PCR and DGGE to study the microbial ecology of the alteration process. Int. J. Food Microbiol. 90, 83–91. Available from: http://dx.doi.org/10.1016/S0168-1605(03)00296-4.

Connor, C.J., Luo, H., Mcspadden, B.B., Wang, H.H., 2005. Development of a real-time PCR-based system targeting the 16S rRNA gene sequence for rapid detection of *Alicyclobacillus* spp. in juice products. Int. J. Food Microbiol. 99, 229–235. Available from: http://dx.doi.org/10.1016/j.ijfoodmicro.2004.08.016.

Coorevits, A., Dinsdale, A.E., Halket, G., Lebbe, L., De Vos, P., Van Landschoot, A., et al., 2012. Taxonomic revision of the genus *Geobacillus*: emendation of *Geobacillus*, *G. stearothermophilus*, *G. jurassicus*, *G. toebii*, *G. thermodenitrificans* and *G. thermoglucosidans* (nom. corrig., formerly "*thermoglucosidasius*"); transfer of *Bacillus thermantarcticus*. Int. J. Syst. Evol. Microbiol. 62, 1470–1485. Available from: http://dx.doi.org/10.1099/ijs.0.030346-0.

Corradini, M.G., Normand, M.D., Peleg, M., Corrandini, M.G., 2010. Stochastic and deterministic model of microbial heat inactivation. J. Food Sci. 75, R59–R70. Available from: http://dx.doi.org/10.1111/j.1750-3841.2009.01494.x.

Coton, M., Denis, C., Cadot, P., Coton, E., 2011. Biodiversity and characterization of aerobic spore-forming bacteria in surimi seafood products. Food Microbiol. 28, 252–260. Available from: http://dx.doi.org/10.1016/j.fm.2010.03.017.

Danyluk, M.D., Friedrich, L.M., Jouquand, C., Goodrich-Schneider, R., Parish, M.E., Rouseff, R., 2011. Prevalence, concentration, spoilage, and mitigation of *Alicyclobacillus* spp. in tropical and subtropical fruit juice concentrates. Food Microbiol. 28, 472–477. Available from: http://dx.doi.org/10.1016/j.fm.2010.10.008.

Dotzauer, C., Ehrmann, M.A., Vogel, R.F., 2002. Occurrence and detection of *Thermoanaerobacterium* and *Thermoanaerobacter* in canned food. Food Technol. Biotechnol. 40, 21–26.

Doyle, C.J., Gleeson, D., Jordan, K., Beresford, T.P., Ross, R.P., Fitzgerald, G.F., et al., 2015. Anaerobic sporeformers and their significance with respect to milk and dairy products. Int. J. Food Microbiol. 197, 77–87. Available from: http://dx.doi.org/10.1016/j.ijfoodmicro.2014.12.022.

Durand, L., Planchon, S., Guinebretiere, M.-H., André, S., Carlin, F., Remize, F., 2015a. Contamination pathways of spore-forming bacteria in a vegetable cannery. Int. J. Food Microbiol. 202, 10–19. Available from: http://dx.doi.org/10.1016/j.ijfoodmicro.2015.02.019.

Durand, L., Planchon, S., Guinebretiere, M.-H., Carlin, F., Remize, F., 2015b. Genotypic and phenotypic characterization of foodborne *Geobacillus stearothermophilus*. Food Microbiol. 45 Pt A, 103–110. Available from: http://dx.doi.org/10.1016/j.fm.2014.01.015.

Eijlander, R.T., Abee, T., Kuipers, O.P., 2011. Bacterial spores in food: how phenotypic variability complicates prediction of spore properties and bacterial behavior. Curr. Opin. Biotechnol. 22, 180–186. Available from: http://dx.doi.org/10.1016/j.copbio.2010.11.009.

Esteban, M.-D., Conesa, R., Huertas, J.-P., Palop, A., 2015. Effect of thymol in heating and recovery media on the isothermal and non-isothermal heat resistance of *Bacillus* spores. Food Microbiol. 48, 35–40. Available from: http://dx.doi.org/10.1016/j.fm.2014.11.016.

Esty, J., Meyer, K., 1922. The heat resistance of the spores of *B. botulinus* and allied anaerobes XI. J. Infect. Dis. 31, 650–663.

Faille, C., Bénézech, T., Midelet-Bourdin, G., Lequette, Y., Clarisse, M., Ronse, G., et al., 2014. Sporulation of *Bacillus* spp. within biofilms: a potential source of contamination in food processing environments. Food Microbiol. 40, 64–74. Available from: http://dx.doi.org/10.1016/j.fm.2013.12.004.

Filion, G., Laflamme, C., Turgeon, N., Ho, J., Duchaine, C., 2009. Permeabilization and hybridization protocols for rapid detection of *Bacillus* spores using fluorescence in situ hybridization. J. Microbiol. Methods 77, 29–36. Available from: http://dx.doi.org/10.1016/j.mimet.2008.12.009.

Garcia, D., van der Voort, M., Abee, T., 2010. Comparative analysis of *Bacillus weihenstephanensis* KBAB4 spores obtained at different temperatures. Int. J. Food Microbiol. 140, 146–153. Available from: http://dx.doi.org/10.1016/j.ijfoodmicro.2010.04.005.

Geeraerd, A.H., Herremans, C.H., Van Impe, J.F., 2000. Structural model requirements to describe microbial inactivation during a mild heat treatment. Int. J. Food Microbiol. 59, 185–209. Available from: http://dx.doi.org/10.1016/S0168-1605(00)00362-7.

Gómez-Torres, N., Garde, S., Peirotén, Á., Ávila, M., 2015. Impact of *Clostridium* spp. on cheese characteristics: microbiology, color, formation of volatile compounds and off-flavors. Food Control 56, 186–194. Available from: http://dx.doi.org/10.1016/j.foodcont.2015.03.025.

Gounina-Allouane, R., Broussolle, V., Carlin, F., 2008. Influence of the sporulation temperature on the impact of the nutrients inosine and l-alanine on *Bacillus cereus* spore germination. Food Microbiol. 25, 202–206. Available from: http://dx.doi.org/10.1016/j.fm.2007.08.002.

Guinebretiere, M.H., Girardin, H., Dargaignaratz, C., Carlin, F., 2003. Contamination flows of *Bacillus cereus* and spore-forming aerobic bacteria in a cooked, pasteurized and chilled zucchini purée processing line. Int. J. Food Microbiol. 82, 223–232.

Hayrapetyan, H., Abee, T., Nierop Groot, M., 2016. Sporulation dynamics and spore heat resistance in wet and dry biofilms of *Bacillus cereus*. Food Control 60, 493–499. Available from: http://dx.doi.org/10.1016/j.foodcont.2015.08.027.

Herman, L.M.F., De Block, J.H.G.E., Waes, G.M.A.V.J., 1995. A direct PCR detection method for *Clostridium tyrobutyricum* spores in up to 100 milliliters of raw milk. Appl. Environ. Microbiol. 61, 4141–4146.

Ho, L., 2009. Characterization of *Clostridium* spp. from "blown-pack", Chill-Stored, Vacuum Packaged Beef. University of Alberta Libraries, Edmonton, Canada.

Hornstra, L.M., Ter Beek, A., Smelt, J.P., Kallemeijn, W.W., Brul, S., 2009. On the origin of heterogeneity in (preservation) resistance of *Bacillus* spores: Input for a "systems" analysis approach of bacterial spore outgrowth. Int. J. Food Microbiol. 134, 9–15. Available from: http://dx.doi.org/10.1016/j.ijfoodmicro.2009.02.011.

Iurlina, M.O., Saiz, A.I., Fuselli, S.R., Fritz, R., 2006. Prevalence of *Bacillus* spp. in different food products collected in Argentina. LWT - Food Sci. Technol. 39, 105–110. Available from: http://dx.doi.org/10.1016/j.lwt.2005.01.006.

Jagannath, A., Tsuchido, T., Membré, J., 2005. Comparison of the thermal inactivation of *Bacillus subtilis* spores in foods using the modified Weibull and Bigelow equations. Food Control 22, 233–239. Available from: http://dx.doi.org/10.1016/j.fm.2004.05.004.

Juvonen, R., Haikara, A., 2009. Amplification facilitators and pre-processing methods for PCR detection of strictly anaerobic beer-spoilage bacteria of the class clostridia in brewery samples. J. Inst. Brew. 115, 167–176.

Keijser, B.J.F., Ter Beek, A., Rauwerda, H., Schuren, F., Montijn, R., Spek, H., et al., 2007. Analysis of temporal gene expression during *Bacillus subtilis* spore germination and outgrowth. J. Bacteriol. 189, 3624–3634. Available from: http://dx.doi.org/10.1128/JB.01736-06.

Lavilla, M., Marzo, I., de Luis, R., Perez, M.D., Calvo, M., Sanchez, L., 2010. Detection of *Clostridium tyrobutyricum* spores using polyclonal antibodies and flow cytometry. J. Appl. Microbiol. 108, 488–498. Available from: http://dx.doi.org/10.1111/j.1365-2672.2009.04435.x.

Lawson, P., Dainty, R.H., Kristiansen, N., Berg, J., Collins, M.D., 1994. Characterization of a psychrotrophic *Clostridium* causing spoilage in vacuum-packed cooked pork: description of *Clostridium algidicarnis* sp. nov. Lett. Appl. Microbiol. 19, 153–157.

Le Bourhis, A.-G., Doré, J., Carlier, J.-P., Chamba, J.-F., Popoff, M.-R., Tholozan, J.-L., 2007. Contribution of *C. beijerinckii* and *C. sporogenes* in association with *C. tyrobutyricum* to the butyric fermentation in Emmental type cheese. Int. J. Food Microbiol. 113, 154–163. Available from: http://dx.doi.org/10.1016/j.ijfoodmicro.2006.06.027.

Leguérinel, I., Couvert, O., Mafart, P., 2007. Modelling the influence of the sporulation temperature upon the bacterial spore heat resistance, application to heating process calculation. Int. J. Food Microbiol. 114, 100–104. Available from: http://dx.doi.org/10.1016/j.ijfoodmicro.2006.10.035.

Lenz, C.A., Vogel, R.F., 2015. Differential effects of sporulation temperature on the high pressure resistance of *Clostridium botulinum* type E spores and the interconnection with sporulation medium cation contents. Food Microbiol. 46, 434–442. Available from: http://dx.doi.org/10.1016/j.fm.2014.09.005.

Lücking, G., Stoeckel, M., Atamer, Z., Hinrichs, J., Ehling-Schulz, M., 2013. Characterization of aerobic spore-forming bacteria associated with industrial dairy processing environments and product spoilage. Int. J. Food Microbiol. 166, 270–279. Available from: http://dx.doi.org/10.1016/j.ijfoodmicro.2013.07.004.

Mafart, P., Leguérinel, I., 1997. Modelling the heat stress and the recovery of bacterial spores. Int. J. Food Microbiol. 37, 131–135. Available from: http://dx.doi.org/10.1016/S0168-1605(97)00061-5.

Mafart, P., Leguérinel, I., Couvert, O., Coroller, L., 2010. Quantification of spore resistance for assessment and optimization of heating processes: a never-ending story. Food Microbiol. 27, 568–572. Available from: http://dx.doi.org/10.1016/j.fm.2010.03.002.

Membré, J., van Zuijlen, A., 2011. A probabilistic approach to determine thermal process setting parameters: application for commercial sterility of products. Int. J. Food Microbiol. 144, 413–420.

Miller, R.A., Kent, D.J., Boor, K.J., Martin, N.H., Wiedmann, M., 2015a. Different management practices are associated with mesophilic and thermophilic spore levels in bulk tank raw milk. J. Dairy Sci. 98, 4338–4351. Available from: http://dx.doi.org/10.3168/jds.2015-9406.

Miller, R.A., Kent, D.J., Watterson, M.J., Boor, K.J., Martin, N.H., Wiedmann, M., 2015b. Spore populations among bulk tank raw milk and dairy powders are significantly different. J. Dairy Sci. 98, 8492–8504. Available from: http://dx.doi.org/10.3168/jds.2015-9943.

Mills, J., Donnison, A., Brightwell, G., 2014. Factors affecting microbial spoilage and shelf-life of chilled vacuum-packed lamb transported to distant markets: a review. Meat Sci. 98, 71–80. Available from: http://dx.doi.org/10.1016/j.meatsci.2014.05.002.

Moir, A., 2006. How do spores germinate? J. Appl. Microbiol. 101, 526–530. Available from: http://dx.doi.org/10.1111/j.1365-2672.2006.02885.x.

Molva, C., Baysal, A.H., 2014. Effect of sporulation medium on wet-heat resistance and structure of *Alicyclobacillus acidoterrestris* DSM 3922-type strain spores and modeling of the inactivation kinetics in apple juice. Int. J. Food Microbiol. 189, 82–88. Available from: http://dx.doi.org/10.1016/j.ijfoodmicro.2014.07.033.

Morandi, S., Cremonesi, P., Silvetti, T., Castiglioni, B., Brasca, M., 2015. Development of a triplex real-time PCR assay for the simultaneous detection of *Clostridium beijerinckii*, *Clostridium sporogenes* and *Clostridium tyrobutyricum* in milk. Anaerobe 34, 44–49. Available from: http://dx.doi.org/10.1016/j.anaerobe.2015.04.005.

Mtimet, N., Guégan, S., Durand, L., Mathot, A.-G., Venaille, L., Leguérinel, I., et al., 2016. Effect of pH on *Thermoanaerobacterium thermosaccharolyticum* DSM 571 growth, spore heat resistance and recovery. Food Microbiol. 55, 64–72. Available from: http://dx.doi.org/10.1016/j.fm.2015.11.015.

Nazina, T.N., Tourova, T.P., Poltaraus, A.B., Novikova, E.V., Grigoryan, A.A., Ivanova, A.E., et al., 2001. Taxonomic study of aerobic thermophilic bacilli: descriptions of *Geobacillus subterraneus* gen. nov., sp. nov. and *Geobacillus uzenensis* sp. nov. from petroleum reservoirs and transfer of *Bacillus stearothermophilus*, *Bacillus thermocatenulatus*. Int. J. Syst. Evol. Microbiol. 51, 433–446.

Nguyen Thi Minh, H., Perrier-Cornet, J.-M., Gervais, P., 2008. Effect of the osmotic conditions during sporulation on the subsequent resistance of bacterial spores. Appl. Microbiol. Biotechnol. 80, 107–114. Available from: http://dx.doi.org/10.1007/s00253-008-1519-x.

Olguín-Araneda, V., Banawas, S., Sarker, M.R., Paredes-Sabja, D., 2015. Recent advances in germination of *Clostridium* spores. Res. Microbiol. 166, 236–243. Available from: http://dx.doi.org/10.1016/j.resmic.2014.07.017.

Oomes, S.J.C.M., Van Zuijlen, A.C.M., Hehenkamp, J.O., Witsenboer, H., van der Vossen, J.M. B.M., Brul, S., 2007. The characterisation of *Bacillus* spores occurring in the manufacturing of (low acid) canned products. Int. J. Food Microbiol. 120, 85–94. Available from: http://dx.doi.org/10.1016/j.ijfoodmicro.2007.06.013.

Peck, M.W., Fairbairn, D.A., Lund, B.M., 1992. The effect of recovery medium on the estimated heat-inactivation of spores of non-proteolytic *Clostridium botulinum*. Lett. Appl. Microbiol. 15, 146–151.

Peck, M.W., Evans, R.I., Fairbairn, D.A., Hartley, M.G., Russell, N.J., 1995. Effect of sporulation temperature on some properties spores of non-proteolytic *Clostridium botulinum*. Int. J. Food Microbiol. 28, 289–297.

Pennacchia, C., Breeuwer, P., Meyer, R., 2014. Development of a Multiplex-PCR assay for the rapid identification of *Geobacillus stearothermophilus* and *Anoxybacillus flavithermus*. Food Microbiol. 43, 41–49. Available from: http://dx.doi.org/10.1016/j.fm.2014.05.002.

Pepe, O., Blaiotta, G., Moschetti, G., Greco, T., Villani, F., 2003. Rope-producing strains of *Bacillus* spp. from wheat bread and strategy for their control by lactic acid bacteria. Appl. Environ. Microbiol. 69, 2321–2329. Available from: http://dx.doi.org/10.1128/AEM.69.4.2321.

Piggot, P.J., Hilbert, D.W., 2004. Sporulation of *Bacillus subtilis*. Curr. Opin. Microbiol. 7, 579–586. Available from: http://dx.doi.org/10.1016/j.mib.2004.10.001.

Postollec, F., Bonilla, S., Baron, F., Jan, S., Gautier, M., Mathot, A.G., et al., 2010. A multiparametric PCR-based tool for fast detection and identification of spore-forming bacteria in food. Int. J. Food Microbiol. 142, 78–88. Available from: http://dx.doi.org/10.1016/j.ijfoodmicro.2010.06.005.

Postollec, F., Bernard, M., Divanac'h, M.-L., Pavan, S., Sohier, D., Mathot, A.-G., et al., 2012. Tracking spore-forming bacteria in food: from natural biodiversity to selection by processes. Int. J. Food Microbiol. 158, 1–8. Available from: http://dx.doi.org/10.1016/j.ijfoodmicro.2012.03.004.

Prevost, S., Andre, S., Remize, F., 2010. PCR detection of thermophilic spore-forming bacteria involved in canned food spoilage. Curr. Microbiol. 61, 525–533. Available from: http://dx.doi.org/10.1007/s00284-010-9648-8.

Pujol, L., Albert, I., Magras, C., Johnson, N.B., Membré, J.-M., 2015. Estimation and evaluation of management options to control and/or reduce the risk of not complying with commercial sterility. Int. J. Food Microbiol. 213, 124–129. Available from: http://dx.doi.org/10.1016/j.ijfoodmicro.2015.05.014.

Ratphitagsanti, W., Park, E., Lee, C.S., Amos Wu, R.-Y., Lee, J., 2012. High-throughput detection of spore contamination in food packages and food powders using tiered approach of

ATP bioluminescence and real-time PCR. LWT - Food Sci. Technol. 46, 341–348. Available from: http://dx.doi.org/10.1016/j.lwt.2011.09.003.

Reindl, A., Dzieciol, M., Hein, I., Wagner, M., Zangerl, P., 2014. Enumeration of clostridia in goat milk using an optimized membrane filtration technique. J. Dairy Sci. 97, 6036–6045. Available from: http://dx.doi.org/10.3168/jds.2014-8218.

Rigaux, C., Denis, J.-B., Albert, I., Carlin, F., 2013. A meta-analysis accounting for sources of variability to estimate heat resistance reference parameters of bacteria using hierarchical Bayesian modeling: estimation of D at 121.1°C and pH 7, zT and zpH of *Geobacillus stearothermophilus*. Int. J. Food Microbiol. 161, 112–120. Available from: http://dx.doi.org/10.1016/j.ijfoodmicro.2012.12.001.

Rigaux, C., André, S., Albert, I., Carlin, F., 2014. Quantitative assessment of the risk of microbial spoilage in foods. Prediction of non-stability at 55°C caused by *Geobacillus stearothermophilus* in canned green beans. Int. J. Food Microbiol. 171, 119–128. Available from: http://dx.doi.org/10.1016/j.ijfoodmicro.2013.11.014.

Rosenkvist, H., 1995. Contamination profiles and characterisation of *Bacillus* species in wheat bread and raw materials for bread production. Int. J. Food Microbiol. 26, 353–363. Available from: http://dx.doi.org/10.1016/0168-1605(94)00147-X.

Rückert, A., Ronimus, R.S., Morgan, H.W., Ruckert, A., 2004. A RAPD-based survey of thermophilic bacilli in milk powders from different countries. Int. J. Food Microbiol. 96, 263–272. Available from: http://dx.doi.org/10.1016/j.ijfoodmicro.2004.03.020.

Rueckert, A., Ronimus, R.S., Morgan, H.W., 2005. Development of a rapid detection and enumeration method for thermophilic bacilli in milk powders. J. Microbiol. Methods 60, 155–167. Available from: http://dx.doi.org/10.1016/j.mimet.2004.09.008.

Ryu, J.-H., Beuchat, L.R., 2005. Biofilm formation and sporulation by *Bacillus cereus* on a stainless steel surface and subsequent resistance of vegetative cells and spores to chlorine, chlorine dioxide, and a peroxyacetic acid-based sanitizer. J. Food Prot. 68, 2614–2622.

Sadiq, F.A., Li, Y., Liu, T., Flint, S., Zhang, G., He, G., 2016. A RAPD based study revealing a previously unreported wide range of mesophilic and thermophilic spore formers associated with milk powders in China. Int. J. Food Microbiol. 217, 200–208. Available from: http://dx.doi.org/10.1016/j.ijfoodmicro.2015.10.030.

Saitou, N., Nei, M., 1987. The neighbor-joining method: a new method for reconstructing phylogenetic trees. Mol. Biol. Evol. 4, 406–425.

Samapundo, S., Devlieghere, F., Xhaferi, R., Heyndrickx, M., 2014. Incidence, diversity and characteristics of spores of psychrotolerant spore formers in various REPFEDS produced in Belgium. Food Microbiol. 44, 288–295. Available from: http://dx.doi.org/10.1016/j.fm.2014.06.020.

Sattin, E., Andreani, N.A., Carraro, L., Fasolato, L., Balzan, S., Novelli, E., et al., 2016. Microbial dynamics during shelf-life of industrial *Ricotta* cheese and identification of a *Bacillus* strain as a cause of a pink discolouration. Food Microbiol. 57, 8–15. Available from: http://dx.doi.org/10.1016/j.fm.2015.12.009.

Scheldeman, P., Herman, L., Goris, J., De Vos, P., Heyndrickx, M., 2002. Polymerase chain reaction identification of *Bacillus sporothermodurans* from dairy sources. J. Appl. Microbiol. 92, 983–991.

Setlow, P., 2003. Spore germination. Curr. Opin. Microbiol. 6, 550–556. Available from: http://dx.doi.org/10.1016/j.mib.2003.10.001.

Sevenier, V., Delannoy, S., André, S., Fach, P., Remize, F., 2012. Prevalence of *Clostridium botulinum* and thermophilic heat-resistant spores in raw carrots and green beans used in French canning industry. Int. J. Food Microbiol.. Available from: http://dx.doi.org/10.1016/j.ijfoodmicro.2012.02.009.

Smit, Y., Cameron, M., Venter, P., Witthuhn, R.C., 2011. *Alicyclobacillus* spoilage and isolation--a review. Food Microbiol. 28, 331–349. Available from: http://dx.doi.org/10.1016/j.fm.2010.11.008.

Sonenshein, A.L., 2000. Control of sporulation initiation in *Bacillus subtilis*. Curr. Opin. Microbiol. 3, 561–566.

Stringer, S.C., Webb, M.D., Peck, M.W., 2009. Contrasting effects of heat treatment and incubation temperature on germination and outgrowth of individual spores of nonproteolytic *Clostridium botulinum* bacteria. Appl. Environ. Microbiol. 75, 2712–2719. Available from: http://dx.doi.org/10.1128/AEM.02572-08.

Stringer, S.C., Webb, M.D., Peck, M.W., 2011. Lag time variability in individual spores of *Clostridium botulinum*. Food Microbiol. 28, 228–235. Available from: http://dx.doi.org/10.1016/j.fm.2010.03.003.

Talukdar, P.K., Olguín-Araneda, V., Alnoman, M., Paredes-Sabja, D., Sarker, M.R., 2015. Updates on the sporulation process in *Clostridium* species. Res. Microbiol. 166, 225–235. Available from: http://dx.doi.org/10.1016/j.resmic.2014.12.001.

Ter Beek, A., Hornstra, L.M., Pandey, R., Kallemeijn, W.W., Smelt, J.P.P.M., Manders, E.M.M., et al., 2011. Models of the behaviour of (thermally stressed) microbial spores in foods: tools to study mechanisms of damage and repair. Food Microbiol. 28, 678–684. Available from: http://dx.doi.org/10.1016/j.fm.2010.07.003.

Thompson, J.M., Dodd, C.E.R., Waites, W.M., 1993. Spoilage of bread by *Bacillus*. Int. Biodeterior. Biodegradation 32, 55–66. Available from: http://dx.doi.org/10.1016/0964-8305(93)90039-5.

Thompson, P.J., 1981. Thermophilic organisms involved in food spoilage: aciduric flat-sour sporeforming aerobes. J. Food Prot. 2, 92–163.

Tilsala-Timisjarvi, A., Alatossava, T., 2004. Rapid DNA preparation from milk and dairy process samples for the detection of bacteria by PCR. Food Microbiol. 21, 365–368. Available from: http://dx.doi.org/10.1016/j.fm.2003.08.010.

Tsuda, K., Nagano, H., Ando, A., Shima, J., Ogawa, J., 2015. Isolation and characterization of psychrotolerant endospore-forming *Sporosarcina* species associated with minced fish meat (surimi). Int. J. Food Microbiol. 199, 15–22. Available from: http://dx.doi.org/10.1016/j.ijfoodmicro.2014.12.031.

Valerio, F., De Bellis, P., Di Biase, M., Lonigro, S.L., Giussani, B., Visconti, A., et al., 2012. Diversity of spore-forming bacteria and identification of *Bacillus amyloliquefaciens* as a species frequently associated with the ropy spoilage of bread. Int. J. Food Microbiol. 156, 278–285. Available from: http://dx.doi.org/10.1016/j.ijfoodmicro.2012.04.005.

Vos, P., Garrity, G., Jones, D., Krieg, N.R., Ludwig, W., Rainey, F.A., et al. (Eds.), 2009. Bergey's Manual of Systematic Bacteriology, Volume 3: The Firmicutes. Springer Science & Business Media. Available from: http://dx.doi.org/10.1007/978-0-387-68489-5.

Wagner, I.D., Wiegel, J., 2008. Diversity of thermophilic anaerobes. Ann. N Y Acad. Sci. 1125 (1), 1–43. Available from: http://dx.doi.org/10.1196/annals.1419.029.

Watterson, M.J., Kent, D.J., Boor, K.J., Wiedmann, M., Martin, N.H., 2014. Evaluation of dairy powder products implicates thermophilic sporeformers as the primary organisms of interest. J. Dairy Sci. 97, 2487–2497. Available from: http://dx.doi.org/10.3168/jds.2013-7363.

Xiao, Y., Francke, C., Abee, T., Wells-Bennik, M.H.J., 2011. Clostridial spore germination versus bacilli: genome mining and current insights. Food Microbiol. 28, 266–274. Available from: http://dx.doi.org/10.1016/j.fm.2010.03.016.

Zeigler, D.R., 2014. The *Geobacillus* paradox: why is a thermophilic bacterial genus so prevalent on a mesophilic planet?. Microbiology 160, 1–11. Available from: http://dx.doi.org/10.1099/mic.0.071696-0.

Chapter 5

Yeasts

Marianne Perricone, Mariangela Gallo, Maria Rosaria Corbo, Milena Sinigaglia and Antonio Bevilacqua
University of Foggia, Foggia, FG, Italy

5.1 INTRODUCTION

Yeasts are eukaryotic organisms that are included in a group of organisms that are called "fungi," which also includes molds and mushrooms. They are generally single-celled organisms that are adapted for life in specialized, usually liquid, environments and, unlike some molds and mushrooms, do not produce toxic secondary metabolites (Rawat, 2015). These organisms reproduce by budding by single cells, or less commonly, divide by fission. Furthermore, many yeasts grow very quickly under strictly anaerobic conditions, and thus yeasts are capable of increasing rapidly in number in liquid environments. On the other hand, reproduction as single cells limits the development or penetration into solid surfaces, where filamentous fungi have an advantage.

Yeasts can have both positive and negative effects on fermented products consumed by humans and animals. Yeasts are used as starter cultures in cheeses and bread, as well as wine, beer, and other alcoholic fermentation products, but they can also initiate spoilage in foods, such as yoghurt, fruit juice, salads, and mayonnaise. Indeed, spoilage yeasts have been isolated from many foods and beverages, including bread, wine, beer, fruit juices, mayonnaise, dairy products, and chocolate (Lowes et al., 2000). Yeasts reproduce more slowly than bacteria, and thus do not compete in environments which favor bacteria, i.e., at neutral pH or at very high temperatures. In common with molds, many yeasts are tolerant of acid conditions. In broad terms, yeasts are more likely to be vital in acidic and liquid environments (Pitt and Hocking, 2009). They often colonize foods with a high sugar or salt content and contribute to the spoilage of maple syrup, pickles, and sauerkraut. Fruits and juices with a low pH are another target, and there are some yeasts that grow on the surfaces of meat and cheese (Rawat, 2015).

The Microbiological Quality of Food. DOI: http://dx.doi.org/10.1016/B978-0-08-100502-6.00008-X

Yeasts are a heterogeneous assembly of often quite unrelated fungi. Pitt and Hocking (2009) reported almost 680 species of yeasts, divided into over 90 genera: 57 were classified as Ascomycetes, 36 as Basidiomycetes, and 2 unclassified (Pitt and Hocking, 2009).

Deák and Beuchat (1996) list 99 yeast species which are naturally present in foods including fruit, beverages, wine, beer, meat, dairy products, low a_w products, and low pH products. Many of these yeasts have a limited growth in processed foods, as they are intolerant of reduced a_w, heat processing, or preservatives. But, there are some exceptions: certain species must be classified as spoilage yeasts because they possess one or more undesirable properties (Pitt and Hocking, 2009).

These yeasts are responsible for major losses of processed foods around the world every year. The correct approach, and often the only recourse in such cases, is to pay attention to manufacturing guidelines, so that this kind of problem is positively eliminated. In such cases the following text will be of little value, as will identification of the yeasts concerned (Pitt and Hocking, 2009).

But, what is the yeast spoilage of food and beverages? Food spoilage caused by yeasts consists in the visible or detectable alteration of physical and sensorial properties of food as result of their activity. Deterioration of food and drinks by yeasts may present several effects, more or less evident, according to the type of food (Loureiro and Querol, 1999).

Research from 60 years ago reported that food-borne yeasts had no significant impact on human health, and they are considered harmless for the consumers even when ingested in high amounts through fermented foods. But, Fleet (1992) reassessed the impact of yeasts on food safety. Several studies reported gastroenteritis caused by the ingestion of foods containing yeasts etiological agents. In addition, some studies, with increasing evidence for the development of allergies and negative effects on humans due to yeasts, became present in literature. Other studies reported that the exposure to antigens from *Saccharomyces cerevisiae* may contribute to inflammatory bowel disease. In a review on new and emergent yeast pathogens it was reported that the infections caused by yeasts were increasingly significant, and it was stated that some yeasts that were previously defined as harmless (e.g., *Candida utilis*, *S. cerevisiae*, and *Candida lipolytica*) are capable of damaging the human body (Loureiro and Querol, 1999).

In recent years, the spoilage yeasts of foods and drinks are considered a problem of great interest in food technology, because they are responsible for significant economic losses. These are considerably high and involve all food industry sectors: manufacturers, suppliers of raw materials and packaging, and, often, retailers (Loureiro and Querol, 1999).

Although the cases reported are rare and other factors are involved, it may be affirmed that the presence of yeasts in foods, either spoilers or not, may be an increasing problem in public health and should be a growing concern for microbiologists and food technologists.

5.2 YEASTS SPOILAGE

Yeast spoilage is very predictable, principally occurring in those products where bacterial growth is either retarded or prevented by the intrinsic, or extrinsic factors, and processing. Yeasts can take over and spoil the product: generally, high acid, low pH foods, products with high sugar or high salt content, and products preserved with weak organic acids (e.g., sorbic, benzoic, acetic) are prone to yeast spoilage; moreover, they can grow with or without oxygen.

A group of authors have published a list of yeasts that are frequently associated with the spoilage of foods and beverages: to these lists of "dangerous" spoiling yeasts four other yeast species have also been added, which were the members of the "second-division spoiling yeasts" (Stratford, 2006). This list contains some species, such as *S. cerevisiae*, *Kluyveromyces marxianus*, *Schizosaccharomyces pombe*, and *Debaryomyces hansenii* that are even used as starter cultures or adjuncts in different food fermentations, but under special conditions or in another type of food, these yeasts may cause spoiling. The presence of certain yeasts during storage is the consequence of single or combined abiotic stress factors and their growth causes symptoms of spoilage.

Moreover, under certain environmental conditions, an additional group of yeasts can cause spoiling, which was described as the "second division" of spoiling yeasts; yeast species were also thought to be selected from the contaminating microbiota and well adapted to one or other environmental extremes. Their special physiological features (osmotolerant, acid tolerant, preservative resistant, and highly fermentative) enable them to proliferate in food environments that are unfavorable to the great majority of yeast species.

Generally, spoilage yeasts can be grouped in four groups: *Zygosaccharomyces* spp. and related genera tolerate high sugar and high salt concentrations and are the usual spoilage organisms in foods such as honey, dried fruit, jams, and soy sauce. They usually grow slowly, producing off-odors and flavors, and carbon dioxide that may cause food containers to swell and burst. *D. hansenii* can grow at salt concentrations as high as 24%, accounting for its frequent isolation from salt brines used for cured meats, cheeses, and olives. This group also includes the most important spoilage organisms in salad dressings. *Saccharomyces* spp. are best known for their role in the production of bread and wine but some strains also spoil wines and other alcoholic beverages by producing gassiness, turbidity, and off-flavors associated with hydrogen sulfide and acetic acid. Some species grow on fruits, including yogurt that contains fruit, and some are resistant to heat processing. *Candida* and related genera are a heterogeneous group of yeasts, some of which also cause human infections, and they are involved in spoilage of fruits, some vegetables, and dairy products. *Dekkera*/*Brettanomyces* are principally involved in spoilage of fermented foods, including alcoholic beverages and some dairy products. They can produce volatile phenolic compounds responsible for off-flavors (Maráz and Kovác, 2014; Rawat, 2015).

In general, spoilage is due to yeast growth but it is difficult to find the correlation between the beginning of spoilage and yeast concentration because the spoilage is the result of yeast biomass and the production of yeast metabolites in foods; though symptoms of spoilage appear when yeast cell numbers increase to approximately 5 log CFU/g or mL and become evident if cell numbers increase to about 7 log CFU/g or mL. The production of metabolites that provoked spoilage depends on the type of food and the metabolic characteristic of yeast; thus the same yeast species could cause different symptoms of spoilage in cases of different types of food, as well as different yeast species being able to generate similar symptoms of spoilage (Stratford, 2006; Maráz and Kovác, 2014).

5.3 EVIDENCES OF YEASTS SPOILAGE

Yeast spoilage of food is primarily that which the consumer detects as abnormal and finds objectionable, although the symptoms of spoilage are many and varied, and not all may be obvious to all consumers.

Spoilage accompanied by excess gas (carbon dioxide) production is the most frequent problem; it is caused by strongly and weakly fermenting species such as *Zygosaccharomyces bailii*, *S. cerevisiae*, *Dekkera bruxellensis*, *Saccharomycodes ludwigii*, or *Candida parapsilosis* and *Candida pseudointermedia*, respectively. Visible symptoms of spoilage are very variable and include sediment and haze/cloudy formation, surface film formation and alterations in the texture and color of food, and the surface film formation is characteristic of the majority of spoiling yeast species. Off-taste and off-odor are the consequences of the production of primary and secondary metabolites by yeasts, like ethanol, acetaldehyde, acetic acid, ethyl acetate, hydrogen sulfide, and more characteristic volatile compounds such as 4-ethylphenol produced by *D. bruxellensis*.

5.3.1 Gas Production

The most visible sign of yeast spoilage is the production of excess gas, which causes swelling of containers or Tetra Paks, that in the extreme example of a glass bottle could explode, leading to physical injury.

Excess gas production by yeasts is due to the fermentation of sugars and the range of sugar substrates fermented by yeasts varies considerably. Furthermore, the volume of gas produced, and hence the pressure formed, varies between yeast species. There are few yeast species capable of forming sufficient gas pressure by fermentation to explode bottles; these include *Z. bailii*, *S. cerevisiae*, *D. bruxellensis*, and *S. ludwigii*, which can generate up to 10 atm overpressure. Behind these highly fermentative species there are other spoilage species that ferment, but to a much lesser extent; these host *C. parapsilosis* and *C. pseudointermedia*, which can generate up to 2 atm overpressure, but this is not sufficient to explode containers.

5.3.2 Visible Symptoms of Yeast Spoilage

The second most obvious sign of yeast spoilage is the visible appearance of yeast cells on the surface of foods. In particular, in solid foods the yeast spoilage may also cause surface discoloration, films or mucous slimes, or formation of extracellular polysaccharides. How obvious this is to a customer depends on the yeast species and the appearance of the food itself. Generally, on most solid foods, pink or red yeasts, such as *Sporobolomyces* or *Rhodotorula* spp. are more visible than white or cream yeast colonies. Instead, in beverages, spoilage yeasts can be seen as hazes, clouds, particulates, surface films or colonies, and sediments. In a clear liquid, spoilage is visible as a faint haze if the yeasts concentration is about 10^5 cells/mL, but can become dense clouds at 10^7-10^8 CFU/mL. Among spoilage microorganisms, yeast cells are densest, due to their thick cell wall, and will sediment rapidly in liquid media. Furthermore, yeasts may also aggregate to form structures, such as floccs, particulates, or surface films, that are more easily identifiable to consumers.

Film formation is most frequently caused by some strains of *Candida* spp., *Issatchenkia occidentalis* and *Issatchenkia orientalis*, *Pichia anomala*, *Pichia fluxuum*, and *Pichia membranifaciens*, but can include spoilage yeasts such as *Dekkera* and *Brettanomyces* spp.

In general, surface film formation is characteristic of all spoilage yeasts in high-sugar syrups, because the buoyant density of the sugar. When it is greater than 35% w/w, the syrup is sufficiently great to cause yeast cells to float, rather than sediment.

5.3.3 Off-Flavors and Off-Odors

To the public, off-tastes and smells in food are difficult to detect and attributable to yeast spoilage; it depends on the perspicacity of the tasters and their familiarity with what the food should normally taste like.

In theory, yeast growth and metabolism can influence the flavor and smell of food in one or two ways: either by removal of flavor of some food components or by production of off-flavors or off-odors. In practice, yeast spoilage is caused by the production of small but organoleptically powerful molecules through primary or secondary metabolism; these molecules are volatile and give off-odors as well as off-tastes. Generally, the off-taste produced in the greatest quantity is of course that of ethanol, that has an unusual slightly sweet taste.

Other volatile off-flavors formed by yeasts include acetaldehyde with a taste of pungent apples; acetic acid with a vinegar taste; ethyl acetate (pineapple note); diacetyl; and acetoin (sweet butter).

Spoilage due to certain yeast species may be characterized by distinct off-flavors.

For example, *D. bruxellensis* may have a distinct mouse flavor, described as leathery, sweaty, farmyard-like, in red wine. *Pichia* spp., in particular, *P. membranifaciens*, have a yeasty aldehyde off-flavor, forming high levels of acetic acid, acetaldehyde, and esters. *Saccharomycodes ludwigii* forms high levels of acetoin and acetaldehyde, while *Kloeckera apiculata* produce high levels of esters and volatile acids. The growth of lipolytic yeasts such as *Yarrowia lipolytica*, on substrates such as cheese or meat, may cause off-odors such as rancidity. In an unexpected twist, the presence of preservatives in foods may make them more susceptible to spoilage by off-flavors, because, if the yeast is resistant to preservative added, it is degraded. For example, sorbic acid can be degraded by *Zygosaccharomyces rouxii* and *D. hansenii* to 1,3-pentadiene, giving a petroleum-like off-odor (Stratford, 2006).

5.4 ECOLOGY OF YEASTS AND STRESS FACTORS IN FOODS

Generally, spoilage is the result of microbial activity. Moreover, the microorganisms that will grow in the foods depend on the microbes that come into contact with the ingredients and the product, and on environmental conditions such as the composition of the product and the storage conditions. In particular, yeasts can generally resist extreme conditions better than bacteria; hence the yeasts growth in the food ecosystem are mainly influenced by different stress factor found in foods, such as the lower a_w, high concentration of salt or sugar, low temperature, starvation, carbon dioxide, low redox potential, and antimicrobial added as preservatives. This is the reason why the ability of some yeasts to survive in harsh conditions makes them potent food spoilage organisms responsible for the large economic losses of some food products.

Yeast cells have evolved to be exceptionally proficient at surviving sudden and often harsh changes in their external environment.

The growth under these harsh conditions requires maintenance of the internal system with different mechanisms, that respond to the external challenges that the cell must deal with, this mechanism of defense is known as "Stress Response." Thus, when environmental conditions change abruptly, the cells have the ability to adjust to environmental stress applied with the production of specific stress response proteins. These types of responses require a complex network of sensing and signal transduction leading to adaptations of cell growth and proliferation as well as to adjustments of the gene expression program, metabolic activities, and other features of the cell. Moreover, yeast cells gain cross protection against different stresses, evident in the fact that cells exposed to a mild dose of one stress become resistant to large, normally lethal doses of other stresses. This observation has sparked the idea that yeast cells use a general mechanism of cellular protection that is provoked when cells are exposed to stressful conditions (Gasch, 2003).

5.4.1 Nutrients

Food by its very nature is expected to be nutritious; therefore, food is a rich habitat for microorganisms.

Yeasts are generally associated with the fermentation of sugars such as glucose and sucrose, but they are able to utilize a variety of other compounds, such as alcohols, organic acids, hydrocarbons, and aromatic compounds. Some yeasts are also capable of utilizing certain acid-based preservatives such as benzoic acid, propionic acid, and sorbic acid, and this can make them a major issue in foods and drinks that rely on these preservatives for stability. A relatively few yeasts can utilize complex carbohydrates such as starch; moreover, some yeasts (i.e., *Y. lipolytica*) show lipolytic activity and are able to utilize fats, although the growth is slow. Finally, most yeasts are unable to grow proteins as a nitrogen source, but they prefer small molecules such as amino acids, *ammonium* ions, and nitrate or nitrite; an exception to this may be the degradation of casein by red yeasts (Stratford, 2006).

5.4.2 Water Activity

Water is essential for life and water activity is one of the most important ecological factors affecting the yeasts growth in foods.

Water activity (a_w) represents a physicochemical parameter that quantified the link between moisture in foods and the ability of microorganisms to growth on them (Scott, 1957; Deak, 2007a,b; Pitt and Hocking, 2009). Practically, water activity corresponds with *equilibrium* relative humidity. The majority of yeasts are more tolerant to reduced a_w than bacteria: food spoilage yeasts have minimum a_w values of 0.90–0.95 for growth; moreover, several species, such as *Z. rouxii*, can grow at a_w as low as 0.62 (Deak, 2007a,b).

5.4.3 Temperature

Another important environmental factor that influences growth is temperature; the range of growth temperature of microorganisms in general can be characterized by cardinal temperature, that is, *minimum*, *optimum*, and *maximum* temperature values. Generally, the temperature range of yeast growth extends from several degrees below 0°C to a few degrees below 50°C, even if normally the temperature range of individual species or strains does not exceed 40°C.

Finally, the growth temperature is influenced by other environmental factors; in particular, it was highlighted that in presence of antimicrobial compounds such as ethanol or bicarbonate the *minimum* temperature of growth increased, whereas the optimum temperature increased in solutions with high solute concentration (Deak, 2007a,b; Pitt and Hocking, 2009).

5.4.4 Oxygen

Some yeast species are strict aerobes whilst others also have a fermentative metabolism. For example, *Saccharomyces* spp. and *Zygosaccharomyces* spp. are capable of growth in complete absence of oxygen; indeed *S. cerevisiae* and *Z. bailii* can continue fermentation under several atmospheres pressure of carbon dioxide (CO_2). Moreover, *Z. bailii* is able to continue fermenting at reduced water activities in the presence of high level of preservatives (Deak, 2007a,b).

5.4.5 Acidity and pH

Yeasts prefer the softly acidic *medium* with a *optimum* pH between 4.5 and 5.5, even if it is known that yeasts show a remarkable resistance to pH values. Many species such as *P. membranifaciens*, *Dekkera intermedia*, and *Candida davenportii* are able to grow at pH values a slow as 1.3–1.7 (Deak, 2007a,b).

Generally, yeasts tolerate acidic conditions better than alkaline ones.

5.5 PRESERVATION OF FOODS BY SPOILAGE

Each food puts up specific and particular microflora, which is a function of the microorganisms present in the raw material, processing, preservation, and storage conditions, but also in water, soil, air, and other foods (Gram et al., 2002). Healthful plants and animals may hinder the development of spoilage yeasts, but when raw material is slaughtered or harvested it becomes susceptible to spoilage microbes. Good manufacturing practices, with particular attention to sanitation and hygiene, may prevent the development of many spoilage yeasts. However, microbes are particularly innovative and seem to circumvent the barriers set up against them. For this reason, controlling the growth and activity of spoilage yeasts requires a good understanding of their physiology, biochemistry, and genetic property. Factors affecting the growth and survival of the yeast are particularly important, to assure a high level of quality control of the final product. For most yeasts, the limit values of the growth and survival, such as inactivation kinetics, are the basis of the technological parameters, and are well defined (Fleet, 2006).

A “preservation system” can be considered as a set of: intrinsic factors to the food, physical and chemical preservation, and hygiene and packaging measures. The intrinsic factors intended are: low pH, low water activity, lack of nutrients, the presence of essential oils. Physical approaches include heat and pasteurization, chilled storage, carbonation or modified atmospheres, and chemical preservation include use of the sorbic, acetic, and benzoic acids.

Below will be discussed the treatments that inhibit and inactivate the growth of spoilage yeasts.

5.5.1 Heat Inactivation

Heat is one of the commonest and most effective methods of preserving foods and beverages. Yeasts are relatively sensitive to heat, and temperatures between of 55°C and 65°C rapidly kill almost all yeasts, but this value depends on the species, composition of the growth medium, age, and phase of the sporulation. The ascospores of yeasts are more resistant to heat than vegetative cells. The *D* value indicates the time required to kill 1 log (90%) of the inoculums, and yeast *D* values rarely exceed 1 minute at 55°C (Stratford, 2006).

Sensibility to temperatures varies between yeast species and strains, and the composition of the foods. The yeasts more resistant to heat include *S. cerevisiae*, *P. membranifaciens*, *P. anomala*, and also *Candida albicans* and *S. pombe*. Yeast species that show sensitivity to heat include *D. hansenii*, *Z. rouxii*, which are killed at temperatures of 40°C (Stratford, 2006). The mechanism of the heat inactivation is not well understood; several studies reported that changes in cellular structures may cause cellular death (Deak, 2007a,b).

Heat may be applied to the complete package (food and container), or liquids may be pasteurized by passage through a pasteurizer.

5.5.2 Alternative and Novel Technologies

Nonthermal preservative methods have received good attention for their positive characteristics on the nutritional quality of foods. Some of the nonthermal processes used in food industries are: high intensity pulsed electric field, high hydrostatic pressure (HHP), high-pressure homogenization (HPH), ultraviolet (UV), ultrasound, and irradiation. These technologies have the ability to inactivate microorganisms but flavor, color, and nutrient value of foods (Aneja et al., 2014).

5.5.2.1 High Hydrostatic Pressure

HHP represents the most advanced alternative technology, and it is commercially used for a variety of foods, such as cooked meat, shellfish, fruit, and vegetable juices, sauces, and dips. High pressure of up to 100 MPa damages cell membranes and inactivates vegetative cells, and yeasts are more sensitive than bacteria (Deak, 2007a,b). The HHP treatment has a lethal effect on microorganisms because this technique damages their cell membrane with inactivation of some key enzymes which are involved in DNA replication and transcription processes. This process has great potential to reduce the microbial load of fruit juices and increase the shelf life of fruit juices (Aneja et al., 2014).

5.5.2.2 *High-Pressure Homogenization*

This involves the pumping of liquid through a homogenizing valve at high pressure of over 100 MPa. This produces high turbulence that results in the breakdown and dispersion of particles throughout the product. After homogenization, uniform particles (size in the range from 0.2 to 2 μm) are obtained. In the past, HPH was proposed as a suitable method for the stabilization of dairy products, but in recent decades it has been suggested for its use for prolongation of the shelf life of fruit juices. HPH inactivates microorganisms by damaging their structural integrity coupled with the sudden rise in temperature produced in this process (Aneja et al., 2014).

5.5.2.3 *Pulsed Electric Field*

Pulsed electric field involves the application of the electric field pulses of short duration and high intensity. The fluid foods are placed between two electrodes in batch and continuous flow treatment (Aneja et al., 2014). This process inactivates microorganisms and enzymes with only a small increase in temperature. Yeasts are killed more easily than bacteria by electric fields, and inactivation increases greatly with field strength and the number of pulses. The technique has potential in the cold pasteurization of liquid foods, such as fruit juices and milk, without causing adverse organoleptic changes (Deak, 2007a,b).

5.5.3 Chemical Inhibition

In the production of foods, several chemical preservatives have been used, some of them traditionally for centuries, others introduced recently. They differ in chemical nature and origin, and can be organic or inorganic, synthetic or natural, biological products (Deak, 2007a,b).

REFERENCES

Aneja, K.R., Dhiman, R., Aggarwal, N.K., Aneja, A., 2014. Emerging preservation techniques for controlling spoilage and pathogenic microorganisms in fruit juices. Int. J. Microbiol. 75, 1–14.

Deak, T., 2007a. Ecology (chapter 3). In: Deak, T. (Ed.), Handbook of Food Spoilage Yeasts-2nd edition. CRC Press Taylor & Francis Group, Boca Raton, FL, pp. 37–58.

Deak, T., 2007b. Preservation: inhibition and inactivation of yeasts (chapter 6). In: Deak, T. (Ed.), Handbook of Food Spoilage Yeasts-2nd edition. Taylor & Francis Group, Boca Raton, FL, pp. 87–115.

Fleet, G.H., 1992. Spoilage yeast. Crit. Rev. Biotechnol. 12, 1–44.

Fleet, G.H., 2006. The commercial and community significance of yeasts in food and beverage production (chapter 1). In: Querol, A., Fleet, G.H. (Eds.), Yeast in Food and Beverages. Springer, Berlin, Germany, pp. 1–12.

Gasch, A.P., 2003. The environmental stress response: a common yeast response to diverse environmental stresses (chapter 2). In: Hohmann, S., Mager, P.W.H. (Eds.), Yeasts Stress Response. Springer, Berlin, Germany, pp. 11–70.

Gram, L., Ravn, L., Rasch, M., Bruhn, J.B., Christensen, A.B., Givskov, M., 2002. Food spoilage-interactions between food spoilage bacteria. Int. J. Food Microbiol. 78, 79–97.

Loureiro, V., Querol, A., 1999. The prevalence and control of spoilage yeasts in foods and beverages. Trends Food Sci. Technol. 10, 356–365.

Lowes, K.F., Shearman, C.A., Payne, J., MacKenzie, D., Archer, D.B., Merry, R.J., et al., 2000. Prevention of yeast spoilage in feed and food by the yeast mycocin HMK. Appl. Environ. Microbiol. 66, 1066–1076.

Maráz, A., Kovác, M., 2014. Food spoilage by cold-adapted yeasts (chapter 23). In: Buzzini, P., Margesin, R. (Eds.), Cold-Adapted Yeasts. Springer, Berlin, Germany, pp. 497–532.

Pitt, I.J., Hocking, A.D., 2009. The ecology of fungal food spoilage (chapter 2). In: Pitt, I.J., Hocking, A.D. (Eds.), Fungi and Food Spoilage. Springer Science + Business Media LLC, New York, NY, pp. 3–9.

Rawat, S., 2015. Food spoilage: microorganisms and their prevention. Asian J. Plant Sci. Res. 5, 47–56.

Scott, W.J., 1957. Water relations of food spoilage microorganisms. Adv. Food Res. 7, 83–127.

Stratford, M., 2006. Food and beverage spoilage yeasts (chapter 11). In: Querol, A., Fleet, G.H. (Eds.), Yeasts in Food and Beverages. Springer, Berlin, Germany, pp. 335–379.

Chapter 6

Molds

Chioma Nwakanma[1] and Marian Unachukwu[2]
[1]*Michael Okpara University of Agriculture, Abia, Nigeria,* [2]*Godfrey Okoye University, Enugu, Nigeria*

6.1 INTRODUCTION TO MOLDS

Molds are microscopically small plant-like organisms, composed of long filaments that grow over the surface and inside nearly all substances of plant, food, dry leaf, or animal origin. They are enabled by their filamentous construction. They can be nearly any color—e.g., white, orange, green, or black. Biologists consider them to be separate from the plant kingdom and members of the kingdom of fungi. They are related to the familiar mushrooms, differing only in not having their filaments united into large fruiting substances.

Molds are vital part of the environment and are needed to break down dead materials. Very tiny and lightweight, mold spores travel easily through the air. Most building surfaces can provide adequate nutrients to support the growth of mold. When mold spores land on material that is damp e.g., walls, floors appliances (such as humidifiers or air conditioners), carpets, or furniture—they can begin to multiply. When molds are present in large numbers, they may cause allergic symptoms similar to those caused by plant pollen.

Molds are those dusty little spots that spread over bread, cheese, books, and other things in the home, cause the loss of millions of Naira and dollars to our economy every year. To deal successfully with this menace, we must understand what molds are and exactly what they are doing.

6.2 DESCRIPTION OF MOLDS

The filaments of the mold fungi are called hyphae. When the hyphae are numerous enough to be seen by the naked eye they form what is called a Mycelium. It is the hyphae and resulting mycelium that invade things in our homes and cause them to decay. Molds reproduce by spores. Spores are like seeds and germinate to produce a new mold colony when they land in a

The Microbiological Quality of Food. DOI: http://dx.doi.org/10.1016/B978-0-08-100502-6.00009-1

suitable place. Unlike seeds, they are very simple in structure and never contain an embryo or any sort of preformed offspring. Spores are produced in a variety of ways and occur in a bewildering array of shapes and sizes. In spite of this diversity, the form of spores is quite constant for any given mold, making it one of the most useful features for identification (see diagram).

The most basic difference between spores lies in their method of initiation, which can be either sexual or asexual. Sexually initiated spores result from a mating between two different organisms or hyphae whereas asexual spores result from a simple internal division or external modifications of an individual hypha. The recognition of a mating and subsequent spore formation is often difficult for an observer, and is usually reserved for patient specialists. However, for practical purposes one can learn to recognize certain indications of the sexual process, namely, the four kinds of sexually determined spores that appear in mold fungi:

1. Oospores
2. Zygospores
3. Ascospores
4. Basidiospores

6.2.1 Oospores

They are produced when male gametes (reproductive nucleic) enter a large spherical cell (oogonium) and fertilize the eggs within. The result, as seen in routine examination, is numerous oogonia containing one to several spherical and often brownish eggs. The oogorus are usually penetrated by one or more hyphae (antheridia) that give rise to the male nucleus Fig. 6.1.

6.2.2 Zygospores

They do not occur inside any kind of enclosing structure, but are produced by the direct fusion of two hyphal protrusious (suspensors) from neighboring filaments. Usually zygospores are recognized as large, nearly spherical, often dark brown or black, rough-walled spores with two connecting hyphae, representing the two maturing gametangia. Sometimes the zygospores may be surrounded by several finger-like extensions from the two gametangia Fig. 6.2.

FIGURE 6.1 Examples of oogonia and antheridia of *Saprolegnia* sp.

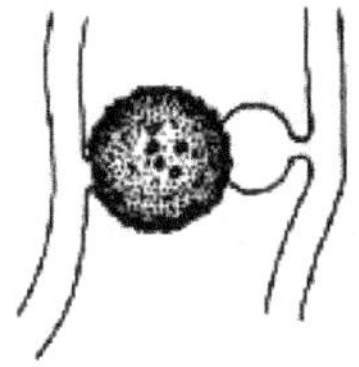

FIGURE 6.2 Examples of zygospores of *Zygorhynchus* sp.

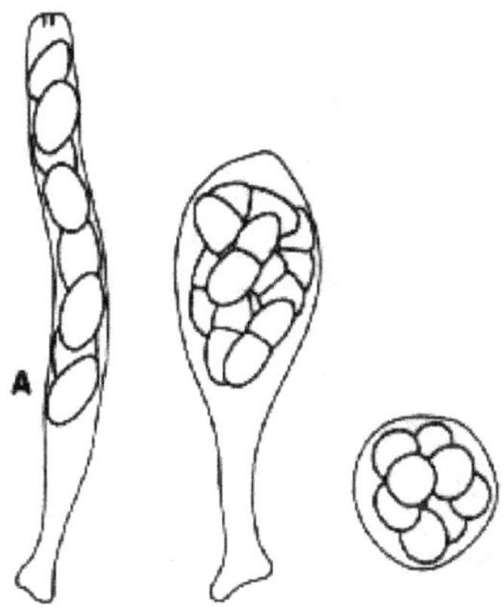

FIGURE 6.3 Examples of ascospores (cylindrical, clavate, and spherical shapes).

6.2.3 Ascospores

They are produced within spherical to cylindrical cells called "asci," most often in groups of four or eight. Usually the asci are produced within some kind of enclosing structure and thus are not found exposed on the hyphae. In a few cases the oogonia with eggs, will never be penetrated by any sort of fertilizing hyphs. Fertilization occurs early in the life cycle and is not evident at the time ascospores are produced Fig. 6.3.

6.2.4 Basidiospore

They are always produced externally on a structure called basidium. Basidia come in a variety of forms, but those commonly encountered on molds will be club-shaped and bear four or eight spores on a sharp protrusion at the apex. At first, it may be difficult to distinguish between a basidiospore and one of the asexually initiated spore types, but one should always suspect the presence of basidia when externally produced spores consistently occur in groups of four or eight. As with ascospores, basidiospores are the result of an early fertilization that is not easily observed Fig. 6.4.

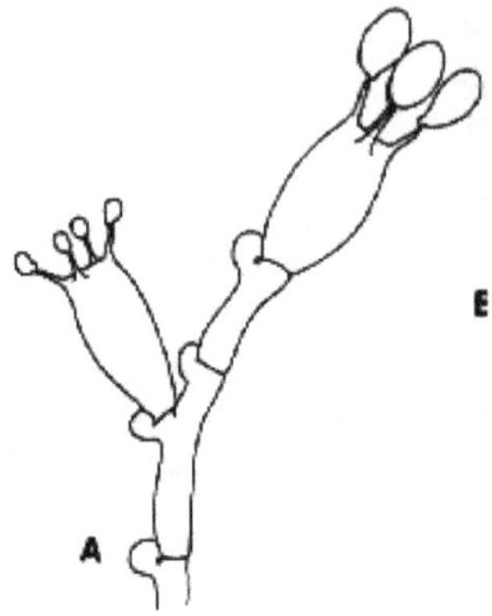

FIGURE 6.4 Examples of basidiomycetes.

Asexual: spores usually occur either in sporangia or as conidia. Sporangia are modified hyphae or cells containing numerous spores (sporangios pores). They never have more than a single connecting hyphae and the spores do not constantly occur in groups of four or eight, as do ascospores.

6.2.5 Conidia

They are the most difficult group to characterize because of their great diversity of form. The only feature that most conidia have in common is that they occur externally on the cells that produce them.

6.3 TYPES OF MOLD

The Centers for Disease Control and Prevention (CDC, United States) states that all molds should be treated the same in terms of health risk and removal.

Some Common Types of Mold

- *Stachybotrys chartarum* (also known as stachybotrysatra)
- *Aspergillus* sp.
- *Penicillium* sp.
- *Fusarium* sp.
- *Trichoderma* sp.
- *Memnoniella* sp.
- *Alternaria* sp.

All fungi strictly speaking should be classified according to their method of sexual reproduction. In many cases this is possible, allowing us to recognize several groups of fungi, of which the following four are of interest as molds.

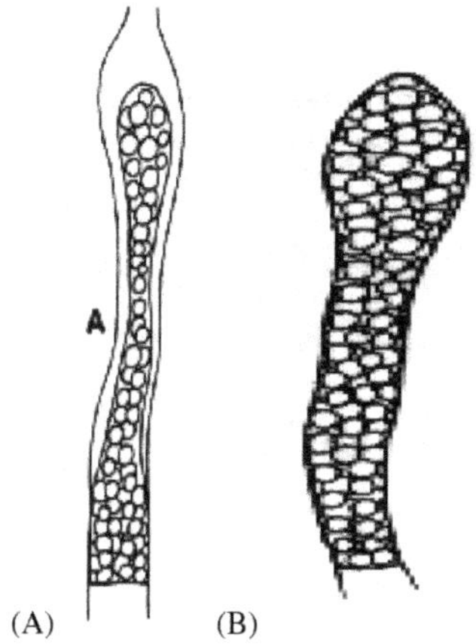

FIGURE 6.5 Examples of (A) zoosporangia of *Sapralegnis* sp. and (B) zoosporangia of *Pythium* sp.

6.3.1 Oomycetes

These members of the group all reproduce by oogonia and eggs. The hyphae have few or no cross-walls (septa) and these appear as long, clear tubes. If a hypha is broken, most of the contents run out. Many oomycetes reproduce asexually by zoospores, which are motile and can swim quite rapidly. Because of their motile zoospores, oomycetes commonly require water for reproduction and are often encountered in water or wet soil (Fig. 6.5).

6.3.2 Zygomycetes

As their name implies, these fungi all produce zygospores. They resemble the oomycetes in having hyphae that usually lack cross-walls or septa, but differ in lacking white spores. Asexual reproduction is by sporangia or conidia. The members of this group are usually terrestrial and will be encountered only occasionally in aquatic conditions (Fig. 6.6).

6.3.3 Ascomycetes

All ascomycetes have ascospores borne inside asci. The hyphae always bear numerous septa. Asexual reproduction is by conidia that always lack motility. Although most ascomycetes are terrestrial some occur in fresh water or marine habitat (Fig. 6.7).

6.3.4 Basidiomycetes

This large group, which includes mushrooms and puff-balls, is characterized by the presence of basidia and basidiospores. Like ascomycetes, to which

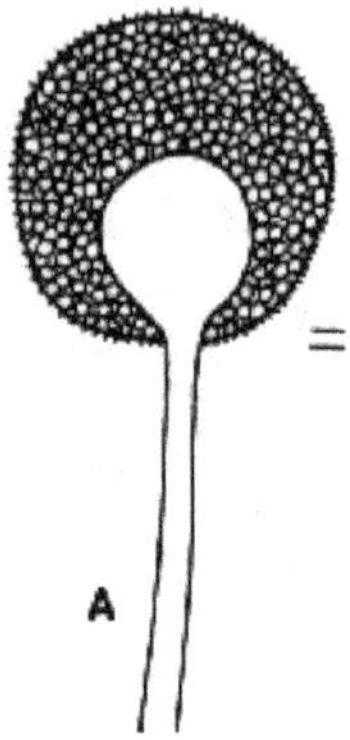

FIGURE 6.6 Examples of zygomycetes. A sporangia of *Mucor* sp.

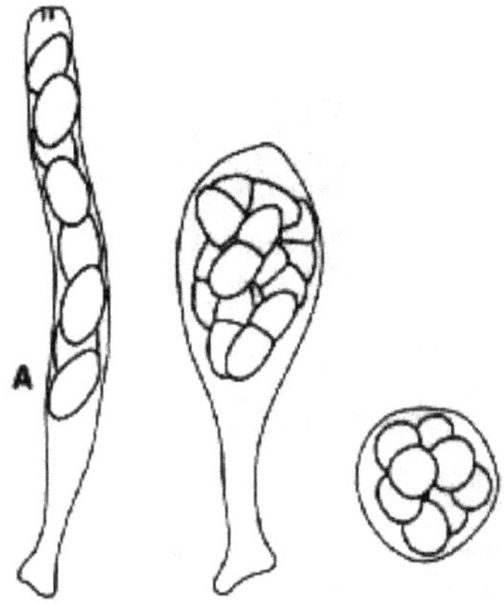

FIGURE 6.7 Examples of ascomycetes asci (cylindrical, clavate, and spherical).

they are related, basidiomycetes have hyphae with septa and lack motile spores. The hyphae of many basidiomycetes bear characterization swellings—called clamp connections—that play a specialized role in nuclear migration. Asexual spores when formed are produced as conidia. Most basidiomycetes are terrestrial.

6.3.5 Some Mold-Like Organisms

Often in routine work we encounter organisms that are similar to molds but which do not fit a strict definition of the term. Either the organism is not filamentous or it is filamentous but not strictly a fungus. Four groups of organisms can be put into this category: bacteria, actinomycetes, yeast, and slime molds.

6.3.5.1 *Bacteria*

They represent a very ancient group of organisms, perhaps as old as four billion years. Colonies of bacteria are composed of minute spore-like cells that together form a slimy mass. Such colonies never contain hyphae and are thus easily distinguished from those of true molds. Bacterial cells, rarely more than 1 um in diameter, are difficult to examine, even with a good microscope, and are best seen when stained. Many bacteria are motile and swim vigorously.

6.3.5.2 *Actinomycetes*

These organisms are usually classified as bacteria but have filaments like fungi. The filaments are seldom more than 1 μm in diameter, however, and are thus considerably narrower than those of molds. Streptomyces, the only actinomycetes genus commonly encountered as a "mold," produces gray to brightly colored powdery colonies, usually with a soil-like odor.

6.3.5.3 *Yeasts*

These are true fungi, but in lacking hyphae cannot be classed as molds. They resemble bacteria in forming pasty or slimy colonies of spore-like cells, but differ in having these cells much larger, usually 2 μm or more in breadth. Reproduction of yeasts is usually by budding, a process where a smaller cell appears to bubble slowly out of the parent cell.

6.3.5.4 *Slime Molds*

Slime molds normally occur on dead or nearly dead materials and other natural materials but many occasionally occur in the laboratory as "molds." Cellular slime molds, the most likely to appear in the laboratory, have creeping amoeba-like cells during one part of the life cycle and sporangium-like structures in another. They most closely resemble members of the zygomycete genus mucor, but differ in their complete lack of hyphae.

6.4 ENVIRONMENTAL OCCURRENCE

Every natural material one can name, no matter how small, will support an indigenous population of molds. Molds are part of the economy of nature, quickly occupying dead or nearly dead materials and returning them to the basic building components of new organisms. They are essential to what biologists cell nutrient cycling, the process whereby nutrients never leave the realm of living things, but simply get used again and again (forming the basis of the passage, "earth to earth, ashes to ashes, dust to dust").

Molds are known to be everywhere; it is useful to classify their habitats under a number of categories based on nutritional characteristics. Molds have specialized nutritional requirements and are not likely to range very far from their usual habitats. The completion among molds having similar nutritional requirements is intense and leaves no room for a mold adapted to another habitat. Fungi that naturally decay seaweed on the beach will not be expected to occur on moldy bread in the kitchen.

Some factors encourage habitat restriction among molds from nutritional and competitive barriers. Most influential is spore dispersal. Although most molds seem to produce astronomical numbers of spores they are, in fact, practicing a strict economy. There can be little waste, in their highly competitive life every minute bit of energy must be put to good use or a more efficient organism will prevail. Thus mold will produce only enough spores to ensure reproduction of their species from year to year. If a mold were to put a disproportionate amount of energy into spore production, it would have to be at the expense of other activities, such as rapid growth. Economy in spore production is best ensured by the existence of mechanisms of spore dispersal that confer a high probability that spores will encounter suitable places to germinate and grow. Many different mechanisms exist and account for the great variety of spores and spore-producing structures found in molds.

Occasionally, molds living together in particular habitats have similar methods of spore dispersal even when the molds themselves are not closely related, illustrating the idea that many structures in organisms are similar because of common ecological pressures rather than common ancestry.

6.4.1 Dead Plant Materials That Support the Growth of Molds

6.4.1.1 *Plants*

The molds that occur on dead herbaceous plants are often the same species that attack dying ones. Thus, we can again list *Alternaria*, *Cladosporium*, and others, as well as a few new ones such as *Epicoccum*, and *Candida*. In temperate climates, where many plants die in the fall, there is a tremendous flush of mold activity in autumn because the dead plant materials simply release their spores in the wind. The likelihood that a spore of *Cladosporium*, e.g., will hit a dead plant in the fall is very great.

Nutritional specialization dictates that different plants will support different fungi and we find that, e.g., dead grasses will yield a different flora of molds than dead milkweeds or mustard, at least to a partial extent.

6.4.1.2 *Wood*

Dead wood is a good source of molds and provides two distinct habitats, depending on whether it is covered with bark or not. Molds that occur under the bark of fallen and standing trees cannot disperse their spores into the air and most often utilize insects for their transmission. Although as we might expect, many have wet and sticky spores, some have dry spores, and we can speculate that these can become trapped among the insects' bodies for hours.

The surface of wood that is not covered with bark often has a distinct mold flora. Although air dispersal of spores is possible in this situation and apparently often occurs, wet-spored molds are still abundant. Again standing and fallen trees support different fungi. Freshly cut wood is a good source of molds causing a phenomenon called blue stain. These molds, common in lumber-yards as bluish to black discoloration on the wood, produce tall sporulating structures bearing wet drops of insect-transmitted spores.

6.4.1.3 Leaves

Dead or dying leaves of trees support molds that are similar to those from herbaceous plants, with some notable exceptions, however. Tree leaves appear to support fewer *Alternaria* and *Cladosporium* colonies and this often allows a better chance to observe the more slowly growing or rarer fungi.

Leaves that become submerged in water support a number of molds with unusual spore forms. The spores may be needle-like, coiled like a watch spring, or coiled into barred-shaped structures that bob to the surface of the water and float away. Others have three or four narrow arms and look something like jacks with two or three of the arms missing. These molds may be examined by submerging leaves in a bowl of water and stirring them vigorously. After a few minutes the spores will float to the surface, where they can be skimmed off in a drop of water.

6.4.1.4 Animals and Humans

A number of diseases of humans and other animals are caused by mold and yeast-like fungi. Many are known only in this habitat and are quite specialized. Notable are the molds called dermatophytes, the cause of a number of skin diseases such as ringworm and athletes foot. The fungus grows on the outermost layer of skin, causing reddening of the surrounding tissues (zoophilic types) and sometimes scaliness (anthropophilic types). Dermatophytes do not normally attack deeper tissues; the symptoms are usually due to an allergic reaction.

6.4.1.5 Soil

Soil is one of the most commonly studied mold habitats. It is a source of plant disease fungi, some human disease fungi, predacious fungi, and a host of forms that bring about the transformation of dead plant and animal material into soil. Fungi, because they are highly specialized users of organic substances, vary from one soil to another and with depth in the same soil. They are entirely dependent upon the kind of material that falls into the soil; a fungus that specializes in oak leaves will probably not colonize a pure needle. This factor alone will account for many differences between the mold populations of two soils. In addition fungi may be sensitive to moisture levels, pH, competition from other organisms, and many other influences.

As a leaf falls into the soil it gradually becomes buried under newer generations of leaves and thus moves down into the lower layers of soil. At the same time, it undergoes continuous decay by molds and other organisms until it finally disappears (somewhere in the B layer; a mineral layer in the soil containing little organic matter). The fungi that first invade the leaf, often while it is still on the, will grow on it until the particular nutrient they need is exhausted and then die back from the leaf. In this way, a leaf moving down through the layers of soil will have a series of molds on it, each replacing a previous population. This transition of populations as the habitat alters and is modified is called succession and is the object of much study and discussion among ecologists.

Although we tend to think of soil fungi as decomposers of litter, they play a number of other roles as well. Many are specialized for animals and animal products and may trap eel worms and other soil animals, or decay the dead bodies of insects and earth worms. Some invade cast-off feathers and hair, and at least one group specializes in old hooves and horns.

Through their roots, living plants offer an attractive habitat for soil-inhabiting fungi. Some live around the root and decay away the dead root layers or substances exuded by the root, others invade the living root tissues and either cause plant disease or live in harmony with them. The latter situation involves a form of symbiosis called mycorrhizae, where the root and fungi within each benefit the other. Most mycorrhizal fungi and plants cannot live apart. Only a few species of plants are consistently free of this association.

6.4.1.6 Air

Air is not habitat for molds, though many disperse their spores by air currents and are encountered in routine work. Fungi have a variety of mechanisms for getting their spores into the air. Simplest is that of exposing dry spore masses to air currents. Many molds use this method, especially those that colonize exposed leaves and stems. Species of *Alternaria*, *Cladosporium*, and some of the basidiomycetes called smuts are well known for this kind of dispersal. These species usually produce large numbers of spores, a necessary expenditure if at least a few randomly dispersed spores are to land on a suitable place to grow.

Ascomycetes and basidiomycetes are usually capable of actually shooting their spores away into the air. Asci are like popguns: the water pressure inside the ascus builds up more and more until the spores are fired out the end with considerable velocity. With such a mechanism, ascomycetes need not be exposed directly to the wind but may grow in sheltered places which allow entry to the air of their discharged spores. Thus, ascomycetes can often be found growing closer to the ground than other air-dispersal organisms.

Some spores become airborne by a mechanism known as droplet adhesion, a process dependent upon the presence of tiny water droplets in the air.

When one of these droplets, as on a foggy day, encounters a mold spore attached to a leaf e.g., the spore adheres to the droplet by surface tension and is carried off on it. It is then either deposited when the drop comes to earth, or becomes truly airborne when the drop dries out.

6.4.1.7 Dung

People although expect dung to be a rather disgusting material to study, many become so intrigued with the molds and other organisms found there that they soon overcome their initial objections. Many fungi found on dung are highly specialized for growth there and never occur anywhere else. They do not produce large numbers of spores and these require a high probability of success in getting their spores from one dung pile to another.

Other Coprophilous molds present their spores to the environment in wet drops at the ends of a stalk. When one of the many insects (especially flies) that visit dung brushes against these stalked spore-drops they carry with them a few of the sticky spores. Later, perhaps on a new dung pile, the spores rub off and germinate to produce a new colony. Most insect-transmitted fungi lack dormancy mechanisms in their spores and are unlikely to be away from dung for more than a few hours.

6.5 IMPORTANCE OF MOLDS—HUMAN-MADE HABITATS

Aside from their role in plant, animal, and human disease, many molds enter directly into the affairs of humans, in either a harmful or beneficial way. On the plus side is the involvement of molds in the greatest contribution to medicine of all time, that of antibiotics. The discovery of penicillin by Sir Alexander Fleming in 1928 probably resulted in the saving of more lives than all other medical discoveries combined. Penicillin, a product of the common mold *Penicillium chrysogenum,* is still one of the safest and most widely used antibiotics in spite of a nearly 40-year search for others.

The foods we eat are as nutritional to many molds as they are to us, a fact often put to use in the preparation of food products, e.g., several types of cheese, such Roquefort, Danish Blue, Camembert, and Brie, owe their distinctive flavor to the presence of mold growing on them. If the mold were absent, these cheeses would not ripen properly. Yeast, although not strictly molds are among the most important fungi in food preparation. Their value, at least for some Saccharomyces species, lies in their ability to produce carbon dioxide and grain alcohol. In wine-making, where the production of alcohol is valued, yeast is added to the grape juice to bring this about. In bread-making, the important product is carbon dioxide, which is necessary in the rising process, and, again yeast is added to the dough. In the production of beer, both alcohol and carbon dioxide, to produce carbonation, may be necessary, although today the carbon dioxide may be added later, artificially.

In the Far East, a number of molds are used in food preparations that remain unexploited in most of the world. Among these are species of *Aspergillus*, *Monascus*, and *Rhizopus*, used to process various rice, bean, and soybean products.

The negative aspects of molds in foods are more noticeable than the positive. Few of us have failed to notice the pink, black, and green molds growing on bread or the green or blue-green *penicillium* rots of citrus fruits. Molds are one of the reasons that food manufacturers include preservatives in their products. One of the most destructive activities of molds in foodstuffs occurs in stored seeds and grains. Certain species of *Aspergillus*, *Penicillium*, and *Eurotium* are able to grow under particularly dry conditions and attack stored grains. To guard against this, the grain must be dried to very low moisture levels. Keeping grains dry in the humid tropics is particularly difficult, if not at times nearly impossible. Not only do such fungi damage the grains or render them unpalatable, they may also excrete toxins that can cause illness or even death. Most famous of these are the aflatoxins—produced by *Aspergillus flavus* and other molds—which are not only immediately toxic but are also known to be carcinogenic. Aspergillus *flavus* often grows on peanuts and was first discovered in peanut feed given to turkeys in Great Britain. Some of the same species of *Eurotium* that attack stored grain commonly appear on the surface of jams, jellies, and syrups. These very sugary substances tend to prevent water from entering the cells of most fungi and thus create what amounts to a microbiological desert. *Eurotium* species, with their ability to grow under dry conditions, are ideally suited to such habitats. Some products, such as pines and dried apricots, are so dry and so surgery that even *Eurotium* species are unable to cope with them. They still can be decayed by certain specialized molds, however. Probably the most remarkable of these is *Xeromyces bisporus*, which cannot even grow at the sugar concentration used to grow *Eurotium* species.

6.5.1 Other Products

Many other man-made products are susceptible to mold attacks. Painted walls, particularly in humid places such as showers, can become overgrown by certain fungi, notably species of *Phoma* and *Exophiala*. Wallpaper also serves as a source of nutrition for some molds. Commonly cited are the *scopulariopsis* species, which have been reported to grow on wallpapers containing arsenic pigments, and to release very poisonous gases. With modern wall pigments, which contain no arsenic, this cannot happen.

Leather is mainly protein and serves as a convenient source of food for molds. Some of the molds that occur here are related to the dermatophytes that attack the outer layer of human skin; others belong to quite different groups. Some substances that seem to be entirely unlikely to support fungal growth may be found to be well-colonized by these organisms. *Exophiala* species, e.g., are found in syrupy solution of polyvinyl alcohol. *Cladosporium* (*A. morphotheca*) *resiniae* often occurs on the surface of aircraft fuels in their tanks and can damage jet engines. *Penicillium ochrochloron* can be found in electroplating solutions that are extremely acid and contain very high levels of toxic upper salts.

6.6 RISK ASSOCIATED WITH MOLDS

Indoor mold is an early warning sign. Mold has been the victim of bad press and it carries an image of something rotting or in the process of decomposition and yet, were it not for the molds that inspired the discovery of penicillin, millions of lives would have been lost. However, the following questions can be asked:

1. HOW I CAN BE EXPOSED TO MOLDS?

   You are exposed to some mold every day, usually by touching, eating or breathing it. When mold is growing on a surface, spores can be released into the air where they can be easily inhaled. A person who ingests or inhales a large number of spores may suffer adverse health effects.
2. WHAT HEALTH EFFECTS CAN BE CAUSED BY EXPOSURE TO MOLD?

   Some people are more sensitive to molds than others, these include:

   a. Infants and children
   b. Elderly persons
   c. Immune-compromised patient (people with HIV infection, cancer, liver disease, etc. or who are undergoing chemotherapy)
   d. People recovering from surgery are usually more susceptible to health problems from molds.
   e. Individuals with existing respiratory conditions, allergies and asthma.

The same amount of mold may have adverse health effects in one person but not in another. Exposure to molds can cause allergic symptoms such as watery eyes, a runny nose, sneezing, nasal congestion, itching, coughing, wheezing, difficulty breathing, headache, and fatigue. The presence of mold does not always means that health problems will occur. However, for some people the inhalation of the mold, fragments of the molds, or spores can lead to health problems or make certain health conditions worse. In addition, many of these molds make "Mycotoxins" are metabolites or by-products from the molds that have been identified as being toxic to humans. These toxins can slowly wear down the immune system and can lead to allergic or respiratory problems.

6.7 HEALTH EFFECTS OF MOLD

Inhalation exposure to mold indoors can cause health effects in some people. Molds produce:

- Allergies (substances that can cause allergic reactions
- Irritants
- Potentially toxic substances or chemicals (mycotoxins)

Inhaling or touching mold or mold spores may cause allergic reactions in sensitive individuals. Mold does not have to be alive to cause an allergic reaction. Dead or alive, mold can cause allergic reactions in some people.

Allergic reactions to mold are common and can be immediate or delayed. Repeated or single exposure to mold, mold spores, or mold fragments may cause nonsensitive individuals to become sensitive to mold and repeated exposure has the potential to increase sensitivity. Allergic responses include hay fever-like symptoms such as:

- Headache
- Sneezing
- Runny nose
- Red eyes (dermatitis)

Molds can cause asthma attacks in people with asthma who are allergic to mold. In addition, whether or not individuals are allergic to mold, molds can irritate:

- Eyes
- Skin
- Nose
- Throat

Inhaling mold may also cause hypersensitivity pneumonitis, an uncommon disease that resembles bacterial pneumonia. In addition, mold exposure may result in opportunistic infections in persons whose immune systems are weakened or suppressed. When mold grows indoors, the occupants of a building may begin to report odors and a variety of symptoms, including:

- Headaches
- Difficulty breathing
- Skin irritation
- Allergic reactions
- Aggravated asthma symptoms

These and other symptoms may be associated with exposure to mold. But all of these symptoms may be caused by other exposures or conditions unrelated to mold growth. Therefore, it is important not to assume that, whenever any of these symptoms occurs, mold is necessarily the cause.

Although mold is frequently found in damp buildings, it is not the only potential contaminant. Biological contaminants other than mold, as well as nonbiological contaminants, are often present and may also cause health effects. Damp buildings may attract rodents and other pests. Damp or wet building components and furnishings may release chemicals indoors. Potential contaminants in damp or wet buildings include:

- Bacteria
- Dust mites
- Cockroaches and other pests
- Chemicals emitted by damp building materials and furnishings

3. HOW WOULD YOU TELL IF YOU ARE INHALING POTENTIALLY HARMFUL LEVELS OF SPORE CONCENTRATION?

   The way to tell if occupants of a building were inhaling potentially harmful levels of spore concentration is to take air samples. Air may be taken to be analyzed by culture methods or by direct microscopic examination. Culture methods may significantly underestimate the total airborne mold concentration in the air because only the viable spores or fragments of the mold would be detected if the media used for culturing were not suitable for the types of molds present in air. Samples taken for direct microscopic examination allow the analyst to count all the spores and fragments regardless of whether the spores were viable or dead. Since susceptible individuals can also react to dead spores, nonviable samples would be the best to give an idea of whether occupants were inhaling potentially hazardous levels of mold spores.

6.7.1 Mold Mycotoxins and Health Effects

As molds grow, some of them may produce potentially toxic byproducts called mycotoxins under some conditions. Some of these molds are commonly found in moisture-damaged buildings. More than 200 mycotoxins from common molds have been identified and many more remain to be identified. The amount and types of mycotoxins produced by a particular mold depends on many environmental and genetic factors. No one can tell whether a mold is producing mycotoxins just by looking at it.

Some mycotoxins are known to affect people, but for many mycotoxins little health information is available. Research on mycotoxins is ongoing. Exposure to mycotoxins can occur from inhalation, ingestion, and skin contact. It is prudent to avoid unnecessary inhalation exposure to mold.

6.7.2 Moldy Smell

Some compounds produced by molds have a strong smell and are volatile and quickly released into the air. These compounds are known as microbial volatile organic compounds (mVOcs). Because mVOcs often have strong or unpleasant odors, they can be the source of the "moldy odor" or musty smell frequently associated with mold growth. A moldy odor suggests that mold is growing in the building and should be investigated.

The health effects of inhaling mVOcs are largely unknown, although exposure to mVOcs has been linked to symptoms such as:

- Headaches
- Nasal irritation
- Dizziness
- Fatigue
- Nausea

More research is needed to determine whether there are any human health effects from nonoccupational indoor exposure to mVOcs.

6.7.3 Biocides

Biocides are substances that can destroy living organisms. The use of a biocide or a chemical that kills organisms such as molds (Chlorine bleach, for example) is not recommended as a routine practice during mold clean up. There may be instances, however, when professional judgment indicates their use (e.g., when immune-compromised individuals are present).

In most cases, it is not possible or desirable to sterilize an area, and a background level of mold spores will remain, but these spores will not grow if the moisture problem has been resolved. If disinfectants or biocides are used, always ventilate the area and exhaust the air to the outdoors. Never mix chlorine bleach with other cleaning solutions or with detergents that contain ammonia because toxic fumes could be produced. Dead mold is allergenic and may cause allergic reactions and other health effects in some individuals, so it is not enough to simply kill the mold. It must also be removed.

FURTHER READING

Addressing moisture and Mold in your Home Health Canada. Canadian Center for Occupational Health and safety copyright 1997–2010. Modified 8th March, 2010.

Malloch, D., 1981. Molds their Isolation and Cultivation, and Identification. University of Toronto Press, Buffalo London Printed in Canada.

Mold in Indoor air Health Canada. Guidelines on Assessment and Remediation of Fungi in Indoor Environments, New York City. Departmental of Health. 2008.

Mold and Bacteria Consulting Service Copyright @ 2016 (MBS). 1020 Brevik place Unit 1A Mississanya Law 4 N 7 or 2044475 way Burne Drive, Burnby BC V5G.

Institute of Medicine Report, Damp Indoor Spaces and Health, published by the National Academics Press in Washington D.C. 2004.

Part II

A Focus on Food Spoilage

Chapter 7

Spoilage of Milk and Dairy Products

Michael Lu and Nam Sun Wang
University of Maryland, College Park, MD, United States

7.1 INTRODUCTION

Milk and dairy products provide unique conditions for the growth of microorganisms. Effectively controlling these microbes is a critical challenge for the dairy industry. The predominant microorganisms leading to product spoilage can vary widely depending on a combination of factors such as the type of product, method of processing, storage temperature, pH, headspace oxygen concentration, and the source of contamination. This chapter discusses milk's composition and its ability to sustain the growth of microorganisms that cause spoilage. Next, it examines different milk and dairy products and the major microorganisms responsible for their spoilage. This chapter then discusses pathogenic bacteria and methods of processing that destroy or inhibit microbial growth and extend shelf life. And finally, this chapter explains light spoilage and various milk spoilage detection methods.

7.2 MILK COMPOSITION

In general, liquid bovine milk contains approximately 87% water, 3.5% fat, 4.9% carbohydrates, 3.9% protein, and 0.7% ash such as various vitamins and minerals. Milk composition varies widely depending on the species and their immediate environment. Milk from cow, goat, sheep, and other species have unique compositions and differentiate further within various breeds. For example, Guernsey and Jersey cattle are well-known for producing milk with a rich flavor due to its higher fat content, while Holstein-Friesian cattle produce larger quantities at a lower fat content (Jensen, 1995). Other factors, including feeding regimen, dietary fat, intervals between milkings, and season, contribute to the composition of milk (Palmquist et al., 1993; Linn, 1988). The components of milk provide many health advantages, such as promoting proper growth and

The Microbiological Quality of Food. DOI: http://dx.doi.org/10.1016/B978-0-08-100502-6.00010-8

body function, as well as decreasing risk of diseases such as obesity, metabolic syndromes, and hypertension (Griffiths, 2010).

7.2.1 Fat

Compared to other edible fats, milk fat is complex and contains many different types of fatty acids. The majority of these fatty acids are short-chain fatty acids and C_{14}, C_{16}, C_{18}, and $C_{18:1}$ fatty acids (Fox and McSweeney, 1998). Certain fatty acids, such as $C_{14:0}$ and $C_{16:0}$ β-hydroxy fatty acids, are important for the formation of dairy products such as butter. In liquid milk, these fatty acids form milk fat globules (MFG) with a diameter generally between 1 and10 μm. These globules are stabilized by the milk fat globule membrane (MFGM), which is composed of a multitude of phospholipids, proteins, and enzymes. The protein and enzyme components have been shown to be important for child development and maintenance of important cellular processes (Cavaletto et al., 2008). Fat-soluble vitamins, such as vitamins A, D, E, and K, are also found in the lipid components of milk. When milk sits, the larger MFGs rise, initiating a process known as "creaming." As these fat globules rise, they create the distinct "cream line" at the surface and expose the lipids to degradative reactions that shorten shelf life. Creaming can be prevented, however, by homogenizing the milk (Varnam and Sutherland, 2001).

7.2.2 Carbohydrates

Of all of the carbohydrates in milk, lactose predominates in bovine milk, making up approximately 98% of all carbohydrates (Scrimshaw and Murray, 1988). Other sugars, including glucose, galactose, and additional mono- and oligosaccharides, also exist but in trace amounts. Lactose, a disaccharide of glucose and galactose, must be broken down by lactase into its constituent monosaccharides for intestinal absorption. Lactose provides most of the calories in milk. It is estimated that 70% of the world's population suffers from lactase deficiency, which leads to an inability to break down lactose causing symptoms such as osmotic diarrhea. Due to a variety of other factors involved in lactose tolerance, however, many people do not experience symptoms (Matthews et al., 2005).

7.2.3 Proteins

Proteins in milk can be classified into two main groups: caseins and whey proteins. Caseins make up approximately 82% of total milk proteins. They are further subdivided into 5 main groups: α_{S1}, α_{S2}, β, γ, and κ. These subunits bound by calcium phosphate bridges form aggregates with other subunits forming micelles. Caseins mainly exist in milk as micelles and there are approximately 10^{15} casein micelles/L (Jensen et al., 1995). It is these casein micelles that give milk its characteristic white color.

Whey proteins make up the other 18% of total milk proteins. Whey proteins include β-lactoglobulin, α-lactalbumin, blood serum albumin, immunoglobulins, and other proteins. These proteins are also known as "serum proteins" because they remain if fat and casein are removed from milk. Isolated whey products are major by-products formed from cheese production and often are added to other foods such as breads.

Milk proteins are a tremendous source of value to human health. The combination of caseins and whey proteins provide all eight essential amino acids that the body cannot produce on its own. In addition, the lysine-rich amino acid composition complements many diets, including vegetarian-only diets. Caseins and whey proteins confer many health benefits. For example, caseins are readily digestible for growth and tissue development (Munro and Crim, 1988), while whey proteins can help with protein synthesis, muscle hypertrophy, and can help stimulate and improve recovery times after stresses from exercise (Ha and Zemel, 2003). It has also been demonstrated that whey protein can improve insulin levels and fasting lipids in overweight and obese individuals (Pal et al., 2010).

7.2.4 Minerals

Bovine milk also contains a multitude of various ions, minerals, and vitamins essential for human health. The most obvious is calcium, but potassium, sodium, chloride, phosphate, and magnesium also appear in milk. Calcium and phosphate are particularly important due to their involvement in the formation of casein micelles and their role in bone development, cellular signaling, and enzyme activation. Vitamins such as A, B, C, D, E, and K also contribute to human health by providing necessary cofactors for physiological reactions. Milk can also be fortified, particularly in the case of skimmed milk, which must be replenished with vitamins A and D.

7.3 SPOILAGE CHARACTERISTICS

7.3.1 What Is Milk Spoilage?

While milk's components provide many health benefits, the components also provide ideal conditions for microorganisms to cause degeneration and spoiling of milk. The standard for "spoiled" milk is subjective and can be interpreted in many ways. For example, different types of dairy products, such as moldy cheeses and yogurts, may be avoided by one person yet sought after by another. While the term "spoilage" is difficult to normalize, there is a general consensus among consumers and manufacturers that characteristics such as off-flavors, color changes, and loss of consistent texture can signal spoilage. Specific examples include an "acidic aftertaste," "chalky mouth-feel," and "sourness" (Bandler and Barnard, 1984).

7.3.2 Causes of Spoilage

While there are many culprits that contribute to spoilage, the most common cause of milk spoilage is the proliferation of microorganisms, such as bacteria and fungi, which metabolize milk components to form unwanted by-products. Milk spoilage is characteristic for its rapid onset, which is a direct result of exponential microbial growth. The rate of spoilage depends on a number of factors, the most important of which are the number and species of the initial microflora and the storage temperatures and conditions (Schröder et al., 1982). Milk has a unique spoilage process due to its high lactose content and the refrigerator conditions that it is stored in. As microorganism growth increases, the components of milk undergo distinct changes leading to the spoiled by-products that consumers avoid.

7.3.3 Events of Microorganism Spoilage

7.3.3.1 Carbohydrates

In addition, by-products of lactose formed from bacteria include various acids and gases which create "sour" and "gassy" defects (De Jonghe et al., 2010). These bacteria also break down lactose into its constituents, releasing galactose, a monosaccharide that can be used by many other microorganisms. The release of those constituents initiates further bacterial growth and perpetuates the spoilage process.

7.3.3.2 Proteins

In raw or pasteurized milk, lactic acid bacteria (LAB) such as the thermoduric *Streptococcus salivarius* can metabolize lactose for growth. As these bacteria proliferate, they produce lactic acid as a by-product, thus decreasing the pH from the industry standard of 6.6 to 4.5. When milk reaches pH levels below 4.5, the acidic conditions can cause precipitation of casein micelles and lead to curdling (O'Connor, 1995).

In addition, other bacteria produce hydrolytic enzymes including proteases, lipases, and phospholipases, which are able to withstand heat treatment and continue activity even after pasteurization (Chen et al., 2003). These enzymes can lead to significant deformation of UHT and sterilized milk and dairy products, even though the microorganism that produced these enzymes are destroyed. Proteases hydrolyze casein resulting in instability and precipitation of casein micelles. Combined with the acidic conditions caused by lactic acid synthesis, these enzymes can lead to milk gelation and coagulation, also known as "curdling" in stored milk. While this process can provide desirable qualities in cheeses, microbial enzymatic degradation of milk proteins can produce bitter peptides and off-flavors described as "bitter," "fruity," and "metallic" (Marshall, 1982; Muir, 1996).

7.3.3.3 Lipids

Lipases released from microorganisms cleave triglycerides, which make up most of the lipid components of milk, into nonesterified free fatty acids and glycerides. Lipase and phospholipase cleavage of milk fat can lead to the release of different length fatty acids that are associated with intense flavor changes in milk. Released short-chain fatty acids, between C_4 and C_8, coincide with "rancid" flavors and odors, while free medium-chain fatty acids, between C_{10} and C_{12}, create a "soapy" flavor (Chrisope and Marshall, 1976). Elevated free fatty acids that rise to a level over 1.5 mmol/L create enough flavor changes in milk to make it no longer suitable for consumption by most people (Deeth, 2006). Lipases produced by *Pseudomonas fragi* can create "fruity" off-flavors by combining low levels of ethanol in milk with free fatty acids to form ethyl butyrate and ethyl hexanoate esters (Deeth and Fitz-Gerald, 1983).

Bacterial phospholipases have also been shown to interfere with MFGM which can further expose lipids to both bacterial lipases as well as milk's own natural lipases, consequently impairing its emulsification abilities and furthering degradation (Mabbitt, 1981). An interesting observation in both raw and pasteurized milk, known as "bitty cream," occurs when lecithinase or phospholipase C, which are produced by *Bacillus cereus*, breaks down lecithin into phosphorylcholine and diglyceride. Lecithin is found on MFGM, and hydrolysis of this phospholipid leads to destabilization of the fat globule membrane releasing fat emulsion clumps to the surface of milk. These surface fat clumps are commonly seen, especially when milk is added to a hot beverage (Owens, 1978).

7.4 TYPES OF SPOILAGE MICROORGANISMS IN MILK AND OTHER DAIRY PRODUCTS

Among the thousands of bacterial and fungal species and subspecies, the shelf life of milk and other dairy products largely depends on the activity of only a select group of these microorganisms. This section discusses the major microorganisms involved in the spoilage of common dairy products.

7.4.1 Raw Milk

While raw milk is sterile when first released from mammary glands, contact with the cow's udder, teats, and external environments allow for bacteria to contaminate the milk (Chambers, 2002). Each source provides a variety of bacteria flora in milk and thus, each step that brings the milk in contact with a new source requires interventions to reduce contamination.

The udder and teats harbor many bacteria that colonize on their surfaces. Dirt, feces, soil, and feed all contribute to the microflora that may contaminate raw milk during extraction. While only low levels normally contaminate milk, cows with mastitis may shed high levels of *Staphylococcus aureus*, *Streptococcus agalactiae*, and many other bacterial species contributing to undesirable rates of spoilage. These levels can reach up to 10^7/mL (Bramley and McKinnon, 1990). Infection can also lead to increased shedding of somatic cells, mainly leukocytes. The number of somatic cells is a strong indicator of infection and is tightly regulated to control milk quality. Many farmers use antibiotics to treat mastitis. In regulation with the Grade A Pasteurized Milk Ordinance, however, milk cannot be used for human consumption if it contains any detectable level of antibiotics (Food and Drug Administration and US Public Health Service, 2013).

LAB such as *Lactobacillus* contribute to milk spoilage by producing fermentative lactic acid as a by-product of metabolic reactions. LAB spoilage can create "sour" off-flavors and odors due to production of acetic and propionic acids. *Lactococcus lactis* can also lead to the production of various aldehydes and alcohols such as 2-methylpropanal and 3-methylbutanal, which can create a "malty flavor" in milk (Morgan, 1976). The lactic acid produced can sufficiently lower pH levels and induce souring, off-odors, and conformational changes in casein leading to curdling. LAB normally colonize the teat and can be found in the immediate environment around cows such as feed and bedding. Milk spoilage due to LAB is commonly seen in developing countries that do not store raw milk in cold temperatures. LAB are mesophilic and prefer growth in temperatures around 10°C–37°C. It has been demonstrated that milk stored at 4°C for 24 hours shifts bacterial flora toward those that are better able to reproduce under refrigerated conditions and away from mesophilic bacteria such as LAB (Lafarge et al., 2004). In developed countries, milk dominated by LAB indicates milk stored in inadequate refrigeration (Boor and Murphy, 2002).

With the use of refrigeration to store raw milk in most developed countries, psychrotrophic bacteria have become the leading cause of spoilage. This is largely due to their ubiquitous distribution in nature, as well as their ability to grow at low, cold temperatures (less than 7°C). Psychrotrophs are commonly found in soil, water, as well as air (Schröder, 1984). Contamination can also occur due to poor hygiene of milk-handling equipment. Psychrotrophs are notable for their ability to form biofilms on unsanitized equipment, which allows for further contamination. From milk extraction to bottling, psychrotrophs can contaminate milk throughout its processing. In particular, raw milk is stored in refrigerated conditions for extended periods before pasteurization, which allows for psychrotrophs to dominate the microflora.

Psychrotrophs found in milk are generally classified into Gram-negative nonspore-formers and Gram-positive spore-formers. The Gram-negative psychrotrophs make up a large percentage of raw milk's microflora. Among this group of bacteria, *Pseudomonas* spp. represent the largest subgroup, comprising up to 65–75% of psychrotrophs isolated from raw milk (Garcia et al., 1989). Other common Gram-negative psychrotrophic bacteria found in milk include species of *Acinetobacter*, *Bacilli*, and *Actinobacteria* (Hantsis-Zacharov and Halpern, 2007). Milk can also contain Gram-positive psychrotrophs such as *Corynebacterium*, *Micrococi*, *Clostridium*, and *Bacillus* (Cousin, 1982). Psychrotrophic bacteria create significant enzymatic degradation that may occur throughout a product's shelf life. These microorganisms produce a variety of lipases, lecithinases, and proteases that do not denature during pasteurization. Stability of these enzymes continues to persist and degrade components of milk even after the psychrotrophic bacteria producing these enzymes are destroyed (McKellar, 1982).

Soil, bedding, and unclean milk-handling equipment are also sources of spore-forming, thermoduric bacteria. These mostly include the *Paenibacillus*, *Bacillus*, *Clostridium* bacteria (De Jonghe et al., 2010). A characteristic feature is their ability to form endospore structures which are highly protective and resistant to damage from unfavorable conditions such as high heat, extreme pH, and chemical sanitization. Spores can remain dormant for many years and can be found throughout the environment. Spore-formers are common in raw milk, with *B. cereus* found in 80% of raw milk samples. (Johnson et al., 1990); however, their levels are low and generally found below 5000/mL (Mikolajcik and Simon, 1978). Prolonged storage of raw milk can cause spores to germinate and affect milk quality. While entry of these bacteria occurs during the raw milk stage of processing, spoilage activity plays a more significant role postpasteurization (Fromm and Boor, 2004).

7.4.2 Processed Milk

Pasteurization and heat treatment of raw milk eliminates a majority of the bacterial levels to less than 1000 CFU/mL (Boor and Murphy, 2002). However, psychrotrophic spore-forming bacteria that both survive high levels of heat and grow at refrigerated temperatures pose the greatest threat to milk quality among bacteria that exist prepasteurization (Fromm and Boor, 2004) Another more significant cause of spoilage in pasteurized milk is postpasteurization contamination (Champagne et al., 1994). While the initial flora of raw milk can contain high numbers of psychrotrophs, especially after prolonged storage in refrigerated conditions, many psychrotrophs, especially those that are Gram-negative, do not survive the pasteurization process. However, these psychrotrophs can easily contaminate postpasteurized milk through outside sources.

Milk-handling equipment downstream in the processing chain is a principal source of recontamination of milk. While proper sanitization of equipment minimizes recontamination, inadequate aseptic filling and packaging procedures are common routes for bacterial entry (Moseley, 1980). Filling machines, new contacts, and airborne psychrotrophs are all potential sources of postpasteurized contamination (Gruetzmacher and Bradley, 1999). Psychrotrophs can be found on the packaging themselves especially those that contain defects in sealing. During refrigeration, postcontaminant psychrotrophs can once again dominate microflora. Enzymes produced by psychrotrophic bacteria that initially contaminated the raw milk are another cause of spoilage in postpasteurized milk. Degrading enzymes are thermostable and resist denaturation during pasteurization. These enzymes supplement postpasteurization microflora metabolism and lead to decreased shelf life (Moseley, 1980).

Pasteurized milk produced from raw milk containing high levels of psychrotrophic bacteria and stored for longer periods of time leads to faster rates of spoilage compared to pasteurized milk produced by raw milk with lower psychrotroph counts (Vasavada and Cousin, 1993).

Thermoduric bacteria, specifically the spore-forming, Gram-positive psychrotrophs that survive pasteurization, also contribute to degradation of pasteurized milk (Shehata et al., 1971). While postpasteurization contamination is the main cause of spoilage, thermoduric spore-formers are also important to control for, especially in milk that is not aseptically filled and packaged (Champagne et al., 1994). During pasteurization, high heat activates many spores, initiating germination and growth in the postpasteurized product (Cromie et al., 1989). As pasteurized milk is stored under refrigerated temperatures, psychrotrophic spore-formers predominate, especially the *Bacillus* spp. and *Paenibacillus* spp. In controlled, aseptically packaged milk, *Bacillus* spp. dominated the microflora during the early stages periods of shelf life (<7 days). *Paenibacillus* spp. was found to be the dominant spoilage organism during the late stages of shelf life (>10 days) (Ranieri and Boor, 2010). Both families of spore-formers cause a variety of enzymatic-induced defects such as bitty cream and curdling in milk that create off-flavors and odors (Vasavada and Cousin, 1993). "Bitter," "rancid," and "sour" are off-flavors commonly associated with the *Bacillus* spp.

7.4.3 Cheese

Some physical changes, such as casein coagulation, are unwanted processes in fluid milk. However, these "defects" are needed to produce cheeses, and this dairy product provides a very different environment for microorganisms. Lower water content, higher salt levels, removal of whey proteins, lower pH levels, and curd formation are just a few examples of differences between cheese and fluid milks and their impact on different microorganisms responsible for spoilage (Johnson, 2001). Spoilage of cheese products can occur during contamination from processing and ripening, as well as from growth of spoilage thermoduric species that survive heat treatments (Johnson et al., 1990).

Cheeses can generally be classified as soft, semi-soft, or hard, depending on their moisture content. Soft cheeses have a high moisture content (50–80%) and examples include cottage cheese and mozzarella. Because of soft cheeses' higher water content and more neutral pH levels (5.0–6.5), a large variety of microorganisms can proliferate compared to hard cheeses. Psychrotrophic Gram-negative bacteria such as *Pseudomonas* spp. are commonly isolated from these cheeses, and their ability to degrade lipids and fats leads to unwanted flavors and odors (Farkye et al., 2002). Semi-soft and hard cheeses, such as Gorgonzola and Parmesan, respectively, have lower moisture contents and more acidic pH levels that inhibit growth of bacteria more commonly seen in the soft cheese variety.

The *Clostridium* spp., a spore-forming bacteria, causes a common defect known as "late blowing," in which internal gas is generated during ripening and can cause sponge-like defects to form in the cheese. *Clostridium* spp. contamination is due to the use of ensilage in cow diets, and the bacteria that survive pasteurization can metabolize lactate into butyric acid, hydrogen gas, and carbon dioxide, forming the gas bubbles in the cheese (Dasgupta and Hull, 1989). LAB and coliform bacteria, such as *Klebsiella* spp., *Enterococcus* spp., and *Citrobacter* spp., also contribute to the production of a variety of defects in cheese. These bacteria can ferment lactose and produce a variety of acids and gases (Laleye et al., 1987; Massa et al., 1992). Fruity off-flavors in cheeses are also associated with the esterases produced by LAB, which use ethanol to produce ethyl hexanoate and ethyl butyrate (Bills et al., 1965).

The lower pH of these cheeses also facilitates the growth of a variety of yeasts and molds. Yeasts commonly seen in cheese spoilage include *Geotrichum candidum*, *Pichia* spp., and *Candida* spp. (Johnson). Yeasts release alcohol and carbon dioxide contributing to off-flavors and gas production (Horwood et al., 1987). Some strains may additionally produce sulfides, which give cheese an "egg" odor. Molds, including *Penicillium* spp., can often be found in packaged cheeses in refrigerated conditions. These molds require little oxygen and can form characteristic pigmented colonies as well as mycotoxins, which can be dangerous for consumers (Wendorff et al., 1993).

7.4.4 Other Dairy Products

7.4.4.1 Yogurt

Dairy products, such as yogurt and sour cream, are mainly affected by yeast and mold growth, and some bacteria can contribute to spoilage as well.

These products, which are generally refrigerated and acidic, provide conditions that facilitate yeast growth. *Candida*, *Pichia*, *Kluyveromyces*, *Rhodotorula*, *Debaryomyces*, and *Torulopsis* species are all associated with yogurt spoilage (Kosse et al., 1997; Suriyarachchi and Fleet, 1981). Species of *Kluyveromyces* can ferment lactose and are common culprits of yogurt spoilage (Fleet, 1990). Yeast can also ferment sucrose, found in additives to yogurt such as fruit. Yeast fermentation of sucrose frequently creates gas production as a by-product (Fleet, 1990). Certain LAB, such as *Lactobacillus harbinensis*, have been shown to have a powerful antifungal activity and serves as a potential culture starter to extend yogurt shelf life (Delavenne et al., 2013).

7.4.4.2 Butter and Other Dairy Spreads

Butter and dairy spreads are oil-water emulsions that generally contain high levels of fat. Butter contains upwards of 80% fat while dairy spreads have lower fat contents between 30% and 60% (Varnam and Sutherland, 1994). While butter and dairy spreads contain organisms that are initially found in the raw milk, pasteurization, added salt, and preservatives inhibit microbial growth, so spoilage is commonly due to contamination after production.

Among both types of dairy products, a variety of bacteria and fungi present significant danger to their shelf life due to their ability to grow in refrigerated, acidic conditions (Varnam and Sutherland, 1994). Surface bacteria, including *Pseuodmonas* spp. and *Flavobacterium* spp., can proliferate on the surface layers leading to rancid and "cheesy" flavors (Kornacki et al., 2001). Various fungi, such as *Aspergillus*, *Rhizopus*, *Penicillium*, *Geotrichum*, *Mucor*, and *Candida*, are also associated with spoiled butter and dairy spread (Jay et al., 2005). Growth of these fungi can break down fats, produce off-flavors, and cause discoloration. While additives such as sorbate are added to inhibit microorganism growth, species of *Penicillium* are able to metabolize sorbate into by-products that can further reduce shelf life (Marth et al., 1966).

7.4.4.3 Concentrated Milk Products

Condensed and powdered milk may undergo multiple heat treatments during processing which greatly limit microbial spoilage of these products. In condensed milk, canning, proper sanitization, and refrigeration limit only thermoduric organisms, such as *Bacillus* spp., to low levels that maintain longer shelf lives (Kalogridou-Vassiliadou, 1992). Dried milks have extensive shelf lives due to the minute water content that inhibits growth. However, impaired packaging allows dried milk products to absorb water, which can facilitate growth of spore-formers and other microorganisms that persist even after pasteurization and heat treatment.

7.5 PATHOGENIC MICROORGANISMS AND THEIR SIGNIFICANCE IN PUBLIC HEALTH

While there are a large variety of microorganisms that contribute to milk and dairy product spoilage, there are a few specific bacteria that can cause serious illness among consumers. These bacteria, Gram stain classification, and associated symptoms are listed in Table 7.1.

TABLE 7.1 Common Bacteria in Spoiled Milk and Dairy Products

Bacteria	Staining	Clinical Symptoms
B. cereus	Gram-positive	Flu-like illness
		Fever
		Vomiting
		Diarrhea
Brucella spp.	Gram-negative	Undulating fever
		Osteomyelitis
Campylobacter jejuni	Gram-negative	Inflammatory diarrhea
		Arthritis
		Guillain–Barré syndrome
Coxiella burnetii	Gram-negative	Flu-like illness
		Headache
		Fever
		Hepatitis
Escherichia coli 0157: H7	Gram-negative	Inflammatory diarrhea
		Nausea
		Vomiting
		Fever
		Hemolytic uremic syndrome
Listeria monocytogenes	Gram-positive	Meningitis
		Encephalitis
		Septicemia
		Amnionitis in pregnant women
Mycobacterium spp.	Acid fast	Tuberculosis
		Possible association with Crohn's disease

(*Continued*)

TABLE 7.1 (Continued)

Bacteria	Staining	Clinical Symptoms
Salmonella spp.	Gram-negative	Inflammatory diarrhea
		Fever
		Nausea
S. aureus	Gram-positive	Nausea
		Vomiting
		Abdominal pain
		Diarrhea
Yersinia enterocolitica	Gram-negative	Inflammatory diarrhea
		Abdominal pain
		Fever
		Nausea

Generally, pasteurization and heat treatment prevent pathogenic bacteria and fungi from causing disease in consumers; however, improper pasteurization, contamination, and allowance of milk spoilage can lead to human illness. There has been an increase in the demand for raw, unpasteurized milk in the last decade, and consumption of raw milk can pose a significant clinical risk for consumers (Oliver et al., 2009). Consumers of raw milk consumption cite taste, supporting local farms, and health benefits as reasons for drinking raw milk, despite the numerous health warnings against raw milk consumption (Katafiasz and Bartlett, 2012).

According to the Centers for Disease Control and Prevention (CDC), raw-milk related outbreaks continue to increase (Mungai et al., 2015). Comparing 2007–09 and 2010–12, outbreaks in the United States caused by nonpasteurized milk has increased from 30 to 51, respectively, leading to 979 illnesses and 73 hospitalizations. The most common causes of these outbreaks were *Campylobacter* spp., Shiga-toxin producing *Escherichia Coli*, and *Salmonella* spp. During these 6 years, there were 13.5 outbreaks/year. In contrast, from 1993 to 2006, there was an average of only 3.3 outbreaks/year (Langer et al., 2012). This rise in both consumption and, consequently, outbreaks may continue to increase as states become more lenient in regulating raw milk sales and allow more open access to raw milk production. From 2004 to 2011, the number of states in which nonpasteurized milk could legally be sold increased from 22 to 30 (National Association of State Departments of Agriculture, 2011).

In addition to pathogenic bacteria associated with raw milk, there has been increased interest in links between raw milk and the Bovine Leukemia Virus (BLV). There is evidence that BLV, a retrovirus that commonly causes B-cell leukemia/lymphoma in cattle, can also be transmitted to human hosts (Buehring et al., 2014). While the mode of BLV transmission is not well defined, it is speculated that drinking raw milk may be a route of transmission. BLV can be found throughout dairy operations and raw milk. According to a 2007 survey conducted by the US Department of Agriculture, 100% of all large herd dairy operations and 83.9% of all US dairy operations tested positive for BLV antibodies (US Department of Agriculture Animal and Plant Health Services, Veterinary Services, Center for Epidemiology and Animal Health, 2007). Further studies show that BLV in human hosts are associated with increased incidence of breast cancers (Buehring et al., 2015). The study found an odds ratio of 3.07, consistent with other established breast cancer risk factors. In addition, according to the study, "as many as 37% of breast cancer cases may be attributable to BLV exposure."

Toxins produced by fungi may also cause disease in humans. Pathogenic *Aspergillus* spp. can contaminate cattle feeds leading to production of aflatoxins found in milk (Frobish et al., 1986). Aflatoxins are associated with liver damage and cancer in humans (Ryser, 2001). While these toxins are extremely durable and can potentially survive the pasteurization process, aflatoxins ingested by cattle rarely appear in their milk (0.4–2.2%) (Frobish et al., 1986), and levels of contamination are usually very low (Ryser, 2001).

7.6 MILK PROCESSING AND PREVENTION OF SPOILAGE

Because microbial spoilage is the greatest barrier to extending shelf life of milk and other dairy products, it is imperative to practice adequate sanitization during critical stages of milk production and processing to control bacterial and fungal growth.

7.6.1 Milk Collection

Proper sanitization should begin before any milk is collected. Areas around the cow during collection and the cow's udder and teats should be sufficiently cleaned and sterilized. Milk-collecting equipment should also be thoroughly cleaned routinely to reduce bacterial contamination. Sanitization methods for equipment should make sure to prevent formation of biofilms, which are resistant to standard cleaning and provide a constant source of contamination if not properly treated. Any milk from cows with mastitis and high somatic cell counts must be discarded to prevent mixing of infected milk with stored milk. Storage tanks that are reused must be cleaned before each use and all testing of milk should be performed aseptically. All milk-handling equipment throughout the entire process should continually be cleaned to prevent any contamination downstream after collection. Filling machines, nozzles, packaging materials, and other operating machinery are all main sources of contamination during processing and should be carefully cleaned to prevent milk contamination (Gruetzmacher and Bradley, 1999).

7.6.2 Storage of Raw Milk

After collection, raw milk should be immediately stored in refrigerated conditions (<7°C) to limit microbial growth. Temperatures should be kept as low as possible (<4°C) to extend storage life, and it has been shown that the shelf life of processed milk increases by decreasing temperatures during storage (Janzen et al., 1981). Because raw milk is rarely used immediately, various methods are used to reduce psychrotrophic bacterial counts in raw milk before it is processed and pasteurized. One method, known as thermization, is to heat raw milk to 60–65°C for 15 seconds and then rapidly cool the milk to below 7°C. This brief heat treatment can kill and limit psychrotrophic bacteria that are able to proliferate under refrigerated conditions, while the rapid cooling prevents germination of spore-forming bacteria (Sun, 2012).

The addition of carbon dioxide is also commonly used to maintain low levels of psychrotrophs (Devlieghere and Debevere, 2000). Carbon dioxide (CO_2) can limit growth through three mechanisms: (1) Displacement of oxygen needed for growth by aerobic organisms. (2) Formation of carbonic acid and decreased pH in milk (Butler, 1991). (3) Directly inhibiting metabolism by inhibiting enzyme production (Daniels et al., 1985). There have been safety concerns that additional carbon dioxide may elicit growth of *Clostridium botulinum*, the bacteria responsible for Botulism. However, studies have shown that additional carbon dioxide does not lead to any additional risk of *C. botulism* growth (Hotchkiss et al., 2006).

7.6.3 Heat Treatment During Processing and Pasteurization

In order to sufficiently preserve milk quality and free it from pathogenic and most spoilage-causing microorganisms, milk must undergo heat treatments such as pasteurization and ultra-high temperature processing (UHT). During pasteurization, milk is heated to high temperatures for a short period of time. In the United States, the most commonly used pasteurization methods are low temperature long time (LTLT) or high temperature short time (HTST). Milk undergoing LTLT pasteurization is generally used to make other dairy products such as cheese. It is heated to 62.8°C for 30 minutes before being cooled to refrigerated conditions. The more commonly used method for fluid milk production is HTST, which heats the milk to 72°C for 15 seconds before being rapidly cooled. Both methods are designed and based off their ability to destroy *Coxiella burnetii*, which is considered the most thermally stable nonspore-forming bacteria (Enright et al., 1957).

For some milk products, ultra-high temperature processing can extend shelf lives and sterilize milk. UHT can destroy spore-forming bacteria that are generally resistant to LTLT or HTST pasteurization. However, some bacteria such as *Bacillus stearothermophilus* may survive and cause spoilage in UHT-processed milk (Lewis, 1999). During UHT processing, milk is heated to 140–150°C for 2 s. Throughout these heat treatments, milk should be continuously tested and monitored in order to comply with FDA Pasteurized Milk Ordinance guidelines.

7.6.4 Other Milk Prevention Methods

While heat treatment is the main method for limiting microbial growth and milk spoilage, other methods have been developed to further extend the shelf lives of milk such as microfiltration, high-pressure processing (HPP), and the addition of Nissin. Microfiltration uses membrane filtration with pore sizes between 0.2 and 5 μm to remove bacteria from milk that survives the pasteurization process. This process has been shown to greatly increase the shelf life of pasteurized milk and reduce microbial loads (Eckner and Zottola, 1991; Schmidt et al., 2012).

HPP is also a growing method that applies pressure up to 1000 MPa to destroy cell spores, membranes, and other bacterial structures, thus inactivating spoilage-causing bacteria. The combination of HPP with high heat has been shown to significantly inactivate *B. cereus* spores in milk (Silva, 2015).

Nissin, an antimicrobial additive, can also be used to inhibit the growth of a variety of Gram-positive bacteria, such as *C. botulinum* (Muir, 1996). It is approved for use in a variety of dairy products such as cheeses and creams.

7.6.5 Industry Standard Milk

In the United States, Grade A milk is the only milk regulated under federal orders. The Grade A Pasteurized Milk Ordinance dictates the proper standards for Grade A milk from production to distribution. Dairy farms producing Grade A milk must have proper licensure and sanitary conditions for production. During processing, Grade A milk must go through rigorous standards including required bacterial counts, somatic cell counts, cooling checks, and sampling must be followed before being cleared for distribution (Food and Drug Administration and US Public Health Service, 2013). While regulations vary from state to state, the FDA Pasteurized Milk Ordinance dictates specific chemical, physical, bacteriological, and temperature standards seen in Table 7.2.

TABLE 7.2 2013 FDA Grade "A" Pasteurized Milk Ordinance

Milk Grade	Quality Standards	
Grade A pasteurized milk	Temperature	• Cooled to 7°C (45°F) or less and maintained thereat
	Bacterial limits	• Not to exceed 20,000 mL^{-1}, or g
	Coliform	• Not to exceed 10 mL^{-1} • Provided, that in the case of bulk milk transport tank shipments, shall not exceed 100 mL^{-1}
	Phosphatase	• Less than 350 mU/L for fluid products and other milk products
	Drugs	• No positive results on drug residue detection methods
Grade A ultrapasteurized (UP) milk	Temperature	• Cooled to 7°C (45°F) or less and maintained thereat
	Bacterial limits	• Not to exceed 20,000 mL^{-1}, or g
	Coliform	• Not to exceed 10 mL^{-1}. Provided, that in the case of bulk milk transport tank shipments, shall not exceed 100 mL^{-1}
	Drugs	• No positive results on drug residue detection methods
Grade A raw milk and milk products for pasteurization, ultra-pasteurization, or aseptic processing	Temperature	• Cooled to 10°C (50°F) or less within 4 hours or less, of the commencement of the first milking, and to 7°C (45°F) or less within 2 hours after the completion of milking
	Bacterial limits	• Individual producer milk not to exceed 100,000 mL^{-1} prior to commingling with other producer milk • Not to exceed 300,000 mL^{-1} as commingled milk prior to pasteurization
	Drugs	• No positive results on drug residue detection methods
	Somatic cell count	• Individual producer milk not to exceed 750,000 mL^{-1}

Source: Adapted from the 2013 Revised FDA Grade "A" Pasteurized Milk Ordinance.

7.7 LIGHT AND SPOILAGE

7.7.1 Light

Light is a well-known cause of milk deterioration. Milk contains high levels of riboflavin (vitamin B_2), which is a sensitive photosensitizer that absorbs visible and UV light. These vitamins, which persist in milk after pasteurization, can transfer light energy to surrounding oxygen molecules that results in formation of highly reactive oxygen species such as singlet oxygen, and can lead to further cascades of oxidative reactions (Min and Boff, 2002). The rate of light-induced oxidation depends on factors such as light intensity, milk composition, and container composition. Both sunlight as well as artificial light can damage milk properties and these reactions result in vitamin loss, oxidation of lipids and proteins, discoloration, and production of off-flavor and odors.

7.7.2 Vitamin Loss

One of light's most prominent photodegradative effects on milk is the decrease in important vitamin levels. Riboflavin oxidation has a strong inverse correlation with levels of singlet oxygen and other reactive oxygen species and these oxidizers react with many various vitamins in milk.

Vitamin D is a major source of milk nutrition affected by singlet oxygen production. Milk itself does not naturally contain high levels of vitamin D, however, pressures from diseases such as rickets in children and osteomalacia in adults have prompted many countries to fortify milk with supplemental vitamin D, approximately 400–600 IU (International Unit) per quart of milk (Calvo et al., 2004). Vitamin D was shown to only degrade in the presence of riboflavin and exposure to light. Light itself did not directly oxidize vitamin D, but it is able to excite riboflavin to a reactive state that forms reactive oxygen species. Rate of degradation increases with concentrations of both riboflavin and Vitamin D. Similarly, vitamins E, K, and folic acid are also highly degraded by reactive oxygen species. (King and Min, 1998). Like riboflavin, vitamin A is directly affected by light energetics rather than oxidation from reactive oxygen species (Berge et al., 1987). One study found that over 75% of vitamin A was lost when milk was exposed to light with an intensity of 2000 lux for 24 hours (Senyk and Shipe, 1981).

When exposed to light, vitamin A forms oxidation products that create low-threshold odors that contribute to the perception of spoilage (Whited et al., 2002). Milk's vitamin A levels rapidly decrease in a linear fashion with increased exposure time to light. The rate of vitamin A degradation has been shown to decrease with increasing concentrations of milk fats, mostly due to reducing light penetration through the milk. However, higher fat content also causes an increase in light-oxidized off-flavors (Whited et al., 2002).

As an antioxidant, vitamin C (ascorbic acid) antagonizes riboflavin's role as a photosensitizer. Loss of vitamin C occurs due to its role as a protector against further oxidative damage. Vitamin C acts as a target for free radicals, thus preventing further degradation of riboflavin and other components of milk (Lee et al., 1998). Similarly to Vitamin C, β-carotenes and tocopherols exhibit antioxidative properties and contribute to the protection of further oxidative damage (Yang and Min, 1994).

7.7.3 Protein Deterioration

Light exposure can cause protein deterioration leading to the production of a variety of off-flavors, also known as "sunlight" or "burnt-feather" flavors. Reactions between reactive oxygen species and the amino acids in milk form these off-flavors (Marsili, 1999). While none of the essential amino acids are lost due to light degradation, off-flavors can quickly cause milk to be less than acceptable by consumers (Dimick and Kilara, 1983). Due to milk's mild flavor, even small concentrations of off-flavors can significantly impact customer perception of milk quality (Alvarez, 2008).

7.7.4 Lipid Deterioration

The light-induced oxidation of lipids also produces distinct off-flavors. Off-flavors associated with lipid deterioration are described as "cardboard-like" and take more time to appear than off-flavors from light-induced protein degradation. Light can increase the oxidation rate of unsaturated fatty acids producing hydroperoxides and low levels of carbonyl compounds such as hexanal and thiobarbituric acid, but ppb levels are needed to lead to noticeable and undesirable changes to milk (Erickson, 1997).

7.7.5 Methods of Protection From Light Degradation

Milk, from production and processing to consumer refrigeration, constantly faces exposure to light. Light-induced oxidation can significantly damage milk's consumer appeal. One study showed that fresh milk protected from light provides consumers with a positive emotional experience, whereas fresh milk that is exposed to light leads to lower acceptability and increased expression of disgust (Walsh et al., 2015). Due to light's significant and undesirable effects on the quality and state of milk, many methods are being used to prevent light exposure.

The main method is through the use of special light-protective packaging (Mestdagh et al., 2005). Milk is most commonly packaged in various plastics, including high-density polyethylene (HDPE), polyethylene terephthalate (PET), and polystyrene (PS). Unlike paperboard cartons, which reduce the passage of light onto milk, plastic packaging materials transmit high levels of light and fail to protect the milk from photo-oxidation (Senyk and Shipe, 1981). HDPE and PET packaging is most often unpigmented and translucent, allowing for high levels of light transmittance. While plastic materials generally have lower light transmittance than glass, HDPE can still transmit 62–85% of light (Webster et al., 2009). These levels of light transmitted by common plastic packaging are still too high and cannot sufficiently prevent light-induced oxidation reactions during retail storage conditions (Johnson et al., 2015).

To counter this problem, many manufacturers are incorporating additives such as titanium dioxide, metal coatings, and various pigments to scatter and reflect light. Titanium dioxide (TiO_2) is commonly used to scatter light as well as absorb UV light (Yam and Lee, 2012). While low levels of TiO_2 (0–1.3%) do not significantly protect milk, high levels of titanium dioxide (4.3%) incorporated in HDPE packaging has been shown to be effective in maintaining riboflavin concentrations and reducing thiobarbiturate production (Johnson et al., 2015).

The use of coating and lamination also protect against photo-oxidation. While coating and lamination are primarily used as barriers against oxygen permeation, layers (5–40 nm thick) of aluminum can be incorporated to protect against visible light, with a positive correlation between layer thickness and concentration of light blocked (Piergiovanni and Limbo, 2004). Paperboard, a commonly used packaging material, often incorporates many of these elements such as a plastic coating and a foil layer to provide both light and oxygen permeating protection.

The addition of pigments to packaging has also been shown to protect against light-induced oxidation. Carbon black, iron oxide, and blue and green copper phthalocyanines are just a few pigments that are being used to color plastics (Robertson, 2012). Pigments can absorb light in specific wavelengths that may initiate photo-oxidation. Wavelengths between 400 and 570 nm can excite riboflavin into its high-energy state; thus, many pigments in the purple, blue, green, and yellow spectrums help prevent the production of singlet oxygen (Webster et al., 2009; Doyle, 2004). For example, amber-colored PET packaging preserved milk flavor better compared to glass, HDPE, and clear PET (Van Aardt et al., 2001).

Still, it is recommended to block all light in order to sufficiently stop light-induced oxidation reactions and vitamin loss. One study showed that flavors from light-oxidative reactions were present even when all riboflavin wavelengths were blocked (Webster et al., 2009). This is most likely due to the involvement of other photo-sensitizers besides riboflavin, such as porphyrins and chlorophyll, which can also produce reactive oxygen species (Wold et al., 2005).

Other additives in packaging have also been explored such as the use of UV blockers as well as oxygen scavengers. However, none of these additives are as capable of reducing photo-oxidation as complete light block. A study compared three types of packaging: (1) a bottle with a perfect light barrier, (2) a transparent bottle with UV-absorbing additive, and (3) a transparent bottle with an active oxygen-binding inner layer. Out of the three methods of packaging, only the bottle with the perfect light barrier was able to avoid light-induced oxidation reactions, off-flavor production, and vitamin degradation (Mestdagh et al., 2005).

7.7.6 Challenges to Implementation

While many of these light-protective packaging methods have demonstrated high efficacy in protecting milk contents from light, milk packaging still exists in forms such as clear plastic and glass that do not offer adequate light protection. Selection of packaging materials depends on a variety of factors such as labeling requirements, costs, target market, consumer appeal, and shipping methods (Yam and Lee, 2012). These factors are often much higher priorities for manufacturers, producers, and retailers, and prevent the adoption of light-protecting materials even with the risk of shorter shelf lives and lower product quality. For example, many studies have demonstrated the need for customers to be able to see their products within the package (Doyle, 2004; Sattar et al., 1975). However, light barriers reduce product visibility. Glass is also increasing in popularity because of its association with high-end beverage products.

Other obstacles, such as lack of knowledge about new and updated packaging technologies, can prevent manufacturers from implementing protective packaging (Yam and Lee, 2012). Even with increasing product development, adoption is slow in the food-related market. New materials are first applied to other fields of interest with higher profit margins, such as in the medical and military industries, before entering the food industry. In addition, because packaging is a food contact material, application must undergo stringent safety and quality testing which can be a long and resource-intensive process for manufacturers (Yam and Lee, 2012).

7.7.7 Light Source

Interventions to reduce light sources during storage are becoming more popular as methods in maintaining milk quality. It has been estimated that milk receives 7 or more days of light exposure by the time it reaches retail (Senyk and Shipe, 1981). This amount of light exposure is plenty for nutrient loss and off-flavor production which can be detected within 1 to 2 hours by the average consumer (Chapman et al., 2002). From production to retail, milk manufacturers can minimize light exposure by two main methods: (1) decreasing time exposed, and (2) minimizing light intensity and wavelengths that initiate the production of reactive oxygen species. In addition, it is estimated that 50% of milk packages are exposed to 8 or more hours from retail lighted display cases, providing more than enough light exposure time for photo-oxidation (Chapman et al., 2002). Both retailers and manufacturers can minimize light damage by making sure that handling and storage areas are designed to keep milk away from any direct light exposure or covering milk at all times. Display cases should position milk away from light bulbs. Some retailers are using motion-activated lights to minimize any unnecessary light exposure. Consumers should also be aware of the damaging effects of light on milk, and should protect milk by using brown paper bags during transport and keeping milk away from direct exposure to light during refrigeration.

If milk, for a variety of reasons, must be exposed to light, there are various ways to minimize oxidative damage by changing the lights themselves. Display cabinets in the average supermarket have a mean light intensity of 1000 lux, which is high enough to initiate off-flavor production in milk (Griffiths, 2010). Less light-intensive bulbs can be used to reduce the light penetration into milk to slow down the degradative effects. Furthermore, the color of the light is also an important factor. Most storage, processing, and retail lights use "cold" lights, which contain higher emissions in the violet, blue, and green wavelengths; however, these should be avoided. These wavelengths coincide with riboflavin absorption and excitation. Instead, "warm" lights containing mostly yellow, orange, and red wavelengths should be used to avoid overlapping with riboflavin's absorption spectrum (Borle et al., 2001). While this can delay photo-oxidation, the best method is simply to prevent total light exposure to keep milk quality at its highest.

7.8 MILK SPOILAGE DETECTION METHODS

To gage levels of spoilage, there are many methods and tests that measure milk quality. The most commonly used method is through bacterial counts. Standard plate counts, coliform, psychrotrophic, thermoduric, and yeast and mold counts are commonly used to test dairy products. While bacterial counts can describe how clean processing operations are, they may not correlate with shelf life and quality (Boor et al., 2005). These tests also are generally slow and take many days to test milk. Yeasts and molds are also difficult and slow to culture using standard plating times (García et al., 2004). To speed up this process, PCR-based detection methods have been developed to identify specific bacterial levels. For example, a rapid PCR-detection method to detect low counts of *Paenibacillus* in raw milk was developed to determine spoilage potentials of the pasteurized product (Ranieri et al., 2012).

Gas sensor arrays have also been developed to detect specific volatile metabolites of different bacterial species. While gas sensor arrays have been shown to rapidly and accurately identify specific species, metabolites must be volatile for this method to work, which may exclude bacterial populations contributing to product spoilage (Haugen et al., 2006).

Infrared spectroscopy, which analyzes spectral changes to determine specific molecular interactions, has also grown as a way of detecting metabolic by-products produced by spoilage microorganisms. One study looked at the use of matrix-assisted laser desorption/ionization time-of-flight mass spectrometry (MALDI-TOF-MS) and found this technology, which is commercially used in the meat industry, to be rapid and accurate for identifying spoilage bacteria in milk (Nicolaou et al., 2010).

Methylene blue reduction combined with an amperometric sensor can also be used to detect spoilage based on rates of bacterial metabolism. This method was determined to accurately predict coliform bacterial concentrations based on the time needed to reduce methylene blue and the corresponding current changes produced by reduction (Lee et al., 2009). While this method is fast and easy to conduct, it could not distinguish different species of bacteria and excluded bacteria that did not produce any significant current change.

As milk spoils, bacterial counts increase leading to higher levels of metabolism of lipid and protein counts. One study found that when *Pseudomonas* spp. levels are highest, lipid and protein levels are lowest (Yagoub et al., 2008). Therefore, protein and lipid counts can be measured as a correlate for milk spoilage.

7.9 FUTURE TRENDS/CONCLUSION

Milk and dairy-related product spoilage make up one of the most significant sources of food loss in the United States. It is estimated that 20% of HTST pasteurized milk is discarded before consumption ever year and approximately 17.4 billions of pounds of milk are lost each year (Kantor et al., 1997). In addition to economic losses, wasted milk also has major environmental consequences. In the United Kingdom, 360,000 tons of milk are wasted each year which is estimated to produce 100,000 tons of CO_2 (Reay et al., 2012). With such large losses and damage to the environment every year, even a small increase in the average shelf life of milk and dairy products can have a drastic effect in reducing wastage.

To prevent microbial spoilage, emerging technologies such as ultrasound, pulsed-electric field, and a variety of new additives, including lantibiotics, are being tested in the dairy industry (Ross et al., 2002). There is also expanding interest in reducing retail milk spoilage through the use of smart packaging, and providing consumers with better spoilage indicators than expiration dates (Restuccia et al., 2010). Expiration dates fail to measure variance in processing, shipping, and storage conditions and may mislead consumers (Kim et al., 1997). Food producers use conservative estimates for expiration dates to protect consumers from safety hazards associated with spoiled milk. However, these ambiguous expiration dates can cause consumers to throw milk away before actual spoilage. Smart packaging such as

"Fresh-Check" can provide both retailers and consumers with quick and more accurate detection of milk spoilage. Boosting detection and prevention are the keys to reducing wastage, and the dairy industry continues to develop new and improved technologies to reach these goals.

REFERENCES

Alvarez, V.B., 2008. Fluid milk and cream products. The Sensory Evaluation of Dairy Products. Springer, New York, NY, pp. 73–133.

Bandler, D.K., Barnard, S.E., 1984. Milk Flavor and Quality--From Cow to Consumer. Cornell University, Ithaca, NY.

Berge, C., Smith, D.E., Warthesen, J.J., 1987. Effect of two antioxidants on the light stability of vitamin A in 2% fat milk. J. Dairy Sci. 70 (Suppl. 1), 60.

Bills, D.D., Morgan, M.E., Libbey, L.M., Day, E.A., 1965. Identification of compounds responsible for fruity flavor defect of experimental cheddar cheeses 1, 2. J. Dairy Sci. 48 (9), 1168–1173.

Boor, K., Carey, N., Murphy, S., & Zadoks, R., 2005. Current status of commercial fluid milk quality. J. Diary Sci. 88 (Suppl. 1), 75.

Boor, K.J., Murphy, S.C., 2002. Microbiology of market milks. In: Robinson, R. (Ed.), Dairy Microbiology, third ed. John Wiley and Sons Inc., New York, NY, pp. 91–122.

Borle, F., Sieber, R., Bosset, J.O., 2001. Photo-oxidation and photoprotection of foods, with particular reference to dairy products: an update of a review article (1993–2000). Sci. Aliments 21 (6), 571–590.

Bramley, A.J., McKinnon, C.H., 1990. The microbiology of raw milk. Dairy Microbiol. 1 (1), 163–208.

Buehring, G.C., Shen, H.M., Jensen, H.M., Choi, K.Y., Sun, D., Nuovo, G., 2014. Bovine leukemia virus DNA in human breast tissue. Emerg. Infect. Dis. 20 (5), 772.

Buehring, G.C., Shen, H.M., Jensen, H.M., Jin, D.L., Hudes, M., Block, G., 2015. Exposure to bovine leukemia virus is associated with breast cancer: a case-control study. PLoS One 10 (9), e0134304.

Butler, J.N., 1991. Carbon Dioxide Equilibria and Their Applications. CRC Press, Boca Raton, FL.

Calvo, M.S., Whiting, S.J., Barton, C.N., 2004. Vitamin D fortification in the United States and Canada: current status and data needs. Am. J. Clin. Nutr. 80 (6), 1710S–1716S.

Cavaletto, M., Giuffrida, M.G., Conti, A., 2008. Milk fat globule membrane components–a proteomic approach. Bioactive Components of Milk. Springer, New York, NY, pp. 129–141.

Chambers, J.V. 2002. The microbiology of raw milk. In: R.K. Robinson (Ed.), Dairy Microbiology Handbook. 3rd ed. New York: John Wiley & Sons Inc., pp. 39–90.

Champagne, C.P., Laing, R.R., Roy, D., Mafu, A.A., Griffiths, M.W., White, C., 1994. Psychrotrophs in dairy products: their effects and their control. Crit. Rev. Food Sci. Nutr. 34 (1), 1–30.

Chapman, K.W., Whited, L.J., Boor, K.J., 2002. Sensory threshold of light-oxidized flavor defects in milk. J. Food Sci. 67 (7), 2770–2773.

Chen, L.D.R.M., Daniel, R.M., Coolbear, T., 2003. Detection and impact of protease and lipase activities in milk and milk powders. Int. Dairy J. 13 (4), 255–275.

Chrisope, G.L., Marshall, R.T., 1976. Combined action of lipase and microbial phospholipase C on a model fat globule emulsion and raw milk. J. Dairy Sci. 59 (12), 2024–2030.

Cousin, M.A., 1982. Presence and activity of psychrotrophic microorganisms in milk and dairy products: a review. J. Food Prot. 45 (2), 172–207.

Cromie, S.J., Schmidt, D., Dommett, T.W., 1989. Effect of pasteurization and storage conditions on the microbiological, chemical and physical quality of aseptically packaged milk. Aust. J. Dairy Technol. 5, 25–30.

Daniels, J.A., Krishnamurthi, R., Rizvi, S.S., 1985. A review of effects of carbon dioxide on microbial growth and food quality. J. Food Prot. 48 (6), 532–537.

Dasgupta, A.P., Hull, R.R., 1989. Late blowing of Swiss cheese: incidence of *Clostridium tyrobutyricum* in manufacturing milk. Aust. J. Dairy Technol. 44 (2), 82–87.

Deeth, H.C., 2006. Lipoprotein lipase and lipolysis in milk. Int. Dairy J. 16 (6), 555–562.

Deeth, H.C., Fitz-Gerald, C.H., 1983. Lipolytic enzymes and hydrolytic rancidity in milk and milk products. Developments in Dairy Chemistry—2. Springer, The Netherlands, pp. 195–239.

De Jonghe, V., Coorevits, A., De Block, J., Van Coillie, E., Grijspeerdt, K., Herman, L., et al., 2010. Toxinogenic and spoilage potential of aerobic spore-formers isolated from raw milk. Int. J. Food Microbiol. 136 (3), 318–325.

Delavenne, E., Ismail, R., Pawtowski, A., Mounier, J., Barbier, G., Le Blay, G., 2013. Assessment of lactobacilli strains as yogurt bioprotective cultures. Food Control 30 (1), 206–213.

Devlieghere, F., Debevere, J., 2000. Influence of dissolved carbon dioxide on the growth of spoilage bacteria. LWT-Food Sci. Technol. 33 (8), 531–537.

Dimick, P.S., Kilara, A., 1983. Photooxidative changes in milk proteins and amino acids. Kieler Milchwirtschaftliche Forschungsberichte 35, 289–299.

Doyle, M., 2004. Consumers have long list of packaging wishes and pet peeves: new report card focuses on what makes packages sell or sit on the shelf. Food Drug Pack. 68 (8), 24.

Eckner, K.F., Zottola, E.A., 1991. Potential for the low-temperature pasteurization of dairy fluids using membrane processing. J. Food Prot. 54 (10), 793–797.

Enright, J.B., Sadler, W.W., Thomas, E.C., 1957. Thermal inactivation of *Coxiella burnetii* and its relation to pasteurization of milk. Am. J. Public Health 47, 1–30.

Erickson, M.C., 1997. Chemical and microbial stability of fluid milk in response to packaging and dispensing. Int. J. Dairy Technol. 50 (3), 107–111.

Farkye, N.Y., Vedamuthu, E.R., 2002. Microbiology of soft cheeses. In: R.K. Robinson (Ed.), Dairy Microbiology Handbook. 3rd ed. New York: John Wiley & Sons Inc., pp. 479–513.

Fleet, G., 1990. Yeasts in dairy products. J. Appl. Bacteriol. 68, 199–211.

Food and Drug Administration and US Public Health Service, 2013. Grade "A" pasteurized milk ordinance.

Fox, P.F., McSweeney, P.L., 1998. Dairy Chemistry and Biochemistry. Springer Science & Business Media, New York, NY.

Frobish, R.A., Bradley, B.D., Wagner, D.D., Long-Bradley, P.E., Hairston, H., 1986. Aflatoxin residues in milk of dairy cows after ingestion of naturally contaminated grain. J. Food Prot. 49 (10), 781–785.

Fromm, H.I., Boor, K.J., 2004. Characterization of pasteurized fluid milk shelf-life attributes. J. Food Sci. 69 (8), M207–M214.

Garcia, M.L., Sanz, B., Garcia-Collia, P., Ordonez, J.A., 1989. Activity and thermostability of the extracellular lipases and proteinases from pseudomonads isolated from raw milk. Milchwissenschaft 44 (9), 547–550.

García, T., Mayoral, B., González, I., López-Calleja, I., Sanz, A., Hernández, P.E., et al., 2004. Enumeration of yeasts in dairy products: a comparison of immunological and genetic techniques. J. Food Prot. 67 (2), 357–364.

Griffiths, M.W. (Ed.), 2010. Improving the Safety and Quality of Milk: Improving Quality in Milk Products, vol. 2. Elsevier, Cambridge, UK.

Gruetzmacher, T.J., Bradley, R.L., 1999. Identification and control of processing variables that affect the quality and safety of fluid milk. J. Food Prot. 62 (6), 625–631.

Ha, E., Zemel, M.B., 2003. Functional properties of whey, whey components, and essential amino acids: mechanisms underlying health benefits for active people (review). J. Nutr. Biochem. 14 (5), 251–258.

Hantsis-Zacharov, E., Halpern, M., 2007. Culturable psychrotrophic bacterial communities in raw milk and their proteolytic and lipolytic traits. Appl. Environ. Microbiol. 73 (22), 7162–7168.

Haugen, J.E., Rudi, K., Langsrud, S., Bredholt, S., 2006. Application of gas-sensor array technology for detection and monitoring of growth of spoilage bacteria in milk: a model study. Anal. Chim. Acta 565 (1), 10–16.

Horwood, J.F., Stark, W., Hull, R.R., 1987. A "fermented, yeasty" flavour defect in Cheddar cheese. Aust. J. Dairy Technol. 42 (1), 25.

Hotchkiss, J.H., Werner, B.G., Lee, E.Y., 2006. Addition of carbon dioxide to dairy products to improve quality: a comprehensive review. Compr. Rev. Food Sci. Food Saf. 5 (4), 158–168.

Janzen, J.J., Bodine, A.B., Bishop, J.R., 1981. Effects of package temperature and days of storage on the flavor score of processed milk. J. Food Prot. 44 (6), 455–458.

Jay J.M., Loessner M.J., and Golden D.A., 2005. Modern Food Microbiology. 7th ed. Springer Science and Business Media, New York, NY, USA.

Jensen, R.G., 1995. Handbook of Milk Composition. Academic Press, San Diego.

Jensen, R.G., Blanc, B., Patton, S., 1995. Particulate Constituents in Human and Bovine Milks. Academic Press, San Diego, CA, pp. 50–62.

Johnson, E.A., Nelson, J.H., Johnson, M., 1990. Microbiological safety of cheese made from heat-treated milk, Part II. Microbiology. J. Food Prot. 53 (6), 519–540.

Johnson, D.S., Duncan, S.E., Bianchi, L.M., Chang, H.H., Eigel, W.N., O'Keefe, S.F., 2015. Packaging modifications for protecting flavor of extended-shelf-life milk from light. J. Dairy Sci. 98 (4), 2205–2214.

Johnson, M., 2001. Cheese products. In: Marth, E.H., Steele, J. (Eds.), Applied Dairy Microbiology. CRC Press, New York, NY, pp. 345–384.

Kalogridou-Vassiliadou, D., 1992. Biochemical activities of *Bacillus* species isolated from flat sour evaporated milk. J. Dairy Sci. 75 (10), 2681–2686.

Kantor, L.S., Lipton, K., Manchester, A., Oliveira, V., 1997. Estimating and addressing America's food losses. Food Rev. 20 (1), 2–12.

Katafiasz, A.R., Bartlett, P., 2012. Motivation for unpasteurized milk consumption in Michigan, 2011. Food Prot. Trends 32 (3), 124–128.

Kim, J., Twede, D., Lichty, J., 1997. Consumer attitudes about open dating techniques for packaged foods and over-the-counter drugs. J. Food Prod. Market. 4 (1), 17–30.

King, J.M., Min, D.B., 1998. Riboflavin photosensitized singlet oxygen oxidation of vitamin D. J. Food Sci. 63 (1), 31–34.

Kosse, D., Seiler, H., Amann, R., Ludwig, W., Scherer, S., 1997. Identification of yoghurt-spoiling yeasts with 18S rRNA-targeted oligonucleotide probes. Syst. Appl. Microbiol. 20 (3), 468–480.

Kornacki, J.L., Flowers, R.S., Bradley, R.L., 2001. Microbiology of butter and related products. In: Marth, E.H., Steele, J. (Eds.), Applied Dairy Microbiology. CRC Press, New York, NY, pp. 127–150.

Lafarge, V., Ogier, J.C., Girard, V., Maladen, V., Leveau, J.Y., Gruss, A., et al., 2004. Raw cow milk bacterial population shifts attributable to refrigeration. Appl. Environ. Microbiol. 70 (9), 5644–5650.

Laleye, L.C., Simard, R.E., Lee, B.H., Holley, R.A., Giroux, R.N., 1987. Involvement of heterofermentative lactobacilli in development of open texture in cheeses. J. Food Prot. 50 (12), 1009–1016.

Langer, A.J., Ayers, T., Grass, J., Lynch, M., Angulo, F.J., Mahon, B.E., 2012. Nonpasteurized dairy products, disease outbreaks, and state laws—United States, 1993–2006. Emerg. Infect. Dis. 18 (3), 385–391.

Lee, K.H., Jung, M.Y., Kim, S.Y., 1998. Effects of ascorbic acid on the light-induced riboflavin degradation and color changes in milks. J. Agric. Food Chem. 46 (2), 407–410.

Lee, Y.G., Wu, H.Y., Hsu, C.L., Liang, H.J., Yuan, C.J., Jang, H.D., 2009. A rapid and selective method for monitoring the growth of coliforms in milk using the combination of amperometric sensor and reducing of methylene blue. Sens. Actuators B Chem. 141 (2), 575–580.

Lewis, M.J., 1999. Microbiological issues associated with heat treated milks. Int. J. Dairy Technol. 52 (4), 121–125.

Linn, J.G., 1988. Factors affecting the composition of milk from dairy cows. In: NRC (Ed.), Designing Foods. Nationall Academy Press, Washington, DC, p. 224.

Mabbitt, L.A., 1981. Metabolic activity of bacteria in raw milk. Kieler Milchwirtschaftliche Forschungsberichte 33, 273–280.

Marshall, R.T., 1982. Relationship between the bacteriological quality of raw milk and the final products. A review of basic information and practical aspects. Kieler Milchwirtschaftliche Forschungsberichte 34, 149–157.

Marsili, R.T., 1999. Comparison of solid-phase microextraction and dynamic headspace methods for the gas chromatographic—mass spectrometric analysis of light-induced lipid oxidation products in milk. J. Chromatogr. Sci. 37 (1), 17–23.

Marth, E.H., Capp, C.M., Hasenzahl, L., Jackson, H.W., Hussong, R.V., 1966. Degradation of potassium sorbate by *Penicillium* species. J. Dairy Sci. 49 (10), 1197–1205.

Massa, S., Gardini, F., Sinigaglia, M., Guerzoni, M.E., 1992. *Klebsiella pneumoniae* as a spoilage organism in mozzarella cheese. J. Dairy Sci. 75 (6), 1411–1414.

Matthews, S.B., Waud, J.P., Roberts, A.G., Campbell, A.K., 2005. Systemic lactose intolerance: a new perspective on an old problem. Postgrad. Med. J. 81 (953), 167–173.

McKellar, R.C., 1982. Factors influencing the production of extracellular proteinase by *Pseudomonas fluorescens**. J. Appl. Bacteriol. 53 (3), 305–316.

Mestdagh, F., De Meulenaer, B., De Clippeleer, J., Devlieghere, F., Huyghebaert, A., 2005. Protective influence of several packaging materials on light oxidation of milk. J. Dairy Sci. 88 (2), 499–510.

Mikolajcik, E.M., Simon, N.T., 1978. Heat resistant psychrotrophic bacteria in raw milk and their growth at 7 C. J. Food Prot. 41 (2), 93–95.

Min, D.B., Boff, J.M., 2002. Chemistry and reaction of singlet oxygen in foods. Compr. Rev. Food Sci. Food Saf. 1 (2), 58–72.

Morgan, M.E., 1976. The chemistry of some microbially induced flavor defects in milk and dairy foods. Biotechnol. Bioeng. 18 (7), 953–965.

Moseley, W.K., 1980. A field topic: pinpointing post-pasteurization contamination. J. Food Prot. 43 (5), pp. 414–414.

Muir, D.D., 1996. The shelf-life of dairy products: 1. Factors influencing raw milk and fresh products. Int. J. Dairy Technol. 49 (1), 24–32.

Mungai, E.A., Behravesh, C.B., Gould, L.H., 2015. Increased outbreaks associated with nonpasteurized milk, United States, 2007–2012. Emerg. Infect. Dis. 21 (1), 119.

Munro, H.N. and Crim, M.C., 1988. The Proteins and Amino Acids. In: Shils, M.E. and Young, V.R. (Eds.), Modern nutrition in health and disease. 7th ed. Lea & Febiger; Philadelphia, PA, pp. 17–37.

National Association of State Departments of Agriculture, 2011. NASDA releases raw milk survey. Available from: <http://www.nasda.org/file.aspx?id = 3916> (accessed 11.11.15.).

Nicolaou, N., Xu, Y., Goodacre, R., 2010. Fourier transform infrared spectroscopy and multivariate analysis for the detection and quantification of different milk species. J. Dairy Sci. 93 (12), 5651–5660.

O'Connor, C., 1995. Rural dairy technology (No. 1). ILRI (aka ILCA and ILRAD).

Oliver, S.P., Boor, K.J., Murphy, S.C., Murinda, S.E., 2009. Food safety hazards associated with consumption of raw milk. Foodborne Pathog. Dis. 6 (7), 793–806.

Owens, J.J., 1978. Lecithinase positive bacteria in milk. Process Biochemistry 13(1), 13–14, 30.

Pal, S., Ellis, V., Dhaliwal, S., 2010. Effects of whey protein isolate on body composition, lipids, insulin and glucose in overweight and obese individuals. Br. J. Nutr. 104 (05), 716–723.

Palmquist, D.L., Beaulieu, A.D., Barbano, D.M., 1993. Feed and animal factors influencing milk fat composition. J. Dairy Sci. 76 (6), 1753–1771.

Piergiovanni, L., Limbo, S., 2004. The protective effect of film metallization against oxidative deterioration and discoloration of sensitive foods. Pack. Technol. Sci. 17 (3), 155–164.

Ranieri, M.L., Boor, K.J., 2010. Tracking and eliminating sporeformers in dairy systems. Aust. J. Dairy Technol. 65 (2), 74.

Ranieri, M.L., Ivy, R.A., Mitchell, W.R., Call, E., Masiello, S.N., Wiedmann, M., et al., 2012. Real-time PCR detection of *Paenibacillus* spp. in raw milk to predict shelf life performance of pasteurized fluid milk products. Appl. Environ. Microbiol. 78 (16), 5855–5863.

Reay, D.S., Davidson, E.A., Smith, K.A., Smith, P., Melillo, J.M., Dentener, F., et al., 2012. Global agriculture and nitrous oxide emissions. Nat. Clim. Change 2 (6), 410–416.

Restuccia, D., Spizzirri, U.G., Parisi, O.I., Cirillo, G., Curcio, M., Iemma, F., et al., 2010. New EU regulation aspects and global market of active and intelligent packaging for food industry applications. Food Control 21 (11), 1425–1435.

Robertson, G.L., 2012. Food Packaging: Principles and Practice. CRC Press, Boca Raton, FL.

Ross, R.P., Morgan, S., Hill, C., 2002. Preservation and fermentation: past, present and future. Int. J. Food Microbiol. 79 (1), 3–16.

Ryser, E., 2001. Public health concerns. In: Marth, E.H., Steele, J. (Eds.), Applied Dairy Microbiology. CRC Press, New York, NY, pp. 397–546.

Sattar, A., deMan, J.M., Furia, T.E., 1975. Photooxidation of milk and milk products: a review. Crit. Rev. Food Sci. Nutr. 7 (1), 13–37.

Schmidt, V.S., Kaufmann, V., Kulozik, U., Scherer, S., Wenning, M., 2012. Microbial biodiversity, quality and shelf life of microfiltered and pasteurized extended shelf life (ESL) milk from Germany, Austria and Switzerland. Int. J. Food Microbiol. 154 (1), 1–9.

Schröder, M.J., 1984. Origins and levels of post pasteurization contamination of milk in the dairy and their effects on keeping quality. J. Dairy Res. 51 (01), 59–67.

Schröder, M.J., Cousins, C.M., McKinnon, C.H., 1982. Effect of psychrotrophic post-pasteurization contamination on the keeping quality at 11 and 5 C of HTST-pasteurized milk in the UK. J. Dairy Res. 49 (04), 619–630.

Scrimshaw, N.S., Murray, E.B., 1988. The acceptability of milk and milk products in populations with a high prevalence of lactose intolerance. Am. J. Clin. Nutr. 48 (4), 1142–1159.

Senyk, G.F., Shipe, W.F., 1981. Protecting your milk from nutrient losses. Dairy Field 164 (3), 81–83.

Shehata, T.E., Duran, A., Collins, E.B., 1971. Influence of temperature on the growth of psychrophilic strains of *Bacillus*. J. Dairy Sci. 54 (11), 1579–1582.

Silva, F.V., 2015. High pressure processing of milk: modeling the inactivation of psychrotrophic *Bacillus cereus* spores at 38–70°C. J. Food Eng. 165, 141–148.

Sun, D.W. (Ed.), 2012. Thermal Food Processing: New Technologies and Quality Issues. CRC Press, New York, NY.

Suriyarachchi, V.R., Fleet, G.H., 1981. Occurrence and growth of yeasts in yogurts. Appl. Environ. Microbiol. 42 (4), 574–579.

US Department of Agriculture Animal and Plant Health Services, Veterinary Services, Center for Epidemiology and Animal Health, 2007. Info sheet: bovine leukosis virus (BLV) on U.S. dairy operations.

Van Aardt, M., Duncan, S.E., Marcy, J.E., Long, T.E., Hackney, C.R., 2001. Effectiveness of poly (ethylene terephthalate) and high-density polyethylene in protection of milk flavor. J. Dairy Sci. 84 (6), 1341–1347.

Varnam, A.H., Sutherland, J.P., 1994. Dairy protein products. Milk and Milk Products. Springer, New York, NY, pp. 159–182.

Varnam, A., Sutherland, J.P., 2001. Milk and Milk Products: Technology, Chemistry and Microbiology, vol. 1. Springer Science & Business Media, Gaithersburg, MD.

Vasavada, P.C., Cousin, M.A., 1993. Dairy Microbiology and Safety. VCH Publishers, New York, NY, pp. 301–426.

Walsh, A.M., Duncan, S.E., Potts, H., Gallagher, D.L., 2015. Comparing quality and emotional responses as related to acceptability of light-induced oxidation flavor in milk. Food Res. Int.

Webster, J.B., Duncan, S.E., Marcy, J.E., O'keefe, S.F., 2009. Controlling light oxidation flavor in milk by blocking riboflavin excitation wavelengths by interference. J. Food Sci. 74 (9), S390–S398.

Wendorff, W.L., Riha, W.E., Muehlenkamp, E., 1993. Growth of molds on cheese treated with heat or liquid smoke. J. Food Prot. 56 (11), 963–966.

Whited, L.J., Hammond, B.H., Chapman, K.W., Boor, K.J., 2002. Vitamin A degradation and light-oxidized flavor defects in milk. J. Dairy Sci. 85 (2), 351–354.

Wold, J.P., Veberg, A., Nilsen, A., Iani, V., Juzenas, P., Moan, J., 2005. The role of naturally occurring chlorophyll and porphyrins in light-induced oxidation of dairy products. A study based on fluorescence spectroscopy and sensory analysis. Int. Dairy J. 15 (4), 343–353.

Yagoub, S.O., Bellow, F.A., El Zubeir, I., 2008. Effect of temperature and storage period on the constituents of milk inoculated with *Pseudomonas aeruginosa*. Res. J. Microbiol. 3 (1), 30–34.

Yam, K.L., Lee, D.S. (Eds.), 2012. Emerging Food Packaging Technologies: Principles and Practice. Elsevier, Cambridge, UK.

Yang, W.T., Min, D.B., 1994. Chemistry of singlet oxygen oxidation of foods. In: Ho, C.T., Hartmand T.G. (Eds.), Lipids in Food Flavors, American Chemical Society, Washington, DC, pp. 15.

Chapter 8

Spoilage of Meat and Fish

Giuseppe Comi
University of Udine, Udine, UD, Italy

8.1 INTRODUCTION

The shelf life of meat, meat products, fish and fish products is the storage time until spoilage, which is a complex condition in which the combination of biological and physico-chemical activities may interact and make the product unacceptable for human consumption (Borch et al., 1996; Gram et al., 2002). A maximum acceptable microbial level and/or unacceptable off-odor and off-flavor identify the exact point of spoilage, which is strictly dependent on the initial numbers and types of contaminating microorganisms, their growth, lipid oxidation, and autolytic enzymatic reactions. Meat and fish, due to their physico-chemical characteristics, are excellent basic nutrients for microbial activity. Indeed the pH, a_w (activity water) and high moisture values can support the growth of a wide variety of microorganisms (Nychas et al., 2007, 2008; Doulgeraki et al., 2011; Casaburi et al., 2015; Remenant et al., 2015) (Table 8.1).

Usually the initial microbial count of meat and cooked meat products is about 2−3 log CFU/cm^2 or g, and for fresh meat and meat and fish products about 4−5 log CFU/cm^2 or g (Blickstad and Molin, 1983; Jackson et al., 1992). The large variety of microbial species originate from the physiological status of the animal during farming, slaughtering, harvesting, fishing, processing, transportation, preservation, and storage conditions (Nychas et al., 2008; Comi and Cattaneo, 2007a,b, Comi and Iacumin, 2012). The contamination also occurs after the heating process that is often used to prolong the shelf life either of meat or fish products. It is estimated that only 10% of the microorganisms initially present are psychrotolerant and may grow in cold-storage and that the fraction causing spoilage is even lower (Borch et al., 1996). During storage, the temperature, gaseous atmosphere, pH, NaCl, and packaging are important factors affecting the selection, the growth rate, and the activity for certain bacteria (Nychas et al., 2007; Comi and Cattaneo, 2007a,b). The shelf life of refrigerated meat, fish,

The Microbiological Quality of Food. DOI: http://dx.doi.org/10.1016/B978-0-08-100502-6.00011-X

TABLE 8.1 Physico-Chemical Characteristics of Meat and Fish

Parameter	Meat	Fish
Moisture	75–80%	75–85%
Protein	15–20%	7–20%
Lipid	2.5–8.2%	1.5–10.5%
Vitamin (B)	0.8–0.9 mg/100 g	1.6–4.9 mg/100 g
Sugar	0.4–1.2%	0.5–5.6%
a_w	0.98–0.99	0.98–0.99
pH	5.5–5.7	5.0–6.5

and meat and fish products may vary from days up to several months (Gill and Molin, 1991; Blickstad and Molin, 1983). The types of packaging such as air, under vacuum (UV) and modified atmosphere packaging (MAP), and cold storage further increase the microbial selection and affect the microbial growth rate and the metabolism. The growth of one or another strain depends on the availability of oxygen and on the affinity of each species to oxygen. Consequently the level of the spoilage is due to the microorganisms that predominate in the meat and fish matrix and produce the typical spoilage compounds, responsible for off-odor and off-flavor, slime, and discoloration (Huis in't Veld, 1996; Comi and Cattaneo, 2007a,b; Gram et al., 2002). The spoilage is the sum of the interaction between microbial growth and enzyme activities (Nychas et al., 2008). Lactic acid bacteria (LAB), *Brochothrix thermosphacta*, Enterobacteriaceae, Pseudomonads, and other Gram-negative bacteria, and Clostridia grow or cogrow in the meat and fish matrix. They produce molecules responsible for spoilage such as organic acids, volatile fatty acids, ethyl esters, sulfur compounds, ketones, aldehydes, alcohols, ammonia, and other molecules, metabolizing glucose, lactic acid, nitrogenous compounds, and free amino acids (Nychas et al., 1998, 2008; Tsigarida and Nychas, 2001; Skandamis and Nychas, 2002). The concentration, the olfactory thresholds, and the balance of the volatile and nonvolatile compounds determine the level of the spoilage.

8.2 INTERACTIONS BETWEEN MEAT AND FISH SPOILING MICROORGANISMS

The intrinsic and the extrinsic parameters of meat and fish are mainly responsible for the selection of the microbial population. In particular the temperature, atmosphere, pH, a_w, redox potential, and the substrate composition select and guide the microbial activities. However, the food spoilage is also determined by the interaction among the different strains present in each food ecosystem. Each strain influences the growth of the other (Boddy and Wimpenny, 1992) and usually the spoilage is determined by up to 10^8-10^9 CFU/g of different microorganisms. So the spoilage level seems to be characterized by the positive or negative interactions between microorganisms (Gram et al., 2002; Fredrickson, 1977). In meat and fish ecosystems, the metabolism of one microorganism can favor the growth of another one. The antagonism, caused by the competition of such ions or molecules, suppresses the growth of less competitive bacteria (Gram et al., 2002). Some strains, producing siderophores, predominate because of their competition for iron. Iron is used in bacterial respiration (as an electron shuttler) and in redox enzymes. Microorganisms develop specific iron chelating systems such as siderophore, which are iron-chelators secreted by the cell. These systems are produced particularly in food with limited iron concentration and consequently microorganisms can adsorb iron from food despite the fact that it is often insoluble (Gram et al., 2002). Many *Pseudomonas* and *Shewanella* strains predominate in some food spoilage for their high siderophores production, but in coculture *Pseudomonas* spp. prevails because its siderophores inhibit *Shewanella* strains (Gram and Melchiorsen, 1996).

Metabiosis consists in the ability of one organism to produce a favorable environment for another one. The removal of oxygen by aerobic psychrotolerant microorganisms can favor the growth of anaerobic strains such as *Clostridium botulinum* type E and other Clostridia. Indeed some proteolytic microorganisms can enhance the growth of others by the secretion of amino acids. Different compounds produced by spoilage microorganisms such as lactic acid can stimulate the growth of *Clostridium butirricum* or *C. tyrobutirricum*. Some nutrient interdependencies play a fundamental role in food spoilage. The cogrowth of LAB and Enterobacteriaceae results in a high amount of biogenic amines in PV fish or meat (Dainty et al., 1985). LAB and *Hafnia alvei*, growing in coculture, produce off-odors typical of spoilage in PV meats, but they don't do it as single cultures (Borch et al., 1996). Consequently, considering that it is impossible that spoilage depends on a single microorganism, the term "specific spoilage organism" (SSO) must be changed to the term "metabiotic spoilage association" as suggested by Jørgensen et al. (2000).

Some strains can predominate due to their ability to quickly metabolize nutriments or to produce quorum-sensing activity. Different Gram-negative bacteria are recognized as being mainly responsible for spoilage because they can express many hydrolytic enzymes. In particular they can communicate via pheromone production (e.g., *N*-acyl homoserine lactones—AHLs), and the result is a growth increase. Indeed gene regulation and expression are often a function of cell density and are influenced by growth phase, nutrients, and external stresses. Microorganisms, able to communicate by chemical signals such as peptides, can increase their spoilage activity for competition against other microbial populations (Gram et al., 2002).

Moreover, the changing of pH, a_w, and temperature also favors the antagonism between bacteria. The lowering of pH selects the microbial population, permitting the growth of acidophilus strains. LAB predominate either because they ferment sugar producing lactic acid and lowering the pH or because they produce bacteriocins and other peptides with antimicrobial activity. The increase of the pH can also select the spoilage microorganisms. Many spoilage molecules can have an antimicrobial activity. Ammonia compounds (NH_3 and trimethylamine) typical of fish and meat spoilage result in toxicity for a large number of microorganisms. Indeed, many *Pseudomonas* strains can predominate because they liberate siderophores and antibiotic compounds against fungi and bacteria (Ellis and Goodacre, 2001; Gram et al., 2002).

8.3 PARAMETERS DETERMINING MEAT AND FISH SPOILAGE

Meat and fish, due to their physico-chemical composition, are easily spoiled by a large number of microorganisms. The ability to multiply depends on favorable parameters like pH and a_w values, and chemical composition. They select the microbial population that metabolizes the basic substrate and cause the production of spoilage characterized by changes in odor and flavor (Jay et al., 2005). The initial contamination depends on the physiological state of the animal, on the slaughtering or harvesting areas, on the handling, on the level of hygiene of employees, and on the tools and equipment used (Nychas et al., 2008; Serraino et al., 2012). Subsequently, the microorganisms' growth and spoilage depends on the following: intrinsic, extrinsic, processing, and implicit parameters.

8.3.1 Intrinsic Parameters

The intrinsic factors are represented by the physical and chemical properties (e.g., a_w, pH, content of nutrients, structure of the foodstuff) (Huis in't Veld, 1996).

Water is important for microbial growth. The presence of ions or molecules, that ligate water, produces a decrease of the a_w and consequently reduces the microbial activity. Fresh meat and fish present a_w values of about 0.98–0.99 and can be spoiled by a large number of microorganisms. For this reason their shelf life is limited. By adding salt to both fish and meat, the a_w decreases and limits the microbial growth. Adding 15% salt, the a_w is reduced to 0.76 and consequently no bacteria can grow. Cod (bacalao) is a typical salted fish.

The pH value selects the microbial population of food. Meat and fish have a pH between 5.5 and 6.5 units. At these levels, all the main spoilage bacteria can grow. The lowering of the pH reduces or totally limits microbial growth. During the production of sausages the pH can reach values of less than 5.3 unities and consequently spoilage microorganisms are not able to grow. The acidification by organic acid (vinegar) or by fermentation of fish allows for an improved shelf life.

The chemical composition also favors spoilage. Both meat and fish contain all the nutrients useful for microbial growth. Indeed, proteins, sugar, nonproteic nitrogen (NPN), vitamins, fat, and organic acids are important growth factors. In particular, the spoilage depends on a sequential growth of microorganisms that metabolize different molecules present in meat and fish.

The food structure can increase but also reduce the microbial activity. Meat limits spoilage through the presence of a high amount of connective tissue, conversely the lower amount of connective tissue in fish permits faster spoilage.

8.3.2 Extrinsic Parameters

The extrinsic parameters are characterized by storage conditions (e.g., temperature) and atmosphere composition and humidity (Huis in't Veld, 1996). Cold-storage selects and reduces bacteria growth. In particular the refrigeration temperatures increase the microbial lag phase, improving the shelf life of both meat and fish. The growth of some bacterial groups, such as Enterobacteriaceae and LAB, is drastically reduced at −1.5°C (Borch et al., 1996). On the other hand, the abuse temperatures (up to 7°C) increase psychrotolerant and mesophilic bacteria activity either in meat or in fish. The effect of packaging on microbial population is often associated to the storage temperatures. Meat and fish products are usually packaged in air, PV, and MAP. The oxygen concentration influences spoilage, because it selects the microbial population. The oxygen favors the aerobic bacteria (e.g., *Pseudomonas* spp.), meanwhile the absence or the reduction of oxygen favors the facultative anaerobic bacteria (e.g., Enterobacteriaceae), LAB and *B. thermosphacta*. However, the use of packaging with low oxygen content (less than 5%), or without, is associated with cold-storage temperature increases meat and fish shelf life. Finally, the microbial spoilage is also influenced by the relative humidity (RH) of the storage rooms. In particular, high RH (>90%) must be associated with low temperatures. The presence of high humidity increases the microbial spoilage in meat, permitting the predominance of molds and enterobacteria strains.

8.3.3 Implicit Parameters

The implicit parameters are represented by the synergistic or antagonistic behaviors between bacteria. Synergistic effects include the ability of certain microorganisms to produce essential compounds for the development of other microorganisms. Usually the synergistic activity induces a sequential microbial growth, and a primary and a secondary spoilage is observed

(Huis in't Veld, 1996). Positive interactions among the microorganisms produce a faster meat and fish spoilage. The changing of the pH, of the redox potential, the competition of essential nutrients, and the antimicrobial compound productions are the main antagonistic effects. These changes have negative effects on the microbial population, because they can influence the metabolism or the survival of the different strains contaminating fish and meat. The lowering of the pH due to the LAB metabolism reduces or does not enable the competitive growth of microorganisms.

8.4 MEAT AND MEAT PRODUCT SPOILING MICROORGANISMS

The meat of all edible animal species is spoiled in the same way. The number and the strains of microorganisms that spoil fresh meat depend mainly on the intrinsic and extrinsic parameters and on processing and storing technologies (Huis in't Veld, 1996; Casaburi et al., 2015; Remenant et al., 2015). In particular, the combination of the different parameters influences the development of either the natural contaminants or the SSO (Nychas et al., 2008). The shelf life of meat maintained in air at temperatures lower than 5°C is limited by microbial growth. Unpleasant odors develop when the number of microorganisms reaches about 10^7 CFU/cm^2 and a slimy coating appears with the count of 10^8 CFU/cm^2, but sometimes a higher microbial concentration has been observed without spoilage signs.

The major biochemical changes occur in the aqueous phase of the meat, which contains amino acids, nucleotides, and peptides. These nutrients are sufficient for bacterial growth and the metabolism of these compounds leads to meat spoilage.

The spoilage microflora is dominated by psychrotrophic Gram-negative rod-shaped aerobic, facultative, and obligated anaerobic bacteria (Table 8.2), mainly belonging to a few genera, including *Pseudomonas, Shewanella, Acinetobacter, Moraxella, Psychrobacter, Alcaligenes*, Enterobacteriaceae, *Flavobacterium, Staphylococcus, Micrococcus, Clostridium*, LAB, *Brochothrix*, and *Corynebacterium* (Comi and Cattaneo, 2007a,b; Comi and Iacumin, 2012; Doulgeraki et al., 2011,2012). However, the growth and the predominance of the different strains is variable because of packaging, temperature, and time of storage (Ercolini et al., 2011; Doulgeraki et al., 2011; Pennacchia et al., 2011). *Pseudomonas* spp. and *Shewanella* spp. are the most common spoilage microorganisms of fresh meat stored in air. Among *Pseudomonas* spp., *P. fragi, P. fluorescens, P. putida*, and *P. lundensis* are the main isolated strains (Casaburi et al., 2015). Indeed, *P. fragi* and *Shewanella putrefaciens* are the most important spoilage strains. Both represent 50–90% of the total microbial population. *Pseudomonas* and *Shewanella* are ubiquitous, psychrotolerant, and consequently easily contaminate meat and grow on it. Other Gram-negative bacteria may also grow with

TABLE 8.2 Spoilage Microorganisms of Meat

Meat Spoilage Microorganisms
Aeromonas
Alcaligenes
Acinetobacter
Bacillus
Brochothrix
Clostridium
Citrobacter
Corynebacterium
Enterobacter/Pantoea
E. coli
Flavobacterium
Lactobacillus
Micrococcus/Staphylococcus
Moraxella
Pseudomonas
Proteus
Psycrobacter
Salmonella
Shewanella
Streptococcus
Aspergillus
Penicillium
Cladosporium
Candida
Rhodotorula
Trichosporon
Geotrichum
Thamnidium
Mucor
Sporotrichum
Rhizopus

Pseudomonas. Aeromonas, Photobacterium are also involved in meat spoilage. Either *Pseudomonas* or the other Gram-negative strains are advantaged in respect to other Gram-negative or Gram-positive strains, because they rapidly grow at chill temperature in aerobic conditions and metabolize the NPN fraction. This is the first substrate of growth for many spoilage microorganisms (Huis in't Veld, 1996). Other molecules are metabolized after the production of hydrolytic enzymes that liberate fatty acids, sugars, and amino acids. Subsequently the basic substrate becomes functional for the SSO. More specifically, it seems that *P. fragi, P. putida*, and *P. fluorescens* are also advantaged by their ability to metabolize glucose to gluconate and 2-oxogluconate, that, accumulating outside the bacterial cell, are then metabolized by the same species or by other *Pseudomonas* strains and by *S. putrefaciens*, but not by microorganisms competitors (Casaburi et al., 2015; Remenant et al., 2015).

Other bacteria are present in small numbers and occasionally constitute an important part of the meat microbial spoilage population. However the spoilage is never due to a single strain but a succession of spoilage-related microbial groups (Table 8.3). Enterobacteriaceae are also involved in meat spoilage. They may grow either in aerobic or in anaerobic conditions. Indeed some Enterobacteriaceae strains are psychrotolerant and consequently can play a secondary role in meat spoilage. They predominate only in temperatures above 7°C. Enterobacteriaceae, generally, dominate over *Pseudomonas* and are responsible for the production of sulfur compounds. Like *Pseudomonas*, Enterobacteriaceae have an environmental origin even though they can originate from fecal contamination (e.g., *Enterobacter* spp. or *Escherichia coli*). *Hafnia alvei, Serratia liquefaciens, Enterobacter agglomerans* are present in low concentrations in respect to *Pseudomonas/Shewanella* group and rarely compete with them at temperatures below 7°C. They begin to develop only when *Pseudomonas* is in the stationary phase, nevertheless *Pseudomonas* remains the main genus.

Also in storage conditions with high levels of RH, the spoilage of animal carcasses appears to be due to Gram-negative bacteria, especially *Pseudomonas* species. However, *Aeromonas, Alcaligenes, Acinetobacter, Moraxella, Flavobacterium, Enterobacter, Microbacterium,* and *Proteus* may occasionally develop and produce slime and off-odor (Comi and Iacumin, 2012).

The Gram-positive bacteria are involved in the spoilage of fresh meat following the activities of the Gram-negative bacteria. Despite the fact that a lot of Gram-positive strains are psychrotolerant, rarely do they predominate over the *Pseudomonas/Shewanella* group.

Brochothrix thermosphacta, a rod Gram-positive bacteria, is present in particular on fresh pork, lamb, and beef and its growth is stimulated by cover fat, by storage temperature above 5°C and pH above 6.5 units, in particular in dark firm dry (DFD) meat. Usually on lamb carcasses, *Pseudomonas* spp.

TABLE 8.3 Main Psychrotolerant Microorganism Strains of Meat and Fish

Strains
P. fluorescens
P. fragi
P. putida
S. putrefaciens
Burkholderia cepacia
Pseudomonas syringae
Photobacterium phosporeum
B. thermosphacta
E. agglomerans
H. alvei
L. sakei
L. curvatus
L. carnosum
L. mesenteroides
W. viridescens
C. divergens
C. piscicola
C. gasigenes
C. putrefaciens,
C. algidixylanolyticum
C. estertheticum
C. frigidicarnis

predominate with *B. thermosphacta*, and Coagulase Negative Catalase Positive Cocci (CNCPC), such as *Staphylococcus xylosus*. Micrococci and staphylococci are natural contaminants of animal carcasses, originating from skin, but their number is limited (less than 10^4 CFU/cm^2), and they rarely play a real role in spoilage (Comi and Iacumin, 2012).

The molds are important spoilers of carcasses stored for a long time just above the freezing point, but the modern technologies of cooling and freezing have eliminated the mold problems. However, they produce colored spots on animal carcasses kept at a storage temperature in high R.H. (>90%).

The refrigeration is not the only technology that permits to improve the shelf life of meat. The type of packaging may affect the microbial activity, too. The oxygen concentration is fundamental for the microbial growth and metabolism (Doulgeraki et al., 2011) in meat. The use of MAP or PV has the aim of inhibiting or at least slowing down the microbial growth. The concentrations of different gases (CO_2, N_2, O_2) are studied according to the type of meat and the effects to be obtained: to extend the shelf life and/or to improve the presentation. Different gases have specific effects on microorganisms. The CO_2 concentrations of about 25–30% inhibit many microorganisms and have a selective action on the surface bacterial population (Gram-negative bacteria, *Pseudomonas*, and other psychrotrophic spoilage microorganisms). Consequently, in meat stored in MAP at 3°C, lactobacilli and streptococci prevail over Gram-negative bacteria, meanwhile at −1°C/0°C over *B. thermosphacta*. It seems that its action consists in penetrating into the cell, interfering in the metabolism and enzymatic systems, in lowering the pH of both microorganisms and meat cytoplasm, in favoring the activity of the LAB that inhibit pathogenic and spoilage microorganisms for the production of H_2O_2 and lactic acid. Oxygen is toxic in high concentrations for all living cells because they form peroxide radicals, but it also reacts with myoglobin to give MbO_2 (bright red). Mixtures of O_2/CO_2 (70%/30%) are used to enhance the color stability and to reduce bacterial growth. The spoilage strains of fresh meat and meat product in PV and in MAP are distinctly different from those that spoil aerobically stored meat. The predominant species are highly dependent on the type of atmosphere and the permeability of the film. In fact, meats wrapped in oxygen-permeable films are spoiled by psychrotrophic aerobic microorganisms such as *Pseudomonas, Acinetobacter*, and *Psychrobacter* spp. The LAB (*Lactobacillus, Carnobacterium*, and *Leuconostoc* spp.) and *B. thermosphacta* prevail in PV or in MAP with low oxygen concentration ($<1\%$). Concentrations above 10^7 CFU/g are necessary to spoil meat in MAP or in PV, while in air the spoilage occurs at lower concentration. In fresh meats, the oxygen absence and the cold temperature allow the development of LAB (*Lactobacillus, Leuconostoc*, and *Carnobacterium* spp.) and *B. thermosphacta*, which induces modification of the color, the flavor, and the smell. In meat packaged with high oxygen concentration (70–80%) and cold-stored, *Lactobacillus* spp., *Carnobacterium* spp., *Leuconostoc* spp., *B. thermosphacta, Pseudomonas* spp., *E. agglomerans, S. liquefaciens*, and *S. putrefaciens* can grow up to high concentrations (above 10^8 CFU/cm^2 or g) and produce H_2S and greening. In 100% CO_2 at −1.5°C only the homofermentative LAB completely dominates the microbial population because neither *B. thermosphacta* nor Enterobacteriaceae are able to grow (Erichsen and Molin, 1981; Jeremiah et al., 1995). Increasing the concentration of CO_2 in MAP, the aerobic microorganisms stop developing. It seems that the microbial growth is affected more by temperature than gas concentration. Roast beef slices packaged in 100% CO_2 have a longer shelf life when stored at −1.5°C compared to those stored at 3 or 10°C. Enterobacteriaceae, such as *S. liquefaciens, E. agglomerans*, and *H. alvei*, are common spoilers of fresh meat stored in MAP or PV (Doulgeraki et al., 2011; Samelis, 2006), conversely *Rahnella* spp. was rarely isolated in beef (Ercolini et al., 2006, 2009; Pennacchia et al., 2011).

The "bone taint", or deep decay, is a fresh meat spoilage that is due to facultative or obligatory anaerobic microorganisms, frequently observed in beef, horse, and pork. Up to now it has been rare. The involved microorganisms include *Clostridium* spp., *C. perfringens, C. bifermentans, C. hystoliticum, C. algidicarnis, C. putrefaciens, C. sporogenes*, which develop in fresh meat kept at temperatures above 15°C. Psychrotolerant *C. gasigenes, C. putrefaciens, C. algidixylanolyticum, C. estertheticum*, and *C. frigidicarnis* also spoil fresh meat in PV and stored at 4 ± 2°C (Cavill et al., 2011; Silva et al., 2011).

The microbial spoilage of meat products depends particularly on the extrinsic parameters such as cooking, pasteurization, sterilization, and storage temperature, even though the product composition could be important. Cooked meat products are cooked, packaged in PV or in MAP and then pasteurized. Only some artisanal meat products are distributed unpacked. The cooking and the pasteurization temperatures decrease the contaminating microorganisms. The thermoduric strains, such as LAB, *Carnobacterium* spp., *B. thermosphacta*, *Leuconostoc* spp., *Weissella* spp., *Streptococcus* spp., and *Micrococcus* spp., are only partially reduced and consequently they grow during storage at 4 ± 2°C. The storage temperature, the types of packaging, the presence of curing salt and nitrite/nitrate are not sufficient to completely inhibit them. Either in PV or in MAP the oxygen concentration remains lower than 0.1% for the whole shelf life, permitting the growth of psychrotrophic LAB and Leuconostocs. Conversely *B. thermosphacta* grows in meat products packaged in film with high oxygen permeability (Nielsen, 1983). It seems that the LAB growth is favored by CO_2 plus N_2, even though some LAB needs oxygen (less than 5%) to grow. *Lactobacillus sakei, Lactobacillus curvatus, Leuconostoc carnosum, Leuconostoc mesenteroides, Weissella viridescens, Carnobacterium divergens*, and *Carnobacterium piscicola* predominate either in PV or in MAP meat products (Dykes et al., 1994; Collins et al., 1987). Finally, *Bacillus* spp. and *Micrococcus* spp. spoil cooked meat products stored in air. Canned meat products are sterilized (F_0 = up to 2.5 min), so either the sporogenous or the asporogenous bacteria are killed. Indeed, the presence of nitrite and nitrate can contribute to preventing the spore germination. Consequently the canned meats are stable and have a long shelf life. Spoilage by sporogenous bacteria (*Clostridium* spp. or *Bacillus* spp.) can be present when the sterilization time/temperature is inadequate and/or unsuitable.

8.4.1 Metabolomics of Spoilage (Off-Odor and Off-Flavor)

Spoilage microorganisms grow and produce different molecules that are involved in off-odor and off-flavor of foods. The spoilage depends on the microorganisms' ability to adapt to the food ecosystem conditions and to metabolize the matrix. In particular they produce different compounds such as esters, ketones, aldehydes, hydrocarbons, alcohols, benzenoids, terpenoids, nitrogen, and sulfur compounds, amines, and volatile fatty acids (Dainty et al., 1985). All these compounds can characterize either the aroma or the spoilage (Tables 8.4 and 8.5). According to Casaburi et al. (2015), "volatile fatty acids and ketones contribute with fatty/gamy/cheesy/dairy notes; aldehydes with their fatty/grassy odor considered fresh and agreeable at low levels, but unpleasant and rancid when concentrations rise; alcohols and esters with their ethereal/fruity/sweet nuances; benzene, sulfur, and terpene compounds with reminiscent plastic, cabbage, and floral/citrus odor, respectively." The equilibrium among these compounds generates pleasant and typical meat and meat product aromas; on the other hand a disequilibrium

TABLE 8.4 Main Meat Spoilage Metabolites

Spoilage Metabolites
Organic acids (lactic, acetic, butyric, propionic, formic)
Hydrogen sulfide
Thiols
Mercaptans
Ammonia compounds
Indole
Scatole
Biogenic amines
α-Gluconate
2-Oxogluconate
Ketones
Aldehydes
Hydrocarbons
Alcohols
Esters
Geosmin (moldy)

TABLE 8.5 Spoilage in Normal and DFD Meat

Meat	Packaging	Spoilage	Microorganisms	Shelf Life
Normal (pH 5.6–6.0)	Air	Slime	*Pseudomonas* spp.	3 weeks
		Off-odor	*Acinetobacter* spp.	
		Off-flavor	*Moraxella* spp.	
	Under vacuum	Off-odor	*Lactobacillus* spp.	10–12 weeks
		Sour	*B. thermosphacta*	
Dark firm dry (up to pH 6.5)	Air	Fast	*Pseudomonas* spp.	1–4 days
		Spoilage	Enterobacteriaceae	
	Under vacuum	Hydrogen sulfide	Enterobacteriaceae *Alteromonas* spp.	1–4 days
		Green color	*Pseudomonas* spp.	

Source: From Comi, G., Cattaneo, P., 2007b. I prodotti carnei, pollame e pesce. In: Cocolin, L.S., Comi, G. (Eds.), Microbiologia Applicata Alle Produzioni Alimentari. Aracne Editrice, Roma, Italy, pp. 79–287, modified.

creates off-odor and off-flavor. In particular in these activities either SSO or other microorganisms are involved (Huis in't Veld, 1996).

The major biochemical changes occur in the meat aqueous phase, which contains sugars, amino acids, nucleotides, and peptides. These nutrients are sufficient for bacterial growth and the metabolism of these compounds leads to the formation of organic acids, neutral metabolites, H_2S, NH_3, indole, skatole, biogenic amines (i.e., cadaverine), and other substances characteristic of meat spoilage.

The residual glucose and the NPN fractions are the first molecules used by spoiler microorganisms. In aerobic conditions both the compounds are oxydated, whereas in anaerobic ones glucose is fermented and depleted in lactate, gluconate, glucose-6-phospate, pyruvate, propionate, formate, ethanol, acetate, acetoin, diacetyl, acetic, isobutyric, isovaleric, and 2-methylbutyric acids, 3-methylbutanol, 2-methylpropanol, independently of the storage and of any packaging conditions (Nychas et al., 1998, 2007; Drosinos and Board, 1995). Acid production, and consequently the changing of the meat pH, must be used to value the spoilage degree. In aerobic conditions, *P. fragi* is the main, and perhaps the only, ethyl esters producer among the aerobic microorganisms and is responsible for a sweet and fruity meat smell in the early stages of spoilage, even though *Pseudomonas* spp. accumulate D-gluconate and gluconate-6-P (Nychas and Arkoudelos, 1990; Casaburi et al., 2015). *Pseudomonas* may also oxidize lactate, pyruvate, and gluconate and utilize pyruvate, gluconate, acetate, and amino acids in anaerobic conditions (Nychas et al., 2007, 2008). Amino acids and NPN fractions are the major source of spoilage metabolites when the glucose is exhausted, with a count greater than 10^7 CFU/cm^2 for normal meats, and 10^6 CFU/cm^2 in meat with high pH (e.g., DFD). *Pseudomonas* are saprophytic bacteria and not nutrition demanding. They use amino acid metabolism as an energy source or as a pathway to produce molecules responsible for off-odor and off-flavor. Indeed, when their load reaches approximately 10^6 to 10^8 CFU/cm^2 or g, *Pseudomonas* mainly produces sulfur compounds, esters, and amines, followed immediately by viscosity and a fruity, putrid, sulfuryl, and cheesy smell (Samelis, 2006), which characterizes the off-odor of the aerobically cold-stored meat (Dainty and Mackey, 1992). In poultry meat stored at 2°C *Pseudomonas* spp. predominates and releases H_2S, methyl mercaptan, dimethyl sulfide, acetone, toluene, *n*-heptane, *n*-heptane, *n*-octane, methyl acetate, ethyl acetate, heptadiene, methanol, and ethanol. *Pseudomonas lundensis*, responsible for the production of unsaturated hydrocarbons, also produces NH_3 and amines from amino acids. The storage temperature influences the type of compounds present on spoiled chicken meat. The glucose and amino acid fermentation is sequential. Enterobacteriaceae initially ferment glucose and then, only after its depletion, they use amino acids, producing sulfur compounds (Gill, 1996). In particular Enterobacteriaceae, *B. thermosphacta* and homofermentative LAB lead to cheesy odors, which are associated with acetoindiacetyl and 3-methylbutanol formation (Borch and Molin, 1989; Dainty and Mackey, 1992). Enterobacteriaceae and *B. thermosphacta* play a secondary role in meat spoilage, producing compounds such as residues of branched chain esters, acetoin, 1-octen-3-ol, 3-methyl-1-butanol, and diacetyl (Casaburi et al., 2015). In beef, it perceived as a sour smell (milk, cheese, yeast), which then changes to bitter (liver) for microbial metabolites (volatile fatty acids) accumulation, mainly produced by LAB on lean tissue. This off-odor appears with LAB concentration higher than 10^8 CFU/g. LAB growth depends on the residual sugar concentration. In meat the LAB include obligatory heterofermentative strains, which produce lactic and acetic acid, CO_2, and ethanol or facultatively heterofermentative strains that produce only lactate by exoses and lactate and acetate without gas formation by pentose fermentation (Kandler, 1983; Kandler and Weiss, 1986; Nychas et al., 1998; Borch et al., 1996). However, some LAB produce lactate and acetate also in meat packaged in air (Samelis, 2006). Acetate may be perceived as a vinegar smell. When sugar is depleted, LAB metabolize amino acids producing ammonia and biogenic amines such as histamine, putrescine, spermine, and tyramine (Montel et al., 1991; Samelis et al., 1994; Labadie, 1999).

The sour smell (yeast) and sweet taste perceived in PV pork and lamb meat depend on the LAB and *B. thermosphacta*, which metabolize glucose in aerobic conditions and ribose, glycerol, and amino acids, producing a lot of volatile compounds in anaerobic conditions (Gardner, 1981; Grau, 1983; Blickstad and Molin, 1984; Borch and Molin, 1989; Nychas et al., 2007, 2008). Enterobacteriaceae and some Gram-negative bacteria also grow, and spoil meat and meat products in PV and MAP and cold-storage. *S. putrefaciens, S. liquefaciens, E. agglomerans*, and *H. alvei*, grown at level of 10^6-10^7 CFU/g in anaerobic conditions, produce H_2S, perceived as a rotten egg smell. This kind of spoilage prevails in high pH meat. Conversely, LAB and *H. alvei* produce only gas (CO_2) and no off-odors or greening (H_2S) are observed at the opening of the packages. Even psychrotrophic Clostridia, in combination with *Leuconostoc* spp., can spoil the PV meat, giving a rotten egg smell. In addition, in PV meat, all the above bacteria produce end molecules of their metabolism such as volatile fatty acids and biogenic amines.

Finally, some psychrotrophic clostridia associated with spoilage of PV meats ferment glucose, producing butyric acid, butanol, carbon dioxide, hydrogen, and causing discoloration and swelling of the packaging (Dainty et al., 1989).

Also the meats packaged in high oxygen-MAP are characterized by a cheesy and rancid smell and flavor. Oxygen favors the *Pseudomonas* spp. developing (Ordòñez et al., 1991; Jackson et al., 1992), but *B. thermosphacta* and LAB are also able to grow and contribute to spoilage. The result is the presence of a high amount of acetoin, acetic acid, and short-chain fatty acids (Blickstad and Molin, 1984; Borch and Molin, 1988).

The off-odors of PV and MAP meat and meat products are less offensive than that of aerobically stored meat (Borch et al., 1996). The spoilage consists in the presence of lactic, acetic, and formic acid that confer a characteristic sour and acid odor and flavor (Remenant et al., 2015; Gardner, 1981; Dainty and Mackey, 1992; Borch and Molin, 1988,1989; Borch et al., 1991). Sometimes sulfur compounds produced by LAB and Enterobacteria may also contribute to off-odor of PV and MAP meat products (Edwards and Dainty, 1987; Egan et al., 1989; Borch et al., 1991,1996). Sulfur compounds are often associated with acetoin, acetic and lactic acid, and diacetyl. However, LAB, *Carnobacterium* and *B. thermosphacta* produce acetoin, which is the main compound found in meat and meat products cold-stored in air (Borch and Molin, 1989).

Microorganisms can also grow in frozen meat held at temperatures less than or equal to $-7°C$. This is possible because they exploit the percentage of water present in the liquid state. In particular molds of *Cladosporium, Geotrichum, Mucor, Rhizopus, Sporotrichum*, and *Thamnidium* genera have been described as able to develop in frozen meat, kept at 4°C but also at $-5°C$ or $-10°C$. The spoilage consists of spots of different colors, strictly dependent on the mold species (Table 8.6).

TABLE 8.6 Mold and Defects in Meat and Fish

Mold Strains	Defect
Cladosporium herbarum	Black spots
Sporotrichum carnis	White spots
Thamnidium elegans	Moldy/feathery
Aspergillus ochraceus	Orange/yellow spots
Penicillium spp.	Green and blue spots
Geotrichum candidum	White spots
Penicillium nordicum	White and pale blue spots
Mucor spp.	White spots
Rizhopus nigricans	White and black spots

Source: From Comi, G., Cattaneo, P., 2007b. I prodotti carnei, pollame e pesce. In: Cocolin, L.S., Comi, G. (Eds.), Microbiologia Applicata Alle Produzioni Alimentari. Aracne Editrice, Roma, Italy, pp. 79–287, modified.

8.4.2 Slime and Gas Formation

The spoilage of PV and MAP meat and meat products often consists in a ropy slime formation. Produced in particular by *Leuconostoc*, LAB, *Carnobacterium*, and *Weissella* during the cold-storage. The pasteurization technology, applied after packaging, reduces the initial contamination but does not completely kill the thermoduric bacteria. The initial contamination, after the heat treatment, is reduced at level of <10 CFU/cm^2, but they, being psychrotolerant, may grow during the cold storage of meat products. The spoilage is usually observed some weeks before their shelf life ends. The slime mainly depends on the homofermentative LAB, but in meat products also the heterofermentative LAB can grow and produce either white slime or gas, inducing a swelling of the packaging (Korkeala et al., 1988; von Holy et al., 1991; Dykes et al., 1994). *Leuconostoc mesenteroides* subs. *destranicum, L. mesenteroides* subs. *mesenteroides, L. amelibiosum, L. curnosum*, and *L. gelidum* are the main bacteria implicated. *Weissella viridescens* and *H. alvei* also produce slime, greening, and gas. All the above microorganisms are psychrotolerant, originating from raw meat and processing rooms (Mäkelä et al., 1992) and do not require sugars to produce the ropy slimes, gas, and whitening or greening.

Even psychrotrophic Clostridia, alone or in combination with *Leuconostoc*, can spoil the PV beef meat, giving rotten egg smells and large amounts of gas (H_2 and CO_2) (Dainty et al., 1989).

8.4.3 Discoloration

Meat and meat products can be spoiled by microorganisms that produce surface modifications of the color of both fat and meat. Brown-black spots appear on fat originating by the development of lipolytic yeasts, such as *Saccharomycopsis lipolytica*. On muscle the spots are due to yeasts, molds, and bacteria.

The bacterial production of hydrogen sulfide converts the muscle pigment to green sulfomyoglobin. Hydrogen sulfide derives from cysteine, which is metabolized only when the glucose and oxygen availability is limited (Egan et al., 1989). Sulfomyoglobin is, however, not formed in anaerobic atmospheres (Borch and Agerhem, 1992) and it is typically associated with high pH meat (>6.5), but may also occur in normal pH meat (5.8). Also H_2O_2 is responsible for greening meat color. Indeed LAB, growth in aerobic conditions, synthesizes H_2O_2 that leads to the oiydation of nitrosohaemochrome to cholemyoglobin, producing green spots on meat and meat products. Homofermentative and heterofermentative LAB, *Leuconostoc* spp., *Enterococcus* spp., *Pediococcus* spp., *C. divergens*, and *W. viridescens*, easily surviving the heat processing and/or contaminating the meat products after cooking (Borch et al., 1996), are able to form H_2O_2, producing green spots (Grant et al., 1988; Borch and Molin, 1989).

In ungutted games and poultry, the H_2S produced in the gut reaches the muscles and after O_2 espousing, it combines with myoglobin to give sulfomyoglobin.

8.5 FISH AND FISH PRODUCT SPOILING MICROORGANISMS

Fish, mollusks, and crustaceans arc highly perishable products if they are not processed or packaged immediately after capture. Refrigeration is one of the best technologics for maintaining their nutritional and sensory properties. However, fish and fish products stored at refrigeration temperatures, or kept on ice or in styrofoam boxes with ice, have limited shelf life depending on the species. The high perishability derives from their chemical composition (Table 8.1), by the presence of a low content of connective tissue, the low acidity of the tissues, by the abundance of protein and NPN fractions (Comi and Cattaneo, 2007a,b; Comi and Iacumin, 2012).

The amount of microorganisms present in fish and fish products is variable and it is between 10^3 and 10^5 CFU/cm^2 on the skin and between 10^3 and 10^7 CFU/cm^2 in the meat.

The microbial population of fish of warm temperate waters includes either aerobic and facultative anaerobic Gram-negative (*Pseudomonas, Moraxella/ Acinetobacter, Photobacterium, Flavobacterium/Cytophaga, Xanthomonas,*

Vibrio, Shewanella, Proteus, Aeromonas, Serratia, Hafnia) or Gram-positive (*Bacillus, Corynebacterium, Micrococcus*, and other cocci and lactobacilli) microorganisms. In cold water, Gram-negative predominates on the skin and Gram-positive (*Clostridium, Bacillus*, Coryneforms) in the intestinal contents. They are psychrotrophic bacteria which are responsible for fresh fish spoilage. The exogenous microflora is formed by bacteria, viruses, yeasts, and molds of terrestrial origin. It consists of pathogens for human beings (*Salmonella* spp., *Listeria monocytogenes, Vibrio* spp., *C. botulinum*, enterotoxigenic *E. coli*, enterotoxigenic strains of *Staphylococcus aureus, Bacillus cereus*, and viruses) (Comi and Cattaneo, 2007a,b; Comi and Iacumin, 2012).

It also includes spoilage or fecal pollution microorganisms: Enterobacteriaceae, total and fecal coliforms, Enterococci.

The fecal microorganisms, *Yersinia enterocolitica, Proteus mirabilis, Klebsiella* spp., *L. monocytogenes,* were found mainly in fish that live usually in coastal area waters. Their concentration depends on the amount of industrial or urban waste, and it is particularly high in Spring and Summer, from the increase of urban waste due to tourism. The degree of water pollution also affects the presence of *Vibrio cholerae, V. fluvialis, V. paraemoliticus, V. alginoliticus, V. vulnificus* as well as *Salmonella* spp. and *L. monocytogenes*. *Vibrio* spp. and *Listeria* spp. include psychrotolerant strains and can affect the shelf life of the fish, conversely *Salmonella* spp. and *E. coli*, being mesophilic, can grow only in temperatures up to 10°C (Comi and Iacumin, 2012).

However, the presence of pathogenic microorganisms is also related to manipulation that the fish undergoes after capture.

In addition there is a high prevalence of Gram-negative over Gram-positive bacteria (Table 8.2). In fact, the Gram-negative microflora may contribute up to 80% of the total flora. The psychrotrophic population includes saprophytic and spoilage bacteria, and can reach concentrations of 10^8 and 10^9 CFU/g, after 10–12 days at 4°C.

Pseudomonas spp., *Photobacterium* spp., and *Shewanella* spp. develop during storage and are the main bacteria isolated in spoiled fish. Values of *Shewanella* spp. and H_2S+ producing bacteria at level of 10^6 CFU/cm^2 represent a spoilage index. Among *Shewanella* spp., *S. baltica*, and *S. putrefaciens* predominate and are the typical spoilers of fresh marine fish and smoked fish (Broekaert et al., 2011; Joffraud et al., 2001). *Photobacterium phosphoreum* is a spoiler of cold water fish, but it is also present in Mediterranean species (sea bream and sea bass). It spoils fresh cod, garfish, halibut, saithe, salmon, and shrimp (Dalgaard et al., 1997, 2006; Hansen et al., 2009; Hovda et al., 2007a,b; Jeyasekaran et al., 2006; López-Caballero et al., 2002; Macé et al., 2013; Olafsdottir et al., 2005; Paarup et al., 2002; Reynisson et al., 2009). Indeed *Photobacterium iliopiscarium* seems to be the main spoiler of PV cold-smoked salmon (Olofsson et al., 2007).

Carnobacterium spp. is also involved in fish and fish products spoilage. In particular Carnobacterium *maltaromaticum* and *C. divergens* are able to spoil raw salmon (*Salmo salar*) fillets and sea fish stored in PV and MAP (Macé et al., 2013), and cooked and peeled shrimp, producing cheese/feet, sour/fermented, and milky-boiled off-odors (Jaffrès et al., 2011; Laursen et al., 2005; Leisner et al., 2007; Vasilopoulos et al., 2008; Casaburi et al., 2011).

Among heterofermentative bacteria *Weissella confusa* represents the major spoiler of oyster gills stored at 10°C (Chen et al., 2013).

Moraxella and *Acinetobacter*, although present in high concentrations in fish products, do not seem to be important spoilers.

Brochothrix thermosphacta is recognized as being responsible for the spoilage of fish hamburgers, tuna packed under CO_2 or in MAP (Lannelongue et al., 1982) and frozen cod (Gardner, 1981). On refrigerated fish and fish products the growth of *B. thermosphacta* depends on the O_2 concentration remaining in the packaging (Lopez-Galves et al., 1995). It can predominate at storage temperature in presence of O_2 above 0.2% or in air (Coton et al., 2013). Instead *Lactobacillus* spp. prevails on *B. thermosphacta* under anaerobic conditions.

Lactobacillus spp. and *Leuconostoc* spp. represent the main spoiler bacteria in PV and MAP fish products. *Lactobacillus alimentarius, L. sakei*, and *L. curvatus* spoil marinated herring (Lyhs et al., 2001, Lyhs and Bjorkroth, 2008), meanwhile *L. curvatus* and *L. sakei* spoil cold-smoked salmon (Jørgensen et al., 2000; Stohr et al., 2001; Joffraud et al., 2001). *Leuconostocs*, *L. gelidum*, and *L. gasicomitatum* spoil marinated herrings, producing slime (Lyhs et al., 2004).

Fish are also preserved frozen. Recently a technique is used that coats the product with a layer of ice in order to prevent evaporation and surface oxidation. The frozen fish products are PV in polyethylene film or other waterproof casings to avoid evaporation, rancidity, loss of color, and aroma. The frozen fish are stable if the cold chain ($T \leq 18°C$) is maintained.

Some seafood products are eviscerated, beheaded, washed, and brined, then they are stored after drying. This technology is performed by exposure to air or in drying plants. The final product is stable because it has less than 15% humidity and an a_w below 0.88; however, it must be stored in a cool, dry (RH < 65%), well-ventilated room or packed in polyethylene film. In fact, if it rehydrates, it permits the development of *Halobacterium* spp., *Staphylococcus* spp., molds and yeasts that give dusty and colored slimes.

Some dietary fish products, or some quality fish products for special consumers (children, babies) are stored lyophilized. The lyophilisate, having less than 5% humidity, is PV packaged and therefore stable for long time.

Salting is another technique that increases the shelf life of fish. Fresh fish is brined whole or after eviscerating and beheading. The product (e.g., cod) is packaged PV and consequently is stable without any further treatment. Indeed molds and staphylococci produce slimes and discoloring, when the product is stored aerobically or in the damp (Comi and Cattaneo, 2007a,b).

The whole or filleted fish are also roasted or fried and then marinated in a solution of water/vinegar (10%)/NaCl (12%). The product is stable due to the brine and the pH of less than 4.5 units. But it must be completely immersed in the brine, otherwise the part that emerges can support yeast and mold growth.

Some products (shellfish, shrimp, prawns, and crab) are processed by cooking combined with additional treatments: freezing, brine/cooling, brine/sterilization. The products subjected to cooking/pasteurization/cooling/packaging have a shelf life of up to 90 days at 4°C, while those sterilized or frozen have unlimited shelf life.

Some fish species are brined, cold (25–30°C) or hot (70–80°C) smoked, vacuum packaged, and stored at 4°C. The salting (Water Phase Salt >3.5%), the smoking, the packaging and the temperature of storage permit a shelf life maximum of 30 to 45 days. In case of long-term storage, or thermal abuse, halotolerant (Micrococci) bacteria, LAB, *B. thermosphacta,* or *Pseudomonas/Shewanella* strains, surviving the smoking process, can develop and produce slimes, off-odor, and off-flavor (Comi and Cattaneo, 2007a,b).

Tuna, salmon, herring, mackerel, and other fish products are steam-cooked and covered with preserving liquid (water and salt) or oil, then canned and sterilized. The control of the time/temperature of sterilization allows for the long-term stability of the product.

Surimi is a product made from fresh, barbed, washed, ground, and heat-treated fish. Herring, catfish, white fish, and some cephalopods are the raw material of surimi. It is stable because it is packaged and sterilized or packaged and frozen. In this case it can last from 6 months to one year.

8.5.1 Metabolomics of Spoilage (Off-Odor and Off-Flavor)

The spoilage begins with the death of the animal and is a process that involves native tissue and bacterial enzymes. The tissue and digestive enzymes are the first to carry out the spoilage and produce low molecular weight catabolites, which represent the ideal substrate for the microorganisms.

These catabolites derive from molecules of meat exudates, such as inosine, ribose, lactate, creatine, urea, anserine, carnosine, free amino acids, and trimethylamine-*N*-oxide (TMAO, not proteic nitrogen fraction present in sea fish). The autochthonous spoilage microorganisms are localized on the surface, on the gills and intestines, while the exogenous derive from the sewage water of cities and rivers. In addition, this microbial population derives from handler's contamination (Comi and Cattaneo, 2007a,b; Comi and Iacumin, 2012).

The fish muscles are initially sterile, then, after the death of the animal, the microbial population invades and contaminates them. The fish microflora depends on many factors, such as the environment in which the animal lives, the fishing season and the water temperature, the method of capture, the handling on the ship, and the technological and sale processes (Comi and Iacumin, 2012).

After catching, fish are stored whole or must be beheaded, gutted, filleted. With the fish death, the microorganisms present in the gut, gills, and skin reach the muscles. The gutting and filleting present an opportunity for dissemination for microorganisms from the gut, and from the surface to the muscles. Increase in contamination is also determined by not sufficiently clean equipment. To prevent excessive contamination it is necessary to use fast technological processes, using sanitized tools and equipment and to refrigerate the product immediately after its capture.

The fish spoilage is made up of two phases. The first phase consists in an autolytic process by endogenous enzymes, which produces the substrate for bacterial growth. The second phase consists in the growth of microorganisms, which occurs more rapidly than in the mammal meats, due to the pH of fish flesh (>6.2) which is higher than that of terrestrial animals. In cephalopods, herring, tuna, and cod, autolytic phenomenon predominate. With the fish death the oxygen of the muscles is depleted, consequently the anaerobic glycolysis starts and produces lactic acid. The pH drops slightly, depending on the species, from 6.8 to 6.1 units in cod and mackerel and from 5.8–6.0 to 5.4–5.6 units in tuna and halibut. After glycolysis, proteolysis, induced by endogenous enzymes activated by the pH, starts. The proteolytic activity is responsible for the progressive softening of the fish tissues. Proteases also cause intestinal autolysis of the abdominal area in ungutted fish. The degradation of protein and NPN molecules produces short-chain peptides, amino acids, and nitrogen compounds, which increase the bacterial growth. Nucleic acids are rapidly degraded by enzymes in tissue. In particular, the adenosine-triphosphate (ATP) is metabolized causing the loss of freshness. The reaction is the following:

$$\text{ATP(adenosinthreephosphatc)} \Rightarrow \text{ADP(adenosindiphosphate)} \Rightarrow \text{AMP (adenosinmonophosphate)} \Rightarrow \text{IMP(inosinmonophosphate)} \Rightarrow \text{Inosine} \Rightarrow \text{Hypoxanthine} \Rightarrow \text{Xanthine} \Rightarrow \text{Uric acid.}$$

The pattern is common to most fish, changing the speed of each reaction (Comi and Cattaneo, 2007a,b).

These early stages, due to degradation by endogenous enzymes, produce catabolites, which constitute an excellent source of nitrogen compounds and stimulate the microorganism's growth. The degradation reactions proceed with the predominance of microbial enzymes. The inosine monophosphate, giving an initially pleasant smell of fresh fish, is transformed into hypoxanthine, which represents the classic bitter and unpleasant taste of spoiled fish (Table 8.7). Ammonia is also produced, and it derives from the microbial catabolism of the amino acids. In elasmobranchs, there are high concentrations of urea (it is present in blood with osmoregulating function and in gills and skin mucous), which is broken down by bacteria to NH_3 (Comi and Iacumin, 2012).

Also the phospholipids, first attacked by endogenous lipases, are converted into free fatty acids by bacterial lipases. This process continues with

TABLE 8.7 Specific Spoilage Organisms (SSO) and Relative Spoilage Compounds (Huss/FAO, 1995) of Fish

Microorganisms	Spoilage Compounds
S. putrefaciens	TMA, H_2S, CH_3SH, $(CH_3)_2S$, hypoxanthine
P. phosphoreum	TMA, hypoxanthine
Pseudomonas spp.	Ketones, aldehydes, esters, sulfur compounds
Vibrio spp.	TMA, H_2S
Anaerobes	NH_3, acetic, butyric and propionic acids

H_2S: Hydrogen sulfide; TMA: Trimethylamine; CH_3SH: Methanethiol; $(CH_3)_2S$: dimethyl sulfide.
Source: From Comi, G., Cattaneo, P., 2007b. I prodotti carnei, pollame e pesce. In: Cocolin, L.S., Comi, G. (Eds.), Microbiologia Applicata Alle Produzioni Alimentari. Aracne Editrice, Roma, Italy, pp. 79–287.

the oxidation of polyunsaturated fatty acids and the formation of hydroperoxides. Then a secondary oxidation determines the production of ketones, aldehydes, alcohols, alkanes, and hydrocarbons (ethane and pentane) responsible for the rancid odor of spoiled fish. Hydroperoxides are also produced by lipoxygenase enzymes present in the gills.

The microbial enzymes are involved in the production of short-chain peptides, and amino acids in the TMAO reduction and in the second stage of ATP degradation. The TMAO is the main constituent of the NPN fraction. It seems to have osmoregulating, detoxifying, and redox functions, and it is present in dark muscle of the big pelagic fish. It seems to be synthesized by tissue or algae enzymes. *Pseudomonas* spp. and *S. putrefaciens* are involved in the TMAO degradation. The reaction is the following:

$$\text{Lactic acid} + 2\ \text{TMAO} \Rightarrow \text{acetic acid} + 2\ \text{TMA(trimetylamine)} + CO_2 + H_2O.$$

The reduction is carried out by bacterial enzymes. However, the demethylation of TMAO occurs even by endogenous enzymes: TMAO ⇒ DMA (Dimethylamine) + FA (formaldehyde). In particular *S. putrefaciens* and *S. baltica* seem to be able to use TMAO as electron acceptor under anaerobic growth conditions. The TMAO is reduced to TMA, that, also at low concentration, is perceived as a urine smell, strong fishy and ammonia-like off-odors (Gram and Huss, 1996; Lopez-Caballero et al., 2001). *Photobacterium phosphoreum* produces 30-fold higher TMA than that of *S. putrefaciens* (Dalgaard, 1995).

In fresh fish, the TMA content can vary from 1 to 4 mg/100 g. A typical stale odor is perceived up to 10 mg/100 g, but with 5–10 mg/100 g, fish are still acceptable.

Biogenic amines and total volatile bases nitrogen (TVB-N) compounds represent the main molecules responsible for the spoilage, too. The TVB-N, formed by the biochemical changes postmortem: mainly NH_3, TMA, and DMA, are considered useful quality indexes. The value depends on the fish species. However, levels up to 20 mg N/100 g correspond to a state of freshness, but also the levels of TVB-N from 20 to 30 are acceptable. Conversely, values between 30 and 50 mg N/100 g indicate the beginning of spoilage. Finally, levels up to 50 mg N/100 g represent a real state of spoilage (Comi and Iacumin, 2012).

The TVB-N limits for certain categories of fishery products and the methods of analysis are shown in the Reg. (EC) n. 2073/2005. In particular the limits are the following: for *Sebastes* spp., *Helicolenus dactylopterus*, *Sebastichthys capensis*: 25 mg N/100 g; for species belonging to the *Pleuronectidae* family (with the exception of halibut: *Hippoglossus* spp.): 30 mg N/100 g and for *S. salar*, species belonging to the *Merluccidae* family, species belonging to the *Gadidae* family: 35 mg N/100 g.

The LAB and micrococci, which metabolize carbohydrates producing lactic acid, acetic acid, and ethanol, also produce hydrogen sulfide, thiols, mercaptans, dimethyl sulfide, indole, skatole, and urine smells in fish and in fish products. The smell of rotten fish also comes from the presence of amino valeric acid, aldehyde amino valeric, and piperidine.

The amino acids are decarboxylated and converted into biogenic amines (Table 8.8). These are toxic to humans and are produced by the microorganisms responsible for the spoilage of fish and fish products. *Hafnia* spp., *Proteus* spp., *Pseudomonas* spp., *Morganella morganii*, and *S. putrefaciens* are the strains mainly involved.

The biogenic amines are common in particular in fish such as Scombridae family (tuna, mackerel), Clupeidae (sardines, herrings), Engraulidae (anchovy), coryphaenidae (dolphinfish), Carangidae (horse mackerel, yellowtail), Pomatomidae, because of the presence of high free amino acid amounts in

TABLE 8.8 Biogenic Amine in Meat, Fish and Meat, and Fish Products

Amino Acid Precursors	Biogenic Amines
Histidine	Histamine
Ornithine	Putrescine
Arginine/putrescine	Spermidine
Lysine	Cadaverine
Tyrosine	Tyramine
Arginine	Agmatine

Source: From Comi, G., Cattaneo, P., 2007b. I prodotti carnei, pollame e pesce. In: Cocolin, L.S., Comi, G. (Eds.), Microbiologia Applicata Alle Produzioni Alimentari. Aracne Editrice, Roma, Italy, pp. 79–287.

muscles. The most important biogenic amines include: Histamine (HIS), Putrescine (PUT), Cadaverine (CAD), Tyramine (TYR) Triptamine (TRY), Spermine (SPN), Spermidine (SPD). *Carnobacterium divergens* can produce TYR in cold-smoked salmon (Connil et al., 2002a,b), conversely *S. putrefaciens* produces PUT and CAD in penaeid shrimp or Mediterranean hake (Baixas-Nogueras et al., 2005; Benner et al., 2004). *Photobacterium phosphoreum* is responsible for various histamine fish poisoning incidents (Dalgaard et al., 2006; Emborg et al., 2005; Kanki et al., 2004, 2007).

The biogenic amines concentration could be used to value the hygienic quality of fresh meat and fish (Hernández-Jover et al., 1997). HIS, in particular, is responsible for headaches, facial flushing and bright red rashes. EFSA (2011) declared that a food is safe if it contains less than 50 mg/kg of HIS. Food containing up to 400 mg/kg are absolutely unsafe (Silla Santos, 1996; Ienistea, 1973).

Pseudomonas and *Shewanella* produce sulfur compounds such as hydrogen sulfide, dimethyl sulfide, methyl mercaptan, which derive from sulfur containing amino acids like methionine and cysteine. In particular *S. putrefaciens* or *S. baltica, S. hafniensis, S. morhuae*, *S. glacialipiscicola*, and *S. algidipiscicola* can produce H_2S (Fonnesbech Vogel et al., 2005; Satomi et al., 2006, 2007). The threshold limits of sulfur smell are very low: 40 ppb for hydrogen sulfide, 0.5 ppb for dimethyl sulfide and 0.05 ppb for methyl mercaptan.

These off-flavors and off-odors are mainly produced by *S. putrefaciens*, and occasionally by *P. fluorescens* and *P. fragi*.

Lactobacillus spp., *Pediococcus pentosaceus*, *Micrococcus* spp., and *B. thermosphacta* produce slight acidification in whole or filleted fish products in PV or MAP.

8.5.2 Slime and Gas Formation

Colored slime and swelling of the packages are often observed in fish products rather than in fish. Heterofermentative LAB and *B. thermosphacta* play a main role in the spoilage. Also homofermentative LAB are involved in the production of slime and off-odors.

Dehydrated and brined cod (*Gadidae*) are spoiled by halotolerant or halophilic staphylococci and micrococci, yeasts, spore-forming bacteria, LAB and molds that produce colored slime, "pink" and "dun," when they are stored unpacked in high RH ($>$70%) or in MAP (Comi and Cattaneo, 2007a,b). Halophilic molds (*Eurotium* spp., *Cladosporium* spp., *Aspergillus* spp.) cover salty fish with a whitish powder. In addition to a poor or insufficient penetration of salt, Gram-negative rod-shaped bacteria produce slimes or putty fish.

Moreover stockfish (dehydrated cod) is spoiled and contaminated by pests. Indeed it also gets moldy or dusty by Gram-positive cocci and molds.

Inadequate time/temperature/RH involve microbial and tissue enzymatic activity, which leads to soft tissue, grayish color, unpleasant odor. In case of deep tissue spoilage, the consistency and the color become respectively mucous and yellowish.

8.5.3 Discoloration

Fish and fish products rarely are subjected to microbial discoloration. Indeed it should be pointed out that oxygen plays the main role, by chemical reactions, in rancidity or discoloration. In particular in cold smoked fish, color can change by chemical oxidation, despite the smoke or PV. However microorganisms such as *Flavobacterium* spp., *Pseudomonas* spp., *Photobacterium* spp., *Micrococcus* spp., *Staphylococcus* spp., LAB, *B. thermosphacta,* and mold, growing on fish and fish products packaged in air, in PV and MAP, produce colored slimes and off-odor and off-flavor.

8.6 CONCLUSION

Meat and meat products and fish and fish products represent an excellent growth media for a variety of spoilage microbial populations (Jay et al., 2005). This large and various microbial population depends on different factors regarding the breeding, the fishing, the harvesting, the preslaughter husbandry practices, the age of the animal at the time of slaughtering, the handling during slaughtering, evisceration and processing, the temperature controls during slaughtering, processing and distribution, and the preservation methods (Dave and Ghaly, 2011). In particular the intestinal tract, the skin of the animal, and the environment are the main sources of contamination. Then a selection pressure is created during meat and fish storage that provides an advantage to bacteria whose metabolic activity may lead to spoilage (Kameník et al., 2012). The increase of hygiene standards and the use of new technology, such as new forms of packaging, affect the composition of the microflora that develops in meat and fish. The current knowledge allows for use of technologies that dramatically restrict the microbial growth and extend the shelf life. Indeed new analysis methods enabled the discovery of the action of psychrotrophic Clostridia or new spoiler strains that must be taken into consideration during meat and fish processing and storage. However, the mechanisms and the interactions that stimulate the growth and the spoilage are far from being completely known and must be still studied, elucidated, and interpreted (Huis in't Veld, 1996).

REFERENCES

Baixas-Nogueras, S., Bover-Cid, S., Veciana-Nogues, M.T., Marine-Font, A., Vidal-Carou, M.C., 2005. Biogenic amine index for freshness evaluation in iced Mediterranean hake (*Merluccius merluccius*). J. Food Prot. 68, 2433–2438.

Benner Jr., R.A., Staruszkiewicz, W.F., Otwell, W.S., 2004. Putrescine, cadaverine, and indole production by bacteria isolated from wild and aquacultured penaeid shrimp stored at 0, 12, 24, and 36°C. J. Food Prot. 67, 124–133.

Blickstad, E., Molin, G., 1983. The microbial flora of smoked pork loin and frankfurter sausage stored in different gas atmospheres at 4°C. J. Appl. Bacteriol. 54, 45–56.

Blickstad, E., Molin, G., 1984. Growth and end-product formation in fermenter cultures of *Brochothrix thermosphacta* ATCC 11507^T and two psychrotrophic *Lactobacillus* spp. in different gaseous atmospheres. J. Appl. Bacteriol. 57, 213–220.

Borch, E., Agerhem, H., 1992. Chemical, microbial and sensory changes during the anaerobic cold storage of beef inoculated with a homofermentative *Lactobacillus* sp. or a *Leuconostoc* sp. Int. J. Food Microbiol. 15, 99–108.

Borch, E., Molin, G., 1988. Numerical taxonomy of psychrotrophic lactic acid bacteria from prepacked meat and meat products. Antonie Van Leeuwen 54, 301–323.

Borch, E., Molin, G., 1989. The aerobic growth and product formation of *Lactobacillus, Leuconostoc, Brochothrix* and *Carnobacterium* in batch cultures. Appl. Microbial. Biotechnol. 30, 81–88.

Borch, E., Berg, H., Holst, O., 1991. Heterolactic fermentation by a homofermentative *Lactobacillus* sp. during glucose limitation in anaerobic continuous culture with complete cell recycle. J. Appl. Bacteriol. 71, 265–269.

Borch, E., Kant-Muemans, M.L., Blixt, Y., 1996. Bacterial spoilage of meat and cured meat products. Int. J. Food Microbiol. 33, 103–120.

Boddy, L., Wimpenny, J.W.T., 1992. Ecological concepts in food microbiology. J. Appl. Bacteriol. Symp. Suppl. 73, 23S–38S.

Broekaert, K., Heyndrickx, M., Herman, L., Devlieghere, F., Vlaemynck, G., 2011. Seafood quality analysis: molecular identification of dominant microbiota after ice storage on several general growth media. Food Microbiol. 28, 1162–1169.

Casaburi, A., Nasi, A., Ferrocino, I., Di Monaco, R., Mauriello, G., Villani, F., et al., 2011. Spoilage-related activity of *Carnobacterium maltaromaticum* strains in air-stored and vacuum-packed meat. Appl. Environ. Microbiol. 77, 7382–7393.

Casaburi, A., Piombino, P., Nychas, G.J., Villani, F., Ercolini, D., 2015. Bacterial population and volatilome associated to meat spoilage. Food Microbiol. 45, 83–102.

Cavill, L., Renteria-Monterrubio, A.L., Helps, C.R., Corry, J.E.L., 2011. Detection of cold- tolerant clostridia other than *Clostridium estertheticum* in raw vacuum-packed chill-stored meat. Food Microbiol. 28, 957–963.

Chen, H., Liu, Z., Wang, M., Chen, S., Chen, T., 2013. Characterization of the spoilage bacterial microbiota in oyster gills during storage at different temperatures. J. Sci. Food Agric. 93 (15), 3748–3754.

Collins, M.D., Farow, J.A.E., Philips, B.A., Ferusu, S., Jones, D., 1987. Classification of *Lactobacillus divergens*, *Lactbacillus piscicola* and some catalase-negative asporogenous, rod-shaped bacteria from poultry in new genus *Carnobacterium*. Int. J. System Bacteriol 37, 110–116.

Comi, G., Cattaneo, P., 2007a. Meat and sausages. In: Gigliotti, C. (Ed.), Food Biotechnology. Piccin Editore, Padova, Italy, pp. 111–140.

Comi, G., Cattaneo, P., 2007b. I prodotti carnei, pollame e pesce. In: Cocolin, L.S., Comi, G. (Eds.), Microbiologia Applicata Alle Produzioni Alimentari. Aracne Editrice, Roma, Italy, pp. 79–287.

Comi, G., Iacumin, L., 2012. Meat and fish products. In: Gobetti, G., Vincenzini, M., Farris, G.A., Neviani, E. (Eds.), Food Microbiology. Ambrosiana Editrice, Milano, Italy, pp. 153–196.

Connil, N., Plissoneau, L., Onno, B., Pilet, M.F., Prévost, H., Dousset, X., 2002a. Growth of *Carnobacterium divergens* V41 and production of biogenic amines and divercin V41 in sterile cold-smoked salmon extract at varying temperatures, NaCl levels, and glucose concentrations. J. Food Prot. 65, 333–338.

Connil, N., Prévost, H., Dousset, X., 2002b. Production of biogenic amines and divercin V41 in cold smoked salmon inoculated with *Carnobacterium divergens* V41, and specific detection of this strain by multiplex-PCR. J. Appl. Microbiol. 92, 611–617.

Coton, M., Joffraud, J.J., Mekhtiche, L., Leroi, F., Coton, E., 2013. Biodiversity and dynamics of the bacterial community of packaged king scallop (*Pecten maximus*) meat during cold storage. Food Microbiol. 35, 99–107.

Dalgaard, P., 1995. Qualitative and quantitative characterization of spoilage bacteria from packed fish. Int. J. Food Microbiol. 26, 319–333.

Dainty, R.H., Mackey, B.M., 1992. The relationship between the phenotypic properties of bacteria from chill-stored meat and spoilage processes. J. Appl. Bacteriol.(Suppl. 21), 103–114 (Symposium).

Dainty, R.H., Edwards, R.A., Hibbard, C.M., 1985. Time course of volatile compounds formation during refrigerated storage of naturally contaminated beef in air. J. Appl. Bacteriol. 59, 303–309.

Dainty, R.H., Edwards, R.A., Hibbard, C.M., 1989. Spoilage of vacuum-packed beef by a *Clostridium* sp. J. Sci. Food Agric. 49, 473–486.

Dalgaard, P., Manfio, G.P., Goodfellow, M., 1997. Classification of Photobacteria associated with spoilage of fish products by numerical taxonomy and pyrolysis mass spectrometry. Zent. Bakteriol. 285, 157–168.

Dalgaard, P., Madsen, H.L., Samieian, N., Emborg, J., 2006. Biogenic amine formation and microbial spoilage in chilled garfish (*Belone belone belone*) effect of modified atmosphere packaging and previous frozen storage. J. Appl. Microbiol. 101, 80–95.

Dave, D., Ghaly, A.E., 2011. Meat spoilage mechanisms and preservation techniques: a critical review. Am. J. Agricul. Biol. Sci. 6 (4), 486–510.

Doulgeraki, A.I., Paramithiotis, S., Nychas, G.J.E., 2011. Characterization of the Enterobacteriaceae community that developed during storage of minced beef under aerobic or modified atmosphere packaging conditions. Int. J. Food Microbiol. 145, 77–83.

Doulgeraki, A.L., Ercolini, D., Villani, F., Nychas, G.J.E., 2012. Spoilage microbiota to the storage of raw meat in different conditions. Int. J. Food Microbiol. 157, 130–141.

Drosinos, E.H., Board, R.G., 1995. Microbial and physicochemical attributes of minced lamb: sources of contamination with Pseudomonads. Food Microbiol. 12, 189–197.

Dykes, G.A., Cioete, T.E., von Holy, A., 1994. Identification of *Leuconostoc* species associated with the spoilage of vacuum-packaged Vienna sausage by DNA-DNA hybridization. Food Microbiol. 11, 271–274.

Edwards, R.A., Dainty, R.H., 1987. Volatile compounds associated with the spoilage of normal and high pH vacuum-packed pork. J. Sci. Food Agric. 38, 57–66.

EFSA, 2011. Scientific opinion on risk based control of biogenic amine formation in fermented foods. EFSA J. 9 (10), 2393–2486.

Egan, A.F., Shaq, B.J., Rogers, P.J., 1989. Factors affecting the production of hydrogen sulphide by *Lactococcus sakei* L13 growing on vacuum-packaged beef. J. Appl. Bacteriol. 67, 255–262.

Ellis, D.I., Goodacre, R., 2001. Rapid and quantitative detection of the microbial spoilage of muscle foods: current status and future trends. Trends Food Sci. Technol. 12, 414–424.

Emborg, J., Laursen, B.G., Dalgaard, P., 2005. Significant histamine formation in tuna (*Thunnus albacares*) at 2°C effect of vacuum- and modified atmosphere packaging on psychrotolerant bacteria. Int. J. Food Microbiol. 101, 263–279.

Ercolini, D., Russo, F., Torrieri, E., Masi, P., Villani, F., 2006. Changes in the spoilage-related microbiota of beef during refrigerated storage under different packaging conditions. Appl. Environ. Microbiol. 72, 4663–4671.

Ercolini, D., Russo, F., Nasi, A., Ferranti, P., Villani, F., 2009. Mesophilic and psychrotrophic bacteria from meat and their spoilage potential in vitro and in beef. Appl. Environ. Microbiol. 75, 1990–2001.

Ercolini, D., Ferrocino, I., Nasi, A., Ndagijimana, M., Vernocchi, P., La Storia, A., et al., 2011. Monitoring of microbial metabolites and bacterial diversity in beef stored in different packaging conditions. Appl. Environ. Microbiol. 77, 7372–7381.

Erichsen, I., Molin, G., 1981. Microbial flora of normal and high pH beef stored at 4°C in different gas environments. J. Food Prot. 44, 866–869.

Fonnesbech Vogel, B., Venkateswaran, K., Satomi, M., Gram, L., 2005. Identification of *Shewanella baltica* as the most important H_2S-producing species during iced storage of Danish marine fish. Appl. Environ. Microbiol. 71, 6689–6697.

Fredrickson, A.G., 1977. Behavior of mixed cultures of microorganisms. Ann. Rev. Microbiol. 31, 63–87.

Gardner, G.A., 1981. Psychrotrophic microorganisms in spoilage and pathogenicity. In: Roberts, T.A., Hobbs, G., Christian, J.H.B., Skovgaard, N. (Eds.), *Brochothrix thermosphacta* in Spoilage of Meats. Academic Press, London, pp. 139–173.

Gill, C.O., 1996. Extending the storage life of raw chilled meats. Meat Sci. 43, 99–109.

Gill, C.D., Molin, G., 1991. Modified atmospheres and vacuum packaging. In: Rusell, N.J., Gould, G.W. (Eds.), Food Preservatives. Blackie, Glasgow, pp. 172–199.

Gram, L., Huss, H.H., 1996. Microbiological spoilage of fish and fish products. Int. J. Food Microbiol. 33, 121–137.

Gram, L., Melchiorsen, J., 1996. Interaction between fish spoilage bacteria *Pseudomonas* sp. and *Shewanella putrefaciens* in fish extracts and on fish tissue. J. Appl. Bacteriol. 80, 589–595.

Gram, L., Ravn, L., Rasch, M., Bruhn, J.B., Christensen, A.B., Givskov, M., 2002. Food spoilage – interaction between food spoilage bacteria. Int. J. Food Microbiol. 78, 79–97.

Grant, G.F., McCurdy, A.R., Osborne, A.D., 1988. Bacterial greening in cured meats: a review. Can. Int. Food Sci. Technol. 21, 50–56.

Grau, F.H., 1983. End products of glucose fermentation by *Brochothrix thermosphacta*. Appl. Environ. Microbiol. 45, 84–90.

Hansen, A.A., Morkore, T., Rudi, K., Rodbotten, M., Bjerke, F., Eie, T., 2009. Quality changes of prerigor filleted Atlantic salmon (*Salmo salar* L.) packaged in modified atmosphere using CO_2 emitter, traditional MAP, and vacuum. J. Food Sci. 74, M242–M249.

Hernández-Jover, T., Izquierdo-Pulido, M., Veciana-Nogués, M.T., Mariné-Font, A., Vidal-Carou, M.C., 1997. Biogenic amine and polyamine contents in meat and meat products. J. Agricol. Food Chem. 45 (6), 2098–2102.

Hovda, M.B., Lunestad, B.T., Sivertsvik, M., Rosnes, J.T., 2007a. Characterisation of the bacterial flora of modified atmosphere packaged farmed Atlantic cod (*Gadus morhua*) by PCR-DGGE of conserved 16S rRNA gene regions. Int. J. Food Microbiol. 117, 68e75.

Hovda, M.B., Sivertsvik, M., Lunestad, B.T., Lorentzen, G., Rosnes, J.T., 2007b. Characterisation of the dominant bacterial population in modified atmosphere packaged farmed halibut (*Hippoglossus hippoglossus*) based on 16S rDNA-DGGE. Food Microbiol. 24, 362–371.

Huis in't Veld, J.H.J., 1996. Microbial and biochemical spoilage of foods: an overview. Int. J. Food Microbiol. 33, 1–18.

Huss, H.H., 1995. Quality and quality changes in fresh fish Fao Fisheries Technical Paper – 348. <www.uniprom/uniprom/a_indagini/htm>.

Ienistea, 1973. Significance and detection of histamine in food. In: Hobbs, B.C., Christian, J.H.B. (Eds.), The Microbiological Safety of Food. Academic Press, London, pp. 327–343.

Jackson, T.C., Acuff, G.R., Vanderzant, C., Sharp, T.R., Savell, J.W., 1992. Identification and evaluation of volatile compounds of vacuum and modified atmosphere packaged beef strip loins. Meat Sci. 31, 175–190.

Jaffrès, E., Lalanne, V., Mace, S., Cornet, J., Cardinal, M., Serot, T., et al., 2011. Sensory characteristics of spoilage and volatile compounds associated with bacteria isolated from cooked and peeled tropical shrimps using SPME-GC-MS analysis. Int. J. Food Microbiol. 147, 195–202.

Jeremiah, L.E., Gibson, L.L., Argonosa, G., 1995. The influence of inherent muscle quality upon the storage life of chilled pork stored in CO_2 at −1.5°C. Food Res. Int. 28, 51–59.

Jeyasekaran, G., Ganesan, P., Anandaraj, R., Jeya Shakila, R., Sukumar, D., 2006. Quantitative and qualitative studies on the bacteriological quality of Indian white shrimp (*Penaeus indicus*) stored in dry ice. Food Microbiol. 23, 526–533.

Joffraud, J.J., Leroi, F., Roy, C., Bergagué, J.L., 2001. Characterisation of volatile com- pounds produced by bacteria isolated from the spoilage flora of cold-smoked salmon. Int. J. Food Microbiol. 66, 175–184.

Jørgensen, L.V., Huss, H.H., Dalgaard, P., 2000. The effect of biogenic amine production by single bacterial cultures and metabiosis on cold-smoked salmon. J. Appl. Microbiol. 89, 920–934.

Jay, J.M., Loessner, M.J., Golden, D.A., 2005. Modern Food Microbiology, seventh ed. Springer Science and Business Media, New York, NY, pp. 63–101.

Kameník, J., Barto, J., Saláková, A., 2012. Haltbarkeit von unter Schutz atmosphäre ver packtem Schweine fleisch. Fleischwirt 92, 100–104.

Kandler, O., 1983. Carbohydrate metabolism in lactic acid bacteria. Antonie van Leeuwenhoek 49, 209–224.

Kandler, O., Weiss, N., 1986. Bergey's manual of systematic bacteriology. In: Sneath, P.H.A., Mair, N.S., Sharpe, M.E., Holt, J.G. (Eds.), Genus *Lactobacillus*. Williams and Wilkins Co, Baltimore, MD, pp. 1208–1234.

Kanki, M., Yoda, T., Ishibashi, M., Tsukamoto, T., 2004. *Photobacterium phosphoreum* caused a histamine fish poisoning incident. Int. J. Food Microbiol. 92, 79–87.

Kanki, M., Yoda, T., Tsukamoto, T., Baba, E., 2007. Histidine decarboxylases and their role in accumulation of histamine in tuna and dried saury. Appl. Environ. Microbiol. 73, 1467–1473.

Korkeala, H., Suortti, T., Mäkelä, P., 1988. Ropy slime formation in vacuum-packed cooked meat products caused by homofermentative lactobacilli and a *Leuconostoc* species. Int. J. Food Microbial. 7, 339–347.

Labadie, J., 1999. Consequences of packaging on bacterial growth. Meat is an ecological niche. Meat Sci. 52, 299–305.

Lannelongue, M., Hanna, M.O., Finne, G., Nickelson, R., Vanderzant, C., 1982. Storage characteristics of fin-fish fillets (*Archosargus probatocephalus*) packaged in modified gas atmospheres containing carbon dioxide. J. Food Prot. 45, 440–444.

Laursen, B.G., Bay, L., Cleenwerck, I., Vancanneyt, M., Swings, J., Dalgaard, P., et al., 2005. *Carnobacterium divergens* and *Carnobacterium maltaromaticum* as spoilers or protective cultures in meat and seafood: phenotypic and genotypic characterization. Syst. Appl. Microbiol. 28, 151–164.

Leisner, J.J., Laursen, H., Prevost, H., Drider, D., Dalgaard, P., 2007. *Carnobacterium*: positive and negative effects in the environment and in foods. FEMS Microbiol. Rev. 31, 592–613.

Lopez-Caballero, M.E., Sanchez-Fernandez, J.A., Moral, A., 2001. Growth and meta- bolic activity of *Shewanella putrefaciens* maintained under different CO_2 and O_2 concentrations. Int. J. Food Microbiol. 64, 277–287.

López-Caballero, M.E., Gonçalves, A., Nunes, M.L., 2002. Effect of CO_2/O_2 containing modified atmospheres on packed deep water pink shrimp (*Parapenaeus longirostris*). Eur. Food Res. Technol. 214, 192–197.

Lopez-Galves, D.E., De la Hoz, L., Ordonez, J.A., 1995. Effect of carbon dioxide and oxygen enriched atmospheres on microbiological and chemical changes in refrigerated bina (*Thunnus alalunga*) steaks. J. Agric. Food Chem. 43, 483–490.

Lyhs, U., Bjorkroth, J.K., 2008. *Lactobacillus sakei/curvatus* is the prevailing lactic acid bacterium group in spoiled maatjes herring. Food Microbiol. 25, 529e533.

Lyhs, U., Korkeala, H., Vandamme, P., Bjorkroth, J., 2001. *Lactobacillus alimentarius*: aspecific spoilage organism in marinated herring. Int. J. Food Microbiol. 64, 355–360.

Lyhs, U., Koort, J.M., Lundstrom, H.S., Bjorkroth, K.J., 2004. *Leuconostoc gelidum* and *Leuconostoc gasicomitatum* strains dominated the lactic acid bacterium population associated with strong slime formation in an acetic-acid herring preserve. Int. J. Food Microbiol. 90, 207e218.

Macé, S., Joffraud, J.J., Cardinal, M., Malcheva, M., Cornet, J., Lalanne, V., et al., 2013. Evaluation of the spoilage potential of bacteria isolated from spoiled raw salmon (*Salmo salar*) fillets stored under modified atmosphere packaging. Int. J. Food Microbiol. 160, 227e238.

Mäkelä, P., Schillinger, U., Korkeala, H., Holzapfel, W.H., 1992. Classification of ropy slime-producing lactic acid bacteria based on DNA-DNA homology, and identification of *Lactobacillus sake* and *Leuconostoc amelibiosum* as dominant spoilage organisms in meat products. Int. J. Food Microbial. 16, 167–172.

Montel, M.C., Talon, R., Fournaud, J., Champomier, M.C., 1991. A simplified key for identifying homofermentative *Lactobacillus* and *Carnobacterium* spp. from meat. J. Appl. Bacteriol. 70, 469e472.

Nielsen, H.-J.S., 1983. Influence of temperature and gas permeability of packaging film on development and composition of microbial flora in vacuum-packed Bologna-type sausage. J. Food Prot. 46, 693–698.

Nychas, G.J.E., Arkoudelos, J.S., 1990. Microbiological and physico-chemical changes in minced meat under carbon dioxide, nitrogen or air at 3°C. Int. J. Food Sci. Technol. 25, 389e398.

Nychas, G.J.E., Drosinos, E., Board, R.G., 1998. Chemical changes in stored meat. In: Board, R.G., Davies, A.R. (Eds.), The Microbiology of Meat and Poultry. Blackie Academic and Professional, London, pp. 288–326.

Nychas, G.-J.E., Marshall, D., Sofos, J., 2007. Meat poultry and seafood. In: Doyle, M.P., Beuchat, L.R., Montville, T.J. (Eds.), Food Microbiology Fundamentals and Frontiers, third ed. ASM Press, pp. 105–140.

Nychas, G.-J.E., Skandamis, P.N., Tassou, C.C., Koutsoumanis, K.P., 2008. Meat spoilage during distribution. Meat Sci. 78, 77–89.

Olafsdottir, G., Jonsdottir, R., Lauzon, H.L., Luten, J., Kristbergsson, K., 2005. Characterization of volatile compounds in chilled cod (*Gadus morhua*) fillets by gas chromatography and detection of quality indicators by an electronic nose. J. Agric. Food Chem. 53, 10140–10147.

Olofsson, T.C., Ahrne, S., Molin, G., 2007. Composition of the bacterial population of refrigerated beef, identified with direct 16SrRNA gene analysis and pure culture technique. Int. J. Food Microbiol. 118, 233–240.

Ordòñez, J.A., De Pablo, B., De Pablo, P.B., Asensio, M.A., Sam, B., 1991. Selected chemical and micorbiological changes in refrigerated pork stored in carbon dioxide and oxygen enriched atmo- spheres. J. Agric. Food Chem. 39, 668–672.

Paarup, T., Sanchez, J.A., Pelaez, C., Moral, A., 2002. Sensory, chemical and bacteriological changes in vacuum-packed pressurised squid mantle (*Todaropsis eblanae*) stored at 4°C. Int. J. Food Microbiol. 74, 1–12.

Pennacchia, C., Ercolini, D., Villani, F., 2009. Development of a real-time PCR assay for the specific detection of *Brochothrix thermosphacta* in fresh and spoiled raw meat. Int. J. Food Microbiol. 134, 230–236.

Pennacchia, C., Ercolini, D., Villani, F., 2011. Spoilage-related microbiota associated with chilled beef stored in air or vacuum pack. Food Microbiol. 28, 84–93.

Remenant, B., Jaffès, E., Dousset, X., Pilet, M.F., 2015. Bacterial spoilers of food: behavior, fitness and functional properties. Food Microbiol. 45, 45–53.

Reynisson, E., Lauzon, H.L., Magnússon, H., Jónsdóttir, R., Olafsdóttir, G., Marteinsson, V., et al., 2009. Bacterial composition and succession during storage of North-Atlantic cod (*Gadus morhua*) at super chilled temperatures. BMC Microbiol. 9, 250.

Samelis, J., 2006. Managing microbial spoilage in meat industry. In: Blackburn, C.W. (Ed.), Food Spoilage Microorganisms. Woodhead Publishing Limited, Cambridge, pp. 213–286.

Samelis, J., Maurogenakis, F., Metaxopoulous, J., 1994. Characterisation of lactic acid bacteria isolated from naturally fermented Greek dry salami. Int. J. Food Microbiol. 23, 179–196.

Satomi, M., Vogel, B.F., Gram, L., Venkateswaran, K., 2006. *Shewanella hafniensis* sp. nov. and *Shewanella morhuae* sp. nov., isolated from marine fish of the Baltic Sea. Int. J. Syst. Evol. Microbiol. 56, 243–249.

Satomi, M., Vogel, B.F., Venkateswaran, K., Gram, L., 2007. Description of *Shewanella glacialipiscicola* sp. nov. and *Shewanella algidipiscicola* sp. nov., isolated from marine fish of the Danish Baltic Sea, and proposal that *Shewanella affinis* is a later heterotypic synonym of *Shewanella colwelliana*. Int. J. Syst. Evol. Microbiol. 57, 347–352.

Serraino, A., Bardasi, L., Riu, R., Pizzamiglio, V., Liuzzo, G., Galletti, G., et al., 2012. Visual evaluation of cattle cleanliness and correlation to carcass microbial contamination during slaughtering. Meat Sci. 90, 502–506.

Skandamis, P.N., Nychas, G.J.E., 2002. Preservation of fresh meat with active and modified atmosphere packaging conditions. Int. J. Food Microbiol. 79, 35–45.

Silla Santos, M.H., 1996. Biogenic amines: their importance in food. Int. J. Food Microbiol. 29, 213–231.

Silva, A.R., Paulo, E.N., Sant'Ana, A.S., Chaves, R.D., Massaguer, P.R., 2011. Involvement of *Clostridium gasigenes* and *C. algidicarnis* in "blown pack" spoilage of Brazilian vacuum-packed beef. Int. J. Food Microbiol. 148, 156–163.

Stohr, V., Joffraud, J.J., Cardinal, M., Leroi, F., 2001. Spoilage potential and sensory profile associated with bacteria isolated from cold-smoked salmon. Food Res. Int. 34, 797e806.

Tsigarida, E., Nychas, G.J.E., 2001. Ecophysiological attributes of a *Lactobacillus* sp. and a *Pseudomonas* sp. on sterile beef fillets in relation to storage temperature and film permeability. J. Appl. Microbiol. 90, 696–705.

von Holy, A., Cloete, T.E., Dykes, G.A., 1991. Quantification and characterization of microbial populations associated with spoiled vacuum-packed Vienna sausages. Food Microbiol. 8, 95–104.

Vasilopoulos, C., Ravyts, F., De Maere, H., De Mey, E., Paelinck, H., De Vuyst, L., et al., 2008. Evaluation of the spoilage lactic acid bacteria in modified atmosphere packaged artisan-type cooked ham using culture-dependent and culture independent approaches. J. Appl. Microbiol. 104, 1341–1353.

Chapter 9

Review of Vegetable Fermentations With Particular Emphasis on Processing Modifications, Microbial Ecology, and Spoilage

Eduardo Medina-Pradas[1], Ilenys M. Pérez-Díaz[2], Antonio Garrido-Fernández[1] and Francisco Noé Arroyo-López[1]
[1]*Instituto de la Grasa (CSIC), University Campus Pablo de Olavide, Seville, Spain,*
[2]*North Carolina State University, Raleigh, NC, United States*

9.1 BRIEF HISTORY OF FOOD FERMENTATIONS

Fermentation is a primitive preservation method primarily used to enable the long-term storage of foods. The elaboration of bread, cheese, or wine is a tradition introduced thousands of years ago and developed all around the world from rural areas to villages. It is well documented that fermented drinks were consumed in Babylon 5000 years ago (Dirar, 1993) and that bread was consumed in Egypt around the 1500 BC (Suhigara, 1985). Anthropologists even suggest that the production of alcoholic beverages was the cause of settling down in primitive inhabitants, which indirectly allowed the introduction of agriculture. The origin of the fermentation of mixed vegetables is situated in China and associated with the diet of the workers who built the Great Wall in the 3rd century BC (Anderson et al., 1988). Furthermore, the use of *chu*, a fermented grain product, is described in the book *Shu-Ching* (written about 1121–256 BC) (Yokotsouka, 1985). Sauerkraut production was described as early as the 1st century by Plinius the elder (Buckenhüskes et al., 1990). The diverse preparation forms of table olives are also described by Columela in his book *De Re Rustica* in the 1st century (Columela, 54).

The Microbiological Quality of Food. DOI: http://dx.doi.org/10.1016/B978-0-08-100502-6.00012-1

Although extremely beneficial to humankind, the fermentation process remained largely uncharacterized for centuries. Deprived of an understanding of the microbiology behind successful and desirable fermentations, antique generations use the cover brines or doughs from fermentations with desirable attributes to initiate fresh ones, a technique known as back slopping. Processors frequently passed the "high-quality" inocula or "yeast paste" to subsequent generations in an attempt to perpetuate specific organoleptic attributes in the desired fermented foods. The possibility to further control the outcome of a fermentation emerged with the discovery of yeasts as living cells by Anton van Leeuwenhoek in 1680 and Cagnard-Latour in 1838 (Nanniga, 2010). The early perception of yeasts as an inanimate paste was transformed into living cells that convert extracted malt sugars to ethanol and CO_2 in 1838 (Nanniga, 2010). Subsequent studies by Louis Pasteur of a faulty alcoholic fermentation in Lille, France catalyzed the discovery of the role of lactic acid bacteria (LAB) in food fermentations (Brock, 1961). A few years later in 1873, Joseph Lister, a leading surgeon in antiseptic surgery, prepared the very first pure starter culture of a LAB species, by diluting fermented milk (Brock, 1961). Defined fermentative cultures were introduced commercially in New Zealand in 1934 (Cogan and Hill, 1993), beginning the era of "controlled" fermentations.

9.2 PRODUCTION OF FERMENTED VEGETABLES

Under the denomination of fermented vegetables are usually included both fruits, considered as the edible part of a plant or tree, consisting of the seed and its envelope or pulp, and vegetables defined as the plant or root cultivated for food. Fruits are mainly considered acidic foods while vegetables have a lower acidity, which makes them more prone to spoilage by microorganisms. However, most of the vegetable fermentation processes imply acidification by microorganisms, regardless of the type of substrate.

In spite of the fact that fermentation of vegetables is an old tradition everywhere, the system still retains increasing popularity due to a series of favorable factors. Stamer (1988) defines such factors as: (1) the minimal requirement of technical training or knowledge, enabling its use from households to industrial production in any country, (2) the low cost of the process, with a reduced energy consumption, able to preserve in bulk very different commodities for a gradual commercialization, (3) the long-lasting safety record making fermented products very reliable for consumers, and (4) also produces a series of metabolites which contribute to the flavor, texture, and appearance as well as the nutritional (e.g., detoxification or increase in absorption of components) characteristics of the products. In addition, the consideration of vegetable fermented products is in agreement with the current trend for more natural products with potential beneficial effects on health, such as decrease of cholesterol, risk of certain types of cancer, modulation of the immunologic system, or probiotic effect. Thus, the prospects for the future of fermented vegetables are promising.

The consumption of vegetables is widespread in the world and represents an important component of the human diet. China is the largest producer and the probable origin of pickles (Fleitag, 2012). In the European Union (EU) the canned pickles, olives, and sauerkraut have been valued at approximately €1746, €1123, and €67.0 million, respectively (CBI, 2009). The largest producer in the EU is Germany with values reaching 800 million euros for pickles (mainly cucumbers). Kimchi (cabbage fermented in garlic, chillies, and vinegar) is also very popular in Asia, being South Korea's national dish (per capita consumption around 24 kg); however, production still relies on household or small-scale users (56%). There are also many other fermented vegetable products all around the world, for which production statistics are not detailed in this report.

9.3 GENERAL DESCRIPTION OF THE MICROBIOLOGY OF FERMENTED VEGETABLES

The commonality of fermented vegetables is their dependence on the growth of LAB, which results in the production of acidity and a decrease in pH. These changes, together with the presence of salt, are the essence of the production of stable and safe fermented vegetables. Total aerobic microbial counts in fresh cucumbers and cabbage may reach 5 $\log_{10}$ CFU/mL (Pérez-Díaz et al., 2015a). It is well known that the community of Enterobacteriaceae, LAB, and yeasts dominate the natural microbial population of vegetable fermentations. The natural microbial population in olives is a fraction of that found in cucumbers and cabbage, presumably due its antimicrobial phenolic content. Most microbes colonize the outermost interior of vegetables (Mattos et al., 2005; Samish et al., 1963). Microbial counts for fresh produce vary with seasons, maturity stage, environmental humidity, temperature, and the use of pesticides among other factors. The diversity of the microbial population of vegetables is drastically reduced as the result of fermentation, which supports the safety record of such preserves. The various Gram-positive and Gram-negative bacteria and yeasts and molds naturally present in fresh vegetables compete for dominance. Enterobacteriaceae, aerobic spore-formers, LAB, and other groups of bacteria and yeasts may be active for several days or weeks depending on factors such as temperature, dissolved oxygen, and the salt concentration used in the cover brines. Very often LAB gain predominance due to their ability to produce lactic acid inducing a drop in pH tolerable by them. Six species of lactic acid-producing bacteria may be naturally and actively present in many vegetable fermentations: *Leuconostoc mesenteroides*, *Pediococcus pentosaceus*, *Pediococcus acidilactici*, *Lactobacillus brevis*, *Lactobacillus plantarum*, and *Lactobacillus pentosus*. Recent studies suggest that *Weissella* spp. are also present during the early stages of sauerkraut fermentation (Plengvidhya et al., 2007). Generally, *L. mesenteroides* survives best in vegetable fermentations with less than 2% NaCl, while *L. plantarum* and *L. brevis* are more resistant to the acidic pH (McDonald et al., 1990). The genome sequences for the main LAB present in vegetable fermentations, *L. plantarum* and *L. pentosus*, are available. The first complete genome sequence for a *L. plantarum* isolate became available in 2003 (Kleerebezem et al., 2003). The human saliva isolate was found to contain putative genes encoding the necessary components of a facultative heterofermentative LAB, and several regulatory and transport-related systems. A study of the genetic diversity among 185 *L. plantarum* isolates obtained from various niches identified 2000 genes as the core genome for this species, including 121 unique *L. plantarum*-marker genes absent in other LAB. More than 50 genes found in the reference *L. plantarum* genome published in 2003 were unique to the saliva isolate (Siezen et al., 2009). Further analysis of multiple *L. plantarum* genome sequences showed the existence of a genomic lifestyle islands consisting of numerous functional gene cassettes related to carbohydrate utilization that may be acquired, shuffled, substituted, or deleted in response to extrinsic environmental factors (Siezen and van Hylckama Vlieg, 2011). The genome sequence for *L. pentosus* MP-10 isolated from naturally fermented Aloreña green table olives with potential as probiotic was published in 2011 (Abriouel et al., 2011). The *L. pentosus* genome sequence has significantly more similarity with the *L. plantarum* than the *L. fermentum* counterparts. It was determined that the *L. pentosus* genome is shaped by horizontal gene transfer mediated by mobile genetic elements. Later it was learned that *L. pentosus* IG1 possess one of the biggest lactobacilli genomes (Maldonado-Barragán et al., 2011). This strain was found to encode for 16 two-component regulatory systems suspected to aid in adapting to changing extrinsic conditions such as those characteristics of active fermentations. The genes encoding for the two-component class IIb plantaricin were found to be identical to open reading frames found in the *L. pentosus* IG1 genome sequence. A comparison of the stress response systems among *L. pentosus*, *L. plantarum*, and *L. paraplantarum* revealed that there is a generalized response in stationary phase to increases in oxidative, heat, and starvation stress (Ricciardi et al., 2012).

A limited number of fermented vegetable processors choose to use starter cultures of *P. pentosaceus/acidilactici* or *L. plantarum* to enhance product consistency and the efficiency of the fermentation as recommended by Etchells et al. (1964, 1973) for cucumber fermentations and Vega Leal-Sánchez et al. (2003) for the olive counterpart. Utilization of *L. mesenteroides* as a starter culture for low salt sauerkraut fermentations has also been evaluated (Johanningsmeier et al., 2007). A method for the preparation of *L. plantarum* and other LAB starter cultures that meets kosher requirements for vegetable fermentations was developed by Pérez-Díaz and McFeeters (2011). The performance of specific lactobacilli as starter cultures in selected vegetable fermentation has been evaluated and is described below. Various species of fermentative yeasts are also typically present in selected vegetable fermentations. Yeasts were implicated in cucumber fermentations in 1941. Salt concentrations in the cover brine were found to be responsible for the extent of the yeasts growth lag and logarithmic phases (Etchells, 1941). Yeast cells are primarily found on the skin of the cucumber fruits, exposed mesocarps, and the fermentation cover brines. The size of yeast cells prevents their penetration into the mesocarp through the fruits skin (Daeschel et al., 1985). Predominating yeasts genera in cucumber fermentations include *Torulopsis*, *Brettanomyces*, *Zygosaccharomyces*, *Hansenula*, *Torulaspora*, and *Kloeckera* (Etchells and Bell, 1950). Oxidative yeasts of the genera *Debaryomyces*, *Pichia*, and *Mycoderma* have also been associated with the formation of a film on the surface of cucumber fermentation cover brines, popularly known as a scum by film-forming yeasts (Etchells and Bell, 1950; Mkar and Bonar, 1939). The fermentation of table olives is known to support the proliferation of *Candida*, *Pichia*, and *Saccharomyces* (a review by Botta and Cocolin, 2012). Such yeasts are known to contribute to the definition of the final organoleptic profile of fermented table olives due to their ability to produce volatile compounds. The utilization of selected yeast cultures as adjunct cultures for vegetable fermentations is been currently evaluated.

9.4 DESCRIPTION OF THE PROCESSING AND MICROBIAL ECOLOGY OF THE MAIN TYPES OF FERMENTED VEGETABLES

A concise description of the processing peculiarities and microbiology of the most relevant fermented vegetables are included below.

1. *Cucumbers (Cucumis sativus).* This vegetable is one of the most important fermented vegetables in the United States, with an average production of around 480,000 tons in 2012–14 and an economic value of between 145 and 175 million dollars (USDA, 2015a). Usually, the fermentation is carried out in large open-top fiberglass tanks to prevent the growth of surface yeasts and molds by the ultraviolet sunlight radiation. The cucumbers are covered with fresh (or recycled brine) containing NaCl and $CaCl_2$ in the adequate proportions to reach 6% and 0.1–0.4%, respectively, at equilibrium. The presence of calcium is recommended for the retention of texture and crispness of the fruits. The main carbohydrates used during the fermentation are fructose and glucose (about 1%) and malic acid. After brining, it is imperative to achieve the removal of the CO_2 produced by both the fruit's respiration and, the gradual malolactic metabolism by LAB, to avoid the formation of bloaters. Although air purging is widely utilized at the commercial-scale production to reduce the incidence of bloating, it also introduces oxygen to the system, which leads to the growth of undesired aerobic organisms such as yeasts and molds, among others. Thus, the incorporation of sorbic acid or benzoic acid in the fermentation cover brines is recommended. The

fermentation is usually spontaneous despite the development of specific malic acid decarboxylase deficient starter culture, recommended for the reduction of the incidence of bloaters (Daeschel et al., 1984). After fermentation, the lactic acid concentration may be about 2% and the pH between 3.1 and 3.5. These conditions, together with the salt and other metabolites produced during fermentation, stabilize the fermented fruits for long-term bulk storage for up to 1 year, which is processed into finished products in line with the market demand. A review of the biochemistry of vegetable fermentation and the genomic research on the LAB involved in them, especially for cucumbers, was presented by Breidt et al. (2013b). As deduced from the description of the cucumber fermentation process described above, a large proportion of waste-waters are generated from the fermentation cover brines and the desalting steps are necessary in order to decrease the sodium chloride concentration from 6% in bulk storage to less than 2% in the finished products to achieve the edibility threshold. Modifications to the traditional cucumber fermentation process have been suggested recently and are being currently optimized for commercial-scale production. One of the most promising modifications is the total substitution of the NaCl by $CaCl_2$. The fermentation biochemistry observed in the modified system brined with calcium chloride is similar to that observed in the traditional fermentation described above (McFeeters and Pérez-Díaz, 2010; Pérez-Díaz et al., 2015b). Due to the high amount of calcium Ca^{++}, the texture is also maintained (McFeeters and Pérez-Díaz, 2010). Further studies on the consumer acceptability of cucumber pickles brined with $CaCl_2$ showed that there was no difference if bulk storage happens for 2–8 months (Wilson et al., 2015). However, a significant preference for cucumber pickles brined with NaCl was observed if a 10-month storage period was considered. It was further concluded that the presence of Ca ions was not the cause of such rejection after 10 months of bulk storage, given that the amount of the calcium salt present in the finished product was well below the taste threshold for $CaCl_2$ at 62 mM. Cucumber fermentations brined with $CaCl_2$ instead of NaCl are a viable process at the commercial scale with a single desalting step and reduced environmental impact from waste waters (Wilson et al., 2015).

2. *Sauerkraut (Brassica oleracea)*. The name is a compound German word derived from *sauer* meaning sour and *kraut* meaning cabbage. In the United States, its production during 2012–14 was about 140,000 tons with an economic value of approximately 10 million dollars (USDA, 2015b). In Germany, production of sauerkraut reached around 100 million euros in 2011 (Fleitag, 2012). It is produced from the sweet and white cabbage varieties, specifically developed for the commercial process, due to their large and dense heads with a minimum of green outer leaves. The sugar content of cabbage is especially high (about 5%), and

mainly composed of fructose and glucose. Also, the raw material is characterized by the presence of a series of inhibitory compounds such as glucosinolates that by hydrolysis produce isothiocyanates, nitriles, and thiocyanates. Processing is relatively simple. After removing the outer leaves, the head is sliced into shreds <0.2 cm, depending on available machinery or tradition, although long, fine shreds are preferred. Then, the shreds are transported into the fermentation tanks (usually concrete tanks, covered by fiberglass, with diverse capacities of up to 180 tons), and are then mixed with the appropriate solid salt so that the final equilibrium reached is around 2.5%. The brine is formed with the liquid extracted by the salt. Finally, the tanks are covered with plastic to prevent the presence of oxygen and to keep the shreds submerged. Fermentation by LAB is initiated immediately after the cabbage is brined. The lactic acid producing microbiota is initially composed of *L. mesenteroides*, *L. citreum*, and *Weissella* spp. which decrease the pH to approximately 6.5 (Plengvidhya et al., 2007). These species are progressively substituted with predominantly *L. plantarum*, *L. curvatus*, and other lactobacilli, which continue to drop the pH to about 4.5 (Plengvidhya et al., 2007). *Lactobacillus brevis* ends the production of acids spearheading the final decrease in pH to $\sim$4.0. *Leuconostoc* and *Weissella* spp. convert the cabbage-derived fructose and glucose heterofermentatively to lactic and acetic acids, and CO_2. In general, adequate progression of cabbage primary and secondary fermentations by the various LAB species is sufficient to stabilize and preserve it preventing microbial spoilage. Although the fermentation may be completed in $\sim$20 days, bulk storage usually occurs in the same tanks until the sauerkraut is processed into finished products according to consumer demand (Fleitag, 2012). The use of selected starter culture of *L. plantarum* and *L. mesenteroides* to control the process improved ascorbigen content (x12), antioxidant (x2.0), and nitric oxide production (2.6). The product obtained with the supplementation of fermentations with starter cultures are known to enjoy higher acceptability by consumers and improved nutritional characteristics (Martinez-Villaluenga et al., 2012). The reduction of the nitrite content in Chinese cabbage using selected starter cultures of *Lactobacillus delbrueckii* and *Lactobacillus paracasei* has also been recently documented (Han et al., 2014). Fermentation of cabbage causes a dramatic degradation of glucosinolates between the second and fifth day of the process, while increasing the concentration of potential bioactive components (ascorbigen and indol-3-carbinol) (Palani et al., 2016).

3. *Table Olives (Olea europaea).* World production of table olives was around 2,600,000 tons in the 2013/14 season, with Spain as the main producer with 25% of the total production (IOC, 2015). There are different types of table olives (green, turning color, and naturally black) which can be debittered using an alkaline solution (treated olives) or by dilution of the bitter compound into the brine (natural). All of them are subjected to fermentation processes after brining. However, the so-called ripe olives (Californian style) are stored in brine or acid solutions just while waiting for the final oxidation process, which is the most important processing step. The most popular style is the green Spanish-style table olives (which account for about 60% of the market) followed by natural black olives (Greek-style). For the Spanish-style, olives are harvested at the green maturation stage, usually by hand (although research into the introduction of mechanical harvesting is in progress). After transportation in bulk 1000 kg bins, the fruits are treated with a NaOH solution (~2.5%, w/v) until the alkali reaches 2/3 of the flesh. Then, the lye is removed and the olives are washed with tap water for 18–24 hours to eliminate the excess alkali. After washing, the olives are submerged in a NaCl solution (10–12% w/v) for a few hours and finally are sent to the fermentation fiberglass tanks with a total capacity of about 16,000 L. Acids such as HCl, lactic acid, and acetic acid are occasionally added to the fermentation cover brines. The fermentation begins just after brining. The first microorganisms to grow are Enterobacteriaceae, followed by LAB, specifically and predominantly *L. plantarum* and *L. pentosus* (De Angelis et al., 2015). Yeasts and LAB may coexist in brined olives (Garrido Fernández et al., 1997). It is relevant to mention that the use of bulk tanks for table olive fermentations have indirectly modified the predominant microbiota, inducing the replacement of lactic acid producing cocci such as *Pediococcus* and *Leuconostoc* to the rods *L. plantarum* or *L. pentosus* (Botta and Cocolin, 2012). The use of *L. plantarum* LPCO10 as a starter culture and of *Enterococcus casseliflavus* cc45 and *L. pentosus* 5138A in sequential inoculations has proven effective in accelerating acid production, and the die-off of pathogenic microbes as compared to spontaneous fermentations (De Castro et al., 2002; Leal-Sánchez et al., 2002; Vega Leal-Sánchez et al., 2003). Application of recently-developed molecular methods have shown that the biodiversity of the microbiota in green Spanish-style fermented olives is higher (Lucena-Padrós et al., 2014) than previously estimated. The fermentation of green Spanish-style olives is mainly spontaneous; however, the formation of biofilms on the olive surface (Dominguez-Manzano et al., 2012) may favor the selection and use of appropriate starter cultures for the making of probiotic olives (Rodríguez-Gómez et al., 2014). For preparing natural black olives, the fruits should be completely ripe but not overripe. Traditionally, the olives are placed in cement open tanks protected with polyester resins. However, they are being progressively substituted with fiberglass tanks with 10-ton capacity, similar to those used for green olives. The brine used has 8–10% (w/v) NaCl concentration or lower (~6% NaCl) as is the case in cold areas. As the black olives are not lye treated, the leakage of nutrients into the brine and the debittering of these olives is slow and, therefore, fermentation is rather long. In fact, the presence of residual fermentable material in the brine of these olives may be observed during the entire fermentation. Fermentation of naturally black olives is mainly achieved by yeasts with a weak lactic acid production, although the growth of LAB is also possible in the presence of low salt levels in cultivars with reduced content of polyphenol. In Spain, the traditional anaerobic process has been substituted with an aerobic fermentation to prevent the formation of gas-pockets in some cultivars (Hojiblanca and Manzanilla) prone to such defects. The modification aims at an increase in facultative yeasts (Garrido-Fernández et al., 1997). Starter cultures are also being developed for this type of olive fermentation (Bleve et al., 2015). The adequate selection of mixed cultures of LAB and yeast for starter cultures could lead to the development of probiotic natural black olives, due to the development of biofilms on the fruits (Rodríguez-Gómez et al., 2014). Combinations of native yeasts and *L. plantarum* have been tested as starter cultures for Bella di Cerignola table olives and have been shown to significantly reduce the biodiversity of the microbiota during the active fermentation period, to modify the concentration of free amino acids and phenolic and volatile compounds, and to generate finished products with increased consumer acceptability (De Angelis et al., 2015).

4. *Other vegetable products*. There are numerous fermented vegetable products that are traditional in specific countries and popular in some other areas around the world. Among them are kimchi, sour onion, mixed vegetables, capers (from *Capparis* sp.) and caper berries. *Capers* and *caper berries* are very common in the counties around the Mediterranean basin. The traditional process for the manufacturing of capers consists of a pretreatment in highly concentrated cover brines, followed by packaging in a 6% NaCl (w/v) brine containing 1% acetic acid (Alvarruiz et al., 1990). The initial step for the making of fermented caper berries consists of storage in water for 4 to 7 days. The water is then substituted with a 10–12% cover brine, the concentration of which is gradually increased to 15% to assure product stability until commercialization (Sánchez et al., 1992). The fermentation of these products consists of lactic acid production by LAB such as *L. plantarum*, which predominates, *L. paraplantarum*, *L. pentosus*, *L. brevis*, *L. fermentum*, *P. pentosaceus*, *P. acidilactici*, and *Enterococcus faecium* (Pérez Pulido et al., 2005). The use of a *L. plantarum* starter culture for caper berry fermentation induces a consistent process and faster sugar catabolism (Palomino et al., 2015). The antioxidant compounds quercetin, kaempferol, rutin, kaempferol-3-O-β-rutinoside and N(1),N(5),N(10)-triphenylpropenoyl spermidine amides have been identified as the most relevant in caper buds extract (Wiese et al., 2013). Although the actual origin of *kimchi* is still disputable, it is the traditional fermented Korean vegetable food, in particular when fermented with red paper (Jang et al., 2015). The name is not related to an individual vegetable, but to a range of fermented products. The two main raw materials used in the making of kimchi are selected cultivars of Chinese cabbage and radish. In contrast to the sauerkraut preparation, the making of kimchi requires half-cuts or quartered brined cabbage. The fermentation of kimchi proceeds at 18°C for a few days, followed by a longer incubation period at refrigerated temperatures to promote microbial stability and the development of excess sourness. Such temperature control provides the advantageous proliferation of heterofermentative *Leuconostoc* spp. at the outset, followed by the growth of homofermentative lactobacilli and *Weissella* spp. (Jung et al., 2012). The use of *L. citreum* as a starter culture for kimchi fermentation has proven to prevent over-ripening and growth of yeasts during refrigerated storage (Chang and Chang, 2010). The health benefits derived from the LAB isolated from kimchi fermentations, in particular with respect to the immunomodulatory effects, have been recently described and mimic those identified for bacteria isolated from dairy products (Choi et al., 2015). The application of metabolomic and transcriptomic tools to study the dynamics of kimchi fermentation suggests that the main changes occur in relation to amino acids, organic acids, and sugars (Jung et al., 2013; Park et al., 2016). Other vegetable fermentations, for which popularity is currently increasing, is that of carrots, onions, and garlic. Carrot lactic acid fermentation is achieved when a lye treatment is applied to a system containing 2.5% NaCl and a mixed starter culture of *L. plantarum* and *Saccharomyces oleaginosus* (Montaño et al., 1997). A mixed LAB starter culture is known to lead a more complete sugar consumption with a concurrent higher acidity as compared to spontaneous fermentation of mixed vegetables including carrots, cabbage, beet, and onions (Gardner et al., 2001). Lactic acid garlic fermentation (3% NaCl) is only possible after blanching the vegetables and yields products with a similar flavor to those packed prior to processing (De Castro et al., 1998). Inhibition of LAB in nonblanched garlic may be associated with the antimicrobial activity of thiosulfinates (Yoshida et al., 1999). However, the same thiosulfinates are ineffective against the abundant population of Gram-negative rods heterofermentative LAB found on prepeeled garlic (Shim and Kyung, 1999). Fermented aged black garlic has shown high favorable bioactivity against induced obesity, hyperlipidemia nephropathy, and hepatopathy, apparently potentiated by yeast fermentation (Jung et al., 2011). The garlic fructan has also been identified as an effective prebiotic component, suggesting the beneficial use of garlic to prevent selected gastrointestinal diseases (Zhang et al., 2013). The anaerobic lactic acid fermentation of onions requires supplementation with sauerkraut cover brine, or slices of cabbage as a source of inocula (Roberts and Kidd, 2005). An insufficient or ineffective natural population of LAB in onions seems to preclude spontaneous fermentation. The fermented product known as sour-onion has a tart acidic taste similar to that which is characteristic of sauerkraut, with the basic onion flavor but without the pungency of raw onion (Roberts and Kidd, 2005).

9.5 SPOILAGE OF FERMENTED VEGETABLES

As mentioned previously, microorganisms play an important role in vegetable fermentations, affecting the quality and safety of the final products. The main alterations suffered by fermented vegetables are described in Table 9.1.

TABLE 9.1 Main Microbial Alterations in Fermented Vegetables and Causative Agents

Type of Alteration	Product Affected	Microorganisms Involved	References
Gas pocket or "alambrado"	Table olives	*Enterobacter* sp.	Garrido-Fernández et al. (1997)
		Citrobacter sp.	
		Escherichia sp.	
		Klebsiella sp.	
		Aeromonas sp.	
		Saccharomyces kluyveri	Duran-Quintana et al. (1979)
		S. cerevisiae	
		W. anomalus	
	Cucumbers	*L. brevis*	Fleming et al. (1975)
Putrid fermentation	Table olives	*Desulfovibrio aestuarii*	Levin and Vaughn (1966)
		Clostridium sp.	Gililland and Vaughn (1943)
Butyric fermentation	Table olives	*C. butyricum*	
		Clostridium beijerinckii	
		Clostridium fallax	
	Cucumbers	*C. bifermentans*	Franco et al. (2012)
Zapatería	Table olives	*Propionibacterium* sp.	Kawatomari and Vaughn (1956)
		Clostridium sporogenes	
		C. bifermentans	

(*Continued*)

TABLE 9.1 (Continued)

Type of Alteration	Product Affected	Microorganisms Involved	References
Lactic acid depletion	Cucumbers	*L. buchneri*	Johanningsmeier and McFeeters (2013)
		Acetobacter pasteurianus	Medina et al. (2016)
		Acetobacter sp.	
		Pectinatus sp.	Breidt et al. (2013a)
		Propioni bacterium sp.	
		E. cloacae	Franco et al. (2012)
		P. manshurica	Franco et al. (2012)
		I. occidentalis	
	Table olives	Oxidative yeasts	Ruiz-Cruz and Gonzalez-Cancho (1969)
Softening	Table olives	*R. minuta*	Hernandez et al. (2007)
		W. anomalus	Arroyo-López et al. (2012)
		D. hansenii	
	Table olives and cucumbers	*Aerobacter*	King and Vaughn (1961)
		Aeromonas	
		Achoromobacter	
		Escherichia	
		Paracolobactrum	
	Cucumbers	*Alternaria*	Costilow et al. (1980)
		Fusarium	
		Mucor	Etchells et al. (1958)
		Filamentous fungi	
	Kimchi	*P. kudriavzevii*	Moon et al. (2014)
White spots	Table olives	LAB species	Vaughn et al. (1953)
Pink/red pigmentation	Sauerkraut	Yeast species	Fred and Peterson (1922)
	Cucumbers	LAB species	Díaz-Muñiz et al. (2007)

Besides LAB and yeast, typically found in vegetable fermentation, Gram-negative bacteria of the *Enterobacter*, *Citrobacter*, and *Escherichia* genera are usually present in the first stage of vegetable fermentations. Such microbial communities decrease during the first few days of fermentation due to the production of lactic acid with a concomitant drop in pH (Breidt and Caldwell, 2011; Garrido-Fernández et al., 1997). If the decline in pH is protracted, Gram-negative bacteria such as Enterobacteriaceae may derive energy for growth from the metabolism of sugars, producing CO_2 and compromising the quality and safety of the preserve (Garrido-Fernández et al., 1997; West et al., 1941). Formation of CO_2 in vegetable fermentations leads to defects associated with significant economic losses. CO_2 can accumulate below the epidermis of the fruits, forming superficial or internal gas-pockets (*alambrado*) in table olives (Garrido-Fernandez et al., 1997) (Fig. 9.1) and cucumbers (Fleming et al., 1975). Formation of gas-pockets is

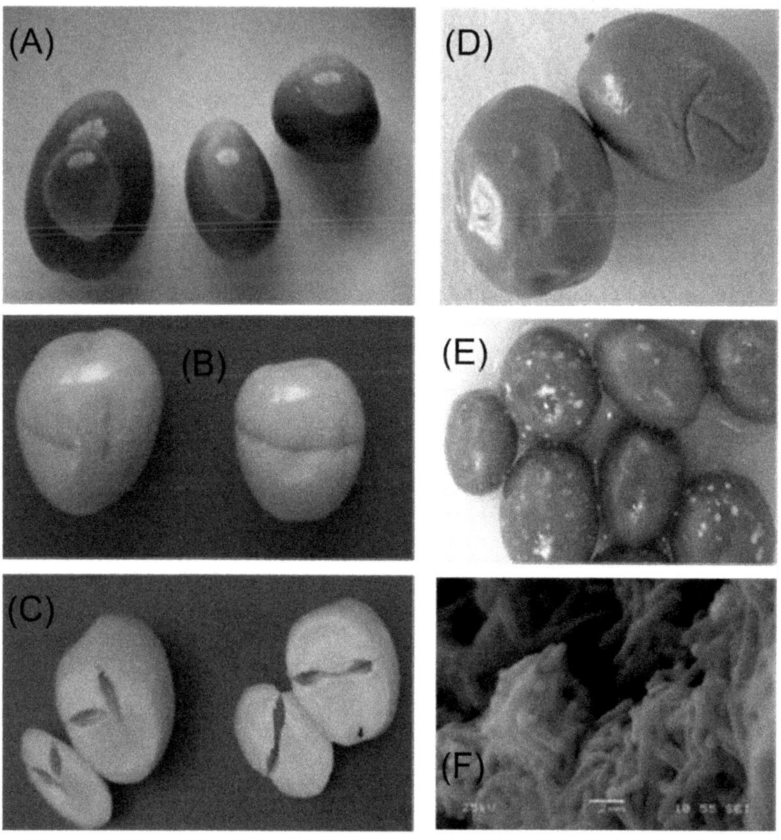

FIGURE 9.1 Olive fruits showing diverse types of spoilage: cuticular (A) and intramesocarpic gas-pockets (B and C), softening (D), presence of white spots on surface (E) and amplified SEM image of the stomatal openings (F). *Pictures courtesy of Dr. Antonio Castro, Dr. Pedro García and Dr. Francisco Rodriguez from Instituto de la Grasa (CSIC).*

the most frequent defect in the production of table olives and cucumber pickles. It may be also induced by the production of CO_2 by yeasts, heterofermentative LAB and homofermentative LAB able to catabolize malic acid, by the respiration of the fresh fruits prior to brining during storage without refrigeration, and by the fruits growing conditions in the field (Etchells et al., 1968, 1975; Duran-Quintana et al., 1979; McFeeters et al., 1984). The use of a nitrogen purging system, a correction of the pH at the beginning of the fermentation process, and the use of preservatives against yeasts would be an effective means with which to control this type of defect (Asehraou et al., 2002; Fleming et al., 1975; Garcia-Garcia et al., 1982).

The richness of nutrients and close to neutral pH existing at the beginning of lactic acid vegetable fermentations may contribute to the development of putrid and butyric fermentations (Gililland and Vaughn, 1943). The main characteristic of putrid fermentation is reminiscent of the smell of decomposing organic matter by the metabolisms of *Clostridium* spp. They are spore-forming anaerobic bacteria able to colonize the bottom of tall fermentation vessels, where anaerobic conditions can develop in the absence of purging. *Clostridium butyricum* in particular has been associated the development of butyric fermentations with a rancid butter smell (Gililland and Vaughn, 1943). *Desulfovibrio* spp. have also been isolated from putrid fermentations characterized by the production of sulfhydric compounds (Levin and Vaughn, 1966). Lack of sanitation practices for fermentation vessels, or the use of contaminated water, are the main causes of this type of spoilage.

Upon completion of a lactic acid vegetable fermentation, physicochemical conditions, high acidity, and low pH ensure the preservation of the product. However, sometimes, an undesired secondary fermentation or spoilage may be initiated by propionic acid bacteria, in particular *Propionibacterium* spp. This group metabolizes sugars or the lactic acid form during the primary fermentation, to produce propionic acid, acetic acid, and CO_2, inducing an increase in pH and volatile acidity (Gonzalez-Cancho et al., 1980). These conditions also encourage the development of *Clostridium* species, which together with *Propionibacterium* can promote *zapatería* spoilage, giving abnormal odors and tastes in table olives (Kawatomari and Vaughn, 1956; Plastourgos and Vaughn, 1957). The resulting higher pH enables the growth of other spoiling microbes such as the spore-forming *Clostridium botulinum*, responsible for the production of the lethal botulinum toxin (Medina-Pradas and Arroyo-López, 2015). Control of pH and salt concentration in brine would prevent the growth of these spoilage microorganisms and their off-odor fermentations, especially when temperatures are warmer during the summer months (Gonzalez-Cancho et al., 1970). In spite of lactic acid depletion, the propionic acid generated by propionic acid bacteria results in desirable attributes in other fermented foods, in particular sauerkraut (Babuchowski et al., 1999). Propionic acid is a potent mold inhibitor, preventing the growth of pathogenic fungi able to produce off-flavors and hazardous substances.

Undesired secondary fermentation or spoilage is also common in cucumber fermentations. The presence of acetic acid bacteria (AAB) and *Lactobacillus* spp. seems to play a relevant role in the initiation of fermented cucumber spoilage (Johanningsmeier and McFeeters, 2013; Medina et al., 2016). AAB can convert lactic acid to acetic acid. *Lactobacillus buchneri* are able to produce acetic acid and 1,2-propanediol from lactic acid during the first stage of the undesired secondary fermentation (Johanningsmeier and McFeeters, 2015). Should the undesired secondary fermentation enable the increase in pH above 4.2, *Propionibacterium* and *Pectinatus* species, and *Clostridium bifermentans* and *Enterobacter cloacae* are able to convert lactic acid to propionic acid (Breidt et al., 2013a) and produce butyric and propionic acids, respectively (Franco and Perez-Diaz, 2013). Production of such organic acids would explain the cheesy and manure-like aroma characteristic of the spoilage of fermented cucumbers.

Yeast populations coexist with the presence of LAB during vegetable fermentation. In fact, yeasts predominate in certain types of elaboration such as directly brined (natural) olives. The roles of yeasts in the processing of fermented vegetables are associated with the development of taste and aroma (Arroyo-López et al., 2012) However, yeasts may be the culprit of certain types of spoilage. *Pichia kudriavzevii* is considered as the main yeast responsible for kimchi spoilage due to superficial film formation, off-odor production, and bad-tastes (Moon et al., 2014). An excessive growth of fermentative yeast, such as *Saccharomyces cerevisiae* and *Wicherhamomyces anomalus*, could lead to a vigorous production of CO_2 that may damage the fruits, due to the formation of gas pockets or *alambrado* (Duran-Quintana et al., 1979; Garrido-Fernandez et al., 1997; Vaughn et al., 1972). Moreover, food spoilage can result by oxidative yeasts which can consume the lactic acids produced during olive fermentation under aerobic conditions (Ruiz-Cruz and Gonzalez-Cancho, 1969). *Pichia manshurica* and *Issatchenkia occidentalis* have the ability to use lactic and acetic acids during aerobic metabolism with the concomitant increases in pH, and have been associated with the initiation of spoilage of fermented cucumbers (Franco et al., 2012).

Another unfavorable property of some yeast strains in vegetable fermentation is the ability of enzyme productions that could cause the softening of fruits, such as proteases, xylanases, and pectinases (Fig. 9.1). Some yeast strains of *Rhodotorula minuta*, *W. anomalus*, and *Debaryomyces hansenii* have been described in green table olive fermentation with these capacities (Hernandez et al., 2007), and *P. kudriavzevii* can lead to softening in kimchi deterioration (Moon et al., 2014). Also, pectinolytic molds belonging to the genera *Alternaria*, *Fusarium*, and *Mucor* are responsible for the softening of brined cucumbers (Costilow et al., 1980). The metabolism of pectinolytic and cellulolytic yeast acts on pectic substances, cellulose, hemicelluloses, and polysaccharides, which damage the cell walls associated with the softening of the fruit (Arroyo-Lopez et al., 2012; Golomb et al., 2013). Also, certain pectinolytic Gram-negative bacteria such as *Aerobacter*, *Aeromonas*, *Achoromobacter*, *Escherichia*, and *Paracolobactrum* produce the softening deterioration and the skin sloughing of the fruit in black olive oxidation process when the temperature is high (Garrido-Fernandez et al., 1997; King and Vaughn, 1961; Vaughn et al., 1969).

Regardless of the habitual low pH and high salt concentration obtained in the final product, packed fermented vegetables can suffer spoilage by microorganisms if residual nutrient concentration is present in the packaging (Garrido-Fernandez et al., 1997). Thermal treatments are an effective method for increasing the microbiological stability of final products. In some products, such as cracked green olives, thermal treatments are not appropriate and result in loss of organoleptic characteristics (Abriouel et al., 2014). Also, residual sugar content can be found in partial lactic acid fermentation, with the subsequent risk of postfermentation. Spoilage microorganisms in packaging are the same as those found in a typical fermentation, LAB and yeasts, and their growth can reach high population levels producing clouding of brines (Arroyo-Lopez et al., 2012). Moreover, yeasts in table olives can also consume the lactic and acetic acids produced during fermentation, under aerobic conditions (Ruiz-Cruz and Gonzalez-Cancho, 1969), suggesting that the development of yeast is stabilized under anaerobic conditions.

Other common alterations in table olive processing is the presence of white spots (see Fig. 9.1). This abnormality is very common in Spanish-style green olives where a vigorous lactic fermentation is carried out by LAB. Vaughn et al. (1953) noticed that the small white spots are colonies of LAB species established in stomatal openings.

Pink sauerkraut is one of the most common defects in the production of that commodity. Although pink sauerkraut has been reported to be edible and is often sold at a lower price, it has been related to changes in texture, flavor, and odor (Fred and Peterson, 1922). Pink sauerkraut is caused by the yeasts naturally present in the cabbage and favored by the concentration and type of sugars present in the cabbage, sodium chloride concentrations above 3%, high acidity, and extrinsic factors such as temperature and the supply of oxygen (Fred and Peterson, 1922). *Lactobacillus brevis* has also been associated with the formation of a water-soluble red pigment in sterile cabbage juice favored at pH values between 5.2 and 6.3 and suppressed by anaerobic conditions (Stamer et al., 1973). Sporadic red-colored fermented cucumber spoilage cases have also been documented within the US cucumber pickling industry. *Lactobacillus casei* and *L. paracasei* have been implicated in such types of spoilage due to their ability to degrade the azo dye tartrazine (FD&C yellow nr 5) used as a yellow coloring in cover brines (Díaz-Muñiz et al., 2007). This type of spoilage can be prevented by the addition of 0.1% sodium benzoate.

9.6 CONTROL MEASURES TO REDUCE ALTERATION OF FERMENTED VEGETABLES

In many industries, vegetable fermentation still occurs spontaneously. Thus, the process is not fully predictable and sometimes can lead to spoilage. Alteration of vegetable packaging can also occur if residual sugar content and microorganisms are present in the finished product. To prevent these problems, the elaboration processes can be controlled through physicochemical or microbiological approaches. Below, we detail the most important factors with easy implementation and control in the industry.

1. *Salting* and *acidification* are two common factors in many vegetable - fermentations, which convert the system into an adverse habitat for spoilage and pathogenic microbes. Salt, mainly sodium chloride, determines the flavor of the final products, reduces the water activity and consequently influences the type and extent of microbial metabolism, indirectly helping to prevent the softening of vegetable tissue, and facilitating the diffusion of components from the fruit tissues into the cover brines. As an example, levels of salt higher than 8% (wt/vol.) are necessary in table olives for long-term bulk storage in order to inhibit spoilage by *Propionibacteria* species (Garrido-Fernández et al., 1997). The pH and/or acidity of a food are generally used to determine processing requirements and applicability of specific regulations. Reduction of pH as the result of fermentation or acidification with organic acids is also a critical preservation factor. Improper acidification to pH above 4.6 may lead to spoilage with consequences of public health significance due to the potential proliferation of toxin-producing *Clostridium* species (Garrido-Fernández et al., 1997; Pérez-Díaz et al., 2015a). The recommended pH and salt concentration varies widely among vegetables (Pérez-Díaz et al., 2015a).
2. A *starter culture* is defined as a preparation of living microorganisms, which are deliberately used to assist the beginning of fermentation, producing specific changes in the chemical composition and the sensorial properties of the substrate to obtain a more homogeneous product. For many years, the search for starter cultures with application in vegetable fermentations was practically focused on the technological activity of LAB. The traditional selection of LAB starter cultures was based on diverse technological criteria, including fast consumption of sugars and acidification rate, homo-fermentative metabolism, pH, and salt tolerance, resistance to bacteriophages, production of bacteriocins, flavor development, wide temperature range for growth and minimum nutritional requirement, among many other characteristics. The starter culture must also dominate the indigenous microbiota by its fast and predominant growth under fermentation conditions. *Lactobacillus pentosus* and *L. plantarum* are among the LAB species with major applications as starter cultures in such

fermented vegetables as cucumber, capers or tables olives, albeit *L. mesenteroides* is also habitually used in low salt fermentations such as sauerkraut, as mentioned previously (Corsetti et al., 2012; Perez-Díaz et al., 2015a). Recently, several publications have emphasized the importance of the role played by yeasts as bioprotective agents when used as starter cultures during vegetable processing. Apart from their technological characteristics (lipase, esterase, β-glucosidase or catalase activities, improvement of LAB growth and the organoleptic profile of raw material), the competitive traits against undesirable microorganisms have attracted increasing attention from scientists, who proposed their application as biocontrol agents in the food industry (Arroyo-López et al., 2012; Muccilli and Restuccia, 2015). These antagonistic activities rely on the production and tolerance of high concentrations of ethanol, competition for nutrients, as well as the synthesis of a large class of antimicrobial compounds (killer toxins, ethyl acetate, ethanol, etc.) which showed clearly a large spectrum of activity against foodborne pathogen and spoilage microorganisms. Thus, latest trends for the development of starters in vegetables are focused on the production of mixed-multifunctional (LAB-yeasts) starters in order to improve and expand the mode of action by the use of two complementary microorganisms with different properties and with the ability to inhibit spoilage microorganisms.

3. *Hazard Analysis and Critical Control Point (HACCP) system.* Fermented vegetables must be prepared and handled in accordance with the appropriate sections of the *General Principles of Food Hygiene* (CAC/RCP 1-1969, Codex Alimentarius Commission, 2003), the *Code of Hygienic Practice for Low-Acid and Acidified Low-Acid Canned Foods* (CAC/RCP 23-1979, Codex Alimentarius Commission, 1993), and the *Code of Hygienic Practice for Canned Fruit and Vegetable Products* (CAC/RCP 2-1969). The product should comply with any microbiological criteria established in accordance with the *Principles for the Establishment and Application of Microbiological Criteria for Foods* (CAC/GL 21-1997, Codex Alimentarius Commission, 1997). HACCP is a valuable tool for complying with all these hygiene criteria and good manufacturing practices in preparation of fermented vegetables. The HACCP is a systematic and preventive procedure, internationally recognized, to tackle biological, physical, and chemical risks by means of applying prevision and prevention principles, instead of final product testing. The preventative approach is based on remarking all essential hygiene controls in each processing step. The industry of fermented vegetables must comply with the requirement for elaborating, applying, and maintaining a permanent procedure based on HACCP principles. Therefore, the HACCP concept permits a systematic approach to the identification of hazards, and an assessment of the likelihood of their occurrence during the manufacture, distribution, and use of the food product, and defines measures for their control in order to ensure food

safety and quality. When being applied, food handlers can acquire knowledge and awareness about food safety and quality, thus improving their handling practices and preventing food alteration. Despite the advantages of HACCP system, food industries have to deal with its implementation across a large variety of food commodities produced. Additionally, for small-scale industries (<20 workers) HACCP implementation is often more difficult given the associated cost and effort. Therefore, HACCP should be adapted for different situations and adjusted to changes in equipment, processing methods or technological achievements.

4. *Preservatives* and *thermal treatment*. To increase shelf life and avoid spoilage of fermented vegetables packaging, the finished products may be pasteurized or sterilized, depending on their characteristics. Thermal treatment considerably reduces the microbial load of finished products, increasing shelf life and stability during storage. Contrarily, it can reduce the texture and produce losses associated with the fresh appearance of the preserved fruits (Garrido-Fernández et al., 1997). Additionally, processed vegetables which undergo thermal treatment may maintain stability at less acidic pH values and reduced salt concentration. Preservatives may substitute thermal treatments if specific processing constraints exist, such as the use of plastic bags or 5-gallon plastic pales for packaging, or the use of heat sensitive seasonings. Such prerogative is applicable in the United States to the stabilization of the products for transportation purposes only. Preservatives such as sorbic and benzoic acids, or their respective salts, are commonly used in finished fermented vegetable products and as processing aids. Potassium sorbate is effectively used to inhibit the growth of yeasts and molds in fermentation cover brines and some processed products. Sodium benzoate is an important preservative used in certain fermented and fresh-pack products to inhibit bacterial growth, especially LAB (Pérez-Díaz et al., 2015a). The effectiveness of these preservatives is dependent on the pH of the medium. The disadvantages in applying such preservatives include their ability to preferably accumulate in the fatty components of vegetables such as olives, limiting the antimicrobial effect, generation of undesirable sensorial notes for consumers, induction of browning of fruits, and degradation by selected microbes compromising the stability of the products. As a result, the table olive sector is demanding research for obtaining more appropriate preservatives (sodium metabisulphite, zinc chloride, or natamicin) as well as other technologies to increase shelf life (e.g., modified atmosphere or high hydrostatic pressure).

9.7 CONCLUDING REMARKS

Vegetable fermentations will continue to be central to the production, availability and supply of foods for a growing population. The use of traditional approaches for spontaneous fermentations is in line with the demand for natural, healthier food products. Safe fermented vegetables with improved quality can be produced provided certain hygienic rules are followed and sufficient knowledge of the underlying microbiology and chemistry is developed (Demarigny, 2012). Consumer trends will likely encourage the emergence of innovative fermented food products made with vegetable materials from tropical and warm countries, using environmentally friendly technologies and applying new scientific knowledge. With the development of DNA sequencing technology and bioinformatics it is currently possible to obtain and translate genomic DNA sequencing information into functional predictions. Microbial genome sequences may be compared in order to improve the selection of starter cultures species that can positively influence the organoleptic attributes of finished products instead of inducing spoilage. The possibility of identifying missing growth factors in vegetable matrixes that could enhance and accelerate the fermentation process and prevent spoilage is around the corner. Identification of specific strains dominating in vegetable fermentations at variable stages is imminent and has been achieved for other environments. Increased knowledge of how to manipulate fermentation systems to increase their bioactivity and retain that which is naturally present awaits.

ACKNOWLEDGMENTS

This paper has received funding from Junta de Andalucía Regional Government through the PrediAlo project (AGR-7755: www.predialo.science.com.es). FNAL wishes to express thanks to the Spanish government for his Ramón y Cajal postdoctoral research contract while EMP would like to thank The Junta de Andalucía for his postdoctoral research contract. Authors also express their gratitude to Drs. Antonio de Castro Gómez-Millán, Pedro García, and Francisco Rodríguez (IG-CSIC) for pictures of spoilage.

REFERENCES

Abriouel, H., Benomar, N., Pulido, R.P., Martínez-Cañamero, M., Gálvez, A., 2011. Annotated genome sequence of *Lactobacillus pentosus* MP-10, which has probiotic potential, from naturally fermented Aloreña green table olives. J. Bacteriol. 193, 4559–4560.

Abriouel, H., Benomar, N., Galvez, A., Pulido, R.P., 2014. Preservation of Manzanilla Aloreña cracked green table olives by high hydrostatic pressure treatments singly or in combination with natural antimicrobials. LWT Food Sci. Technol. 56, 427–431.

Alvarruiz, A., Rodrigo, M., Miquel, J., Giner, V., Feria, A., Vila, A., 1990. Influence of brining and packaging conditions on product quality of capers. J. Food Sci. 55, 196–198.

Anderson, R.E., Daescchel, M.A., Ericksson, C.E., 1988. Controlled lactic acid fermentation of vegetables. In: Durand, G., Babichon, L., Florent, J. (Eds.), Proceedings of the 8th International Symposium. Societé Fraçaise de Microbiologie, Paris (Francia), pp. 855–868.

Arroyo-López, F.N., Romero-Gil, V., Bautista-Gallego, J., Rodriguez-Gómez, F., Jiménez-Díaz, R., García-García, P., et al., 2012. Yeasts in table olive processing: desirable or spoilage microorganisms. Int. J. Food Microbiol. 160, 42–49.

Asehraou, A., Peres, C., Faid, M., Brito, D., 2002. Reducing the bloater spoilage incidence in fermented green olives during storage. Grasas y Aceites 53, 330–334.

Babuchowski, A., Laniewska-Moroz, L., Warminska-Radyko, I., 1999. Propionibacteria in fermented vegetables. Le Lait 79, 113–124.

Bleve, G., Tufariello, M., Durante, M., Grieco, F., Ramires, F.A., Mita, G., et al., 2015. Physico-chemical characterization of natural fermentation process of Conservolea and Kalamata table olives and development of a protocol for the preselection of fermentation starters. Food Microbiol. 46, 368–382.

Breidt, F., Caldwell, J.M., 2011. Survival of *Escherichia coli* O157:H7 in cucumber fermentation brines. J. Food Sci. 76, M198–M203.

Breidt, F., Medina, E., Wafa, D., Perez-Diaz, I., Franco, W., Huang, H.Y., et al., 2013a. Characterization of cucumber fermentation spoilage bacteria by enrichment culture and 16S rDNA cloning. J. Food Sci. 78, 470–476.

Breidt, F., McFeeters, R.F., Díaz-Muñiz, I., Le, C.-H., 2013b. Fermented vegetables. In: Doyle, M.P., Buchanan, R.L. (Eds.), Food Microbiology: Fundamentals and Frontiers, fourth ed. ASM Press, Washington DC.

Brock, T.D., 1961. Milestones in Microbiology. Prentice Hall, Englewood Cliffs, NJ.

Botta, C., Cocolin, L., 2012. Microbial dynamics and biodiversity in table olive fermentation: culture-dependent and independent approaches. Front. Microbiol. Available from: http://dx.doi.org/10.3389/fmicb.2012.00245.

Buckenhüskes, H., Jensen, H.A., Anderson, R., Garrido Fernández, A., Rodrigo, M., 1990. Fermented vegetables. In: Zeuthen, P., Cheftel, J.C., Eriksson, C., Gormley, T.R., Linko, P., Paulus, K. (Eds.), Processing and Quality of Foods. Vol 2. Food Biotechnology: Avenue to Healthy and Nutritious Products. Elsevier Applied Science, London (UK), pp. 2162–2188.

CBI, 2009. Preserved fruit and vegetables. The EU market for canned fruits and vegetable.

Chang, J.Y., Chang, H.C., 2010. Improvements in the quality and shelf-life of kimchi by fermentation with induced bacteriocin-producing strain, *Leuconostoc citreum* GJ7 as a starter. J. Food Sci. 75, M103–M110.

Choi, H.J., Lee, N.K., Paik, H.D., 2015. Health benefits of lactic acid bacteria isolated from kimchi, with respect to immunomodulatory effects. Food Sci. Biotechnol. 24, 783–789.

Codex Alimentarius Commission (CAC), 1993. Recommended code of hygienic practices for low-acid and acidified low-acid canned foods. CAC/RCP 23-1979 (review 1993).

Codex Alimentarius Commission (CAC), 1997. Principles for the establishment and application of microbiological criteria for foods. CAC/GL 21-1997.

Codex Alimentarius Commission (CAC), 2003. Recommended international code of practice general principles of food hygiene. CAC/RCP 1-1969, Rev.4-2003.

Cogan, T.M., Hill, C., 1993. Cheese starter cultures. In: second ed. Fox, P.F. (Ed.), Cheese: Chemistry, Physics, and Microbiology, vol. 1. Chapman and Hall, Great Britain, pp. 193–194.

Columela, L.J.M. (54). De Re Rustica, vol II. Nestlé, A.E.P.A. Santander, Spain.

Corsetti, A., Perpetuini, G., Schirone, M., Tofalo, R., Suzzi, G., 2012. Application of starter cultures to table olive fermentation: an overview on the experimental studies. Front. Microbiol. 3, Art. 248.

Costilow, R.N., Gates, K., Lacy, M.L., 1980. Molds in brine cucumbers. Cause of softening during air-purging of fermentations. Appl. Environ. Microbiol. 40, 417–422.

Daeschel, M.A., McFeeters, R.F., Fleming, H.P., Klaenhamer, T.R., Sanozky, R.B., 1984. Mutation and selection of *Lactobacillus plantarum* strains that do not produce carbon dioxide from malate. Appl. Environ. Microbiol. 47, 419–420.

Daeschel, M.A., Fleming, H.P., Potts, E.A., 1985. Compartmentalization of lactic acid bacteria and yeasts in the fermentation of brined cucumbers. Food Microbiol. 2, 77–84.

De Angelis, M., Campanella, D., Cosmai, L., Summo, C., Rizzello, C.G., Caponio, F., 2015. Microbiota and metabolome of un-started and started Greek-type fermentation of Bella di Cerignola table olives. Food Microbiol. 52, 8–30.

De Castro, A., Montaño, A., Sánchez, A.H., Rejano, L., 1998. Lactic acid fermentation and storage of blanched garlic. Int. J. Food Microbiol. 39, 205–2011.

De Castro, A., Montaño, A., Casado, F.J., Sánchez, A.H., Rejano, L., 2002. Utilization of *Enterococcus casseliflavus* and *Lactobacillus pentosus* as starter culture for Spanish-style green olive fermentation. Food Microbiol. 19, 637–644.

Demarigny, Y., 2012. Fermented food products made with vegetable materials from tropical and warm countries: microbiological and technological considerations. Int. J. Food Sci. Technol. 47, 2469–2476.

Díaz-Muñiz, I., Kelling, R., Hale, S., Breidt, F., McFeeters, R.F., 2007. Lactobacilli and tartrazine as causative agents of a red colored spoilage in cucumber pickle products. J. Food Sci. 72, M240–M245.

Dirar, H., 1993. The Indigenous Fermented Foods in the Sudan: A Study in African Food and Nutrition. CAB International, UK.

Domínguez-Manzano, J., Olmo- Ruiz, C., Bautista-Gallego, J., Arroyo-López, F.N., Garrido-Fernández, A., Jiménez-Díaz, R., 2012. Biofilm formation on abiotic and biotic surfaces during green Spanish-style table olives. Int. J. Food Microbiol. 152, 230–238.

Duran-Quintana, M.C., Gonzalez-Cancho, F., Garrido-Fernandez, A., 1979. Natural black olives in brine. IX. Production of alambrado by some microorganisms isolated from fermentation brines. Grasas y Aceites 30, 361–367.

Etchells, J.L., 1941. Incidence of yeasts in cucumber fermentations. Food Res. 6, 95–104.

Etchells, J.L., Bell, T.A., 1950. Classification of yeasts from the fermentation of commercially brined cucumbers. Farlowia 4, 87–112.

Etchells, J.L., Bell, T.A., Monroe, R.J., Masley, P.M., Demain, A.L., 1958. Populations and softening enzyme activity of filamentous fungi on flowers, ovaries, and fruit of pickling cucumbers. Appl. Microbiol. 6, 427–440.

Etchells, J.L., Costilow, R.N., Anderson, T.E., Bell, T.A., 1964. Pure culture fermentation of brined cucumbers. Appl. Microbiol. 12, 523.

Etchells, J.L., Borg, A.F., Bell, T.A., 1968. Bloater formation by gas-forming lactic acid bacteria in cucumber fermentations. Appl. Microbiol. 16, 1029–1035.

Etchells, J.L., Bell, T.A., Fleming, H.P., Kelling, R.E., Thompson, R.L., 1973. Suggested procedure for the controlled fermentation of commercially brined pickling cucumbers - The use of starter cultures and reduction of carbon dioxide accumulation. Pickle Pak Sci. 3, 4.

Etchells, J.L., Fleming, H.P., Hontz, L.H., Bell, T.A., Monroe, R.J., 1975. Factors influencing bloater formation in brined cucumbers during controlled fermentation. J. Food Sci. 40, 569–575.

Fleitag, C., 2012. Overview about the production and market of sauerkraut. 4th International Kimchi Conference September 13th, Washington, DC.

Fleming, H.P., Etchells, J.L., Thompson, R.L., Bell, T.A., 1975. Purging of CO_2 from cucumber brines to reduce bloater damage. J. Food Sci. 40, 1304–1310.

Franco, W., Perez-Diaz, I.M., 2013. Microbial interactions associated with secondary cucumber fermentation. J. Appl. Microbiol. 114, 161–172.

Franco, W., Perez-Diaz, I.M., Johanningsmeier, S.D., McFeeters, R.F., 2012. Characteristics of spoilage-associated secondary cucumber fermentation. Appl. Environ. Microbiol. 78, 1273–1284.

Fred, E.B., Peterson, W.H., 1922. The production of pink sauerkraut by yeasts. J. Bacteriol. 7, 257–269.

Garcia-Garcia, P., Duran-Quintana, M.C., Garrido-Fernandez, A., 1982. Modifications in the natural fermentation process of black olives in order to avoid spoilage. Grasas y Aceites 33, 9–17.

Gardner, N.C., Savard, T., Obermeier, P., Caldwell, G., Champagne, C.P., 2001. Selection and characterization of mixed starter cultures for lactic acid fermentation of carrot, cabbage, beet, and onion vegetable mixtures. Int. J. Food Microbiol. 64, 261–275.

Garrido-Fernández, A., Fernández-Díez, M.J., Adams, R.M., 1997. Table Olives Production and Processing. Chapman & Hall, London.

Gililland, J.R., Vaughn, R.H., 1943. Characteristics of butyric acid bacteria from olives. J. Bacteriol. 46, 315–322.

Golomb, B.L., Morales, V., Jung, A., Yau, B., Boundy-Mills, K.L., Marco, M.L., 2013. Effects of pectinolytic yeast on the microbial composition and spoilage of olive fermentations. Food Microbiol. 33, 97–106.

Gonzalez-Cancho, F., Nosti-Vega, M., Fernandez-Diez, M.J., Buzcu, N., 1970. Propionibacterium spp. associated with olive spoilage. Factors influencing their growth. Microbiologia Española 23, 233–252.

Gonzalez-Cancho, F., Rejano-Navarro, L., Borbolla y Alcala, J., 1980. Formation of propionic acid during the conservation of green table olives. III. Responsible microorganism. Grasas y Aceites 31, 245–250.

Han, X., Yi, H., Zhang, L., Huang, W., Zhang, Y., Zhang, L., et al., 2014. Improvement of fermented Chinese cabbage characteristics by selected starter cultures. J. Food Sci. 79, M1387–M1392.

Hernandez, A., Martin, A., Aranda, E., Perez-Nevado, F., Cordoba, M.G., 2007. Identification and characterization of yeast isolated from the elaboration of seasoned green table olives. Food Microbiol. 24, 346–351.

IOC, 2015. Table olives. Table 1. Production. <http://www.internationaloliveoil.org/estaticos/view/132-world-table-olive-figures> (last accessd December 2015).

Jang, D.J., Chung, K.R., Yang, H.J., Kim, K.S., Known, D.Y., 2015. Discussion on the origin of *kimchi*, representative of Korean unique fermented vegetables. J. Ethnic. Foods 2, 126–136.

Johanningsmeier, S.D., McFeeters, R.F., 2013. Metabolism of lactic acid in fermented cucumbers by *Lactobacillus buchneri* and related species, potential spoilage organisms in reduced salt fermentations. Food Microbiol. 35, 129–135.

Johanningsmeier, S.D., McFeeters, R.F., 2015. Metabolic footprinting of *Lactobacillus buchneri* strain LA1147 during anaerobic spoilage of fermented cucumbers. Int. J. Food Microbiol. 215, 40–48.

Johanningsmeier, S.D., McFeeters, Fleming, H.P., Thompson, R.L., 2007. Effects of *Leuconostoc mesenteroides* starter culture on fermentation of cabbage with reduced salt concentrations. J. Food Sci. 72, M166.

Jung, Y.-M., Lee, S.-H., Lee, D.-S., You, M.-J., Chung, I.K., Cheon, W.H., et al., 2011. Fermented garlic protects diabetic, obese mice when fed a high-fat diet by antioxidant effects. Nutr. Res. 31, 387–396.

Jung, J.Y., Lee, S.H., Lee, H.J., Seo, H.Y., Park, W.S., Jeon, C.O., 2012. Effects of *Leuconostoc mesenteroides* starter cultures on microbial communities and metabolites during kimchi fermentation. Int. J. Food Microbiol. 153, 378–387.

Jung, J.Y., Lee, S.H., Jin, H.M., Hahn, Y., Madsen, E.L., Jeon, C.O., 2013. Metatranscriptomic analysis of lactic acid bacterial gene expression during kimchi fermentation. Int. J. Food Microbiol. 163, 171–179.

Kawatomari, T., Vaughn, R.H., 1956. Species of *Clostridium* associated with zapatera spoilage of olives. J. Food Sci. 4, 481–490.

King, A.D., Vaughn, R.H., 1961. Media for detecting pectolytic Gram-negative bacteria associated with the softening of cucumbers, olives and other plant tissues. J. Food Sci. 26, 635–643.

Kleerebezem, M., Boekhorst, J., van Kranenburg, R., Molenaar, D., Kuipers, O.P., Leer, R., et al., 2003. Complete genome sequence of *Lactobacillus plantarum* WCFS1. Proc. Natl. Acad. Sci. U. S. A. 100, 1990–1995.

Leal-Sánchez, M.V., Jiménez-Díaz, R., Maldonado-Barragán, A., Garrido-Fernández, A., Ruiz-Barba, J.L., 2002. Optimization of bacteriocin production by batch fermentation of *Lactobacillus plantarum* LPCO10. Appl. Environ. Microbiol. 68, 4465.

Levin, R.E., Vaughn, R.H., 1966. *Desulfovibrio aestuarii*, the causative agent of hydrogen sulfide spoilage of fermenting olive brines. J. Food Sci. 31, 768–772.

Lucena-Padrós, H., Caballero-Guerrero, B., Maldonado-Barragán, A., Ruiz Barba, J.L., 2014. Genetic diversity and dynamics of bacterial and yeast strains associated with Spanish-style green table olive fermentations in large manufacturing companies. Int. J. Food Microbiol. 190, 72–78.

Maldonado-Barragán, A., Caballero-Guerrero, B., Lucena-Pradós, H., Ruiz-Barba, J.L., 2011. Genome sequence of *Lactobacillus pentosus* IG1, a strain isolated from Spanish-style green olive fermentations. J. Bacteriol. 193, 5605.

Martínez-Villaluenga, C., Peñas, E., Sidro, B., Ullate, M., Frias, J., Vidal-Valverde, C., 2012. White cabbage fermentation improves ascorbic content, antioxidant and nitric oxide production inhibitory activity in LPS-induced macrophages. LWT Food Sci. Technol. 46, 77–83.

Mattos, F.R., Fasina, O.O., Reina, L.D., Fleming, H.P., Breidt, F., Damasceno, G.S., et al., 2005. Heat transfer and microbial kinetics modeling to determine the location of microorganisms within cucumber fruit. J. Food Sci. 70, E324.

McDonald, L.C., Fleming, H.P., Hassan, H.M., 1990. Acid tolerance of *Leuconostoc mesenteroides* and *Lactobacillus plantarum*. Appl. Environ. Microbiol. 56, 2120–2124.

McFeeters, R.F., Pérez-Díaz, I., 2010. Fermentation of cucumbers brined with calcium chloride instead of sodium chloride. J. Food Sci. 75, C290–C296.

McFeeters, R.F., Fleming, H.P., Daeschel, M.A., 1984. Malic acid degradation and brined cucumber bloating. J. Food Sci. 49, 999–1002.

Medina, E., Perez-Diaz, I.M., Breidt, F., Hayes, J., Franco, W., Butz, N., et al., 2016. Bacterial ecology of fermented cucumber rising pH spoilage as determined by nonculture-based methods. J. Food Sci. 81, 121–129.

Medina-Pradas, E., Arroyo-Lopez, F.N., 2015. Presence of toxic microbial metabolites in table olives. Front. Microbiol.1–6, Art 873.

Mkar, E.M., Bonar, L., 1939. Film yeasts from pickle brines. Zentr. Bakt. Parasitenk. II 100, 289–294.

Montaño, A., Sánchez, A.H., Rejano, L., de Castro, A., 1997. Processing and storage of lye--treated carrots fermented by a mixed starter culture. Int. J. Food Microbiol. 35, 83–90.

Muccilli, S., Restuccia, C., 2015. Bioprotective role of yeasts. Microorganisms 3, 588–611.

Moon, S.H., Chang, M., Kim, H.Y., Chang, H.C., 2014. *Pichia kudriavzevii* is the major yeast involved in film-formation, off-odor production, and texture-softening in over-ripened Kimchi. Food Sci. Biotechnol. 23, 489–497.

Nanniga, N., 2010. Did van Leeuwenhoek observe yeast cells in 1680? Small Things Considered. American Society for Microbiology, <http://schaechter.asmblog.org/schaechter/2010/04/did-van-leeuwenhoek-observe-yeast-cells-in-1680.html>.

Palani, K., Harbaum-Piayda, B., Meske, D., Keppler, J.K., Bockelmann, W., Heller, K.J., et al., 2016. Influence of fermentation on glucosinolates and glucobrassicin degradation products in sauerkraut. Food Chem. 190, 755–762.

Palomino, J.M., Toledo del Árbol, J., Benomar, N., Abriouel, H., Martínez Cañamero, M., Gálvez, A., et al., 2015. Application of *Lactobacillus plantarum* Lb9 as starter culture in caper berry fermentation. LWT Food Sci. Technol. 60, 788–794.

Park, S.-E., Yoo, A.-A., Seo, S.-H., Lee, K.-I., Na, C.-S., Son, H.-S., 2016. GC-MS based metabolomics approach for the understanding of *Lactobacillus plantarum* fermentation characteristics. LWT Food Sci. Technol. 68, 313–321.

Pérez-Díaz, I.M., McFeeters, R.F., 2011. Preparation of a *Lactobacillus plantarum* starter culture for cucumber fermentations that can meet kosher guidelines. J. Food Sci. 76, M120–M123.

Pérez-Díaz, I.M., Breidt, F., Buescher, R.W., Arroyo-López, F.N., Jiménez-Díaz, R., Garrido-Fernández, A., et al., 2015a. Fermented and acidified vegetables. In: Downes, P., Ito, K. (Eds.), Compendium of Methods for the Microbiological Examination of Foods, fourth ed. American Public Health Association, Washington, DC, pp. 521–532.

Pérez-Díaz, I.M., McFeeters, R.F., Moeller, L., Hayes, J.S., Fornea, D., Johanningsmeier, S.D., et al., 2015b. Commercial scale cucumber fermentations brined with calcium chloride instead of sodium chloride. J. Food Sci. 80, M2827–M2836.

Pérez Pulido, R., Omar, N.B., Abriouel, H., Lucas López, R., Martínez Cañamero, M., Gálvez, A., 2005. Microbiological study of lactic acid fermentation of caper berries by molecular and culture-dependent methods. Appl. Environ. Microbiol. 71, 7872–7879.

Plastourgos, S., Vaughn, R.H., 1957. Species of *Propionibacterium* associated with zapatera spoilage of olives. Appl. Microbiol. 5, 267–271.

Plengvidhya, V., Breidt, F., Lu, Z., Fleming, H.P., 2007. DNA fingerprinting of lactic acid bacteria in sauerkraut fermentations. Appl. Environ. Microbiol. 73, 7697–7702.

Ricciardi, A., Parente, E., Guidone, A., Ianniello, R.G., Zotta, T., Abu Sayem, S.M., et al., 2012. Genotypic diversity of stress response in *Lactobacillus plantarum, Lactobacillus paraplantarum* and *Lactobacillus pentosus*. Int. J. Food Microbiol. 157, 278–285.

Roberts, J.S., Kidd, D.R., 2005. Lactic acid fermentation of onions. LWT Food Sci. Technol. 38, 185–190.

Rodríguez-Gómez, F., Romero Gil, V., Bautista Gallego, J., García García, P., Garrido Fernández, A., Arroyo López, F.N., 2014. Production of potential probiotic Spanish-style green table olives at pilot plant scale using multifunctional starters. Food Microbiol. 44, 278–287.

Ruiz-Cruz, J., Gonzalez-Cancho, F., 1969. The metabolism of yeasts isolated from the brine of pickled Spanish-type green olives. I. The assimilation of lactic, acetic and citric acids. Grasas y Aceites 20, 6–11.

Samish, Z., Etinger-Tulczynska, R., Bick, M., 1963. The microflora within the tissue of fruits and vegetables. J. Food Sci. 28, 259.

Sánchez, A.H., de Castro, A., Rejano, L., 1992. Controlled fermentation of caperberries. J. Food Sci. 57, 675–678.

Shim, S.T., Kyung, K.H., 1999. Natural microflora of pre-peeled garlic and their resistance to garlic antimicrobial activity. Food Microbiol. 16, 165–172.

Siezen, R.J., Tzeneva, V.A., Castioni, A., Wels, M., Phan, H.T., Rademaker, J.L., et al., 2009. Phenotypic and genomic diversity of *Lactobacillus plantarum* strains isolated from various environmental niches. Environ. Microbiol. 12, 758–773.

Siezen, R.J., van Hylckama Vlieg, J.E., 2011. Genomic diversity and versatility of *Lactobacillus plantarum*, a natural metabolic engineer. Microbial. Cell Fact. 10 (Suppl. 1), S3.

Stamer, J.R., 1988. Lactic acid bacteria in fermented vegetables. In: Robinson, R.K. (Ed.), Development in Food Microbiology-3. Elsevier Applied Science., London (UK), pp. 67–85.

Stamer, J.R., Hrazdina, G., Stoyla, B.O., 1973. Induction of red color formation in cabbage juice by *Lactobacillus brevis* and its relationship to pink sauerkraut. Appl. Microbiol. 26, 161–166.

Suhigara, T.F., 1985. Microbiology of bread making. In: Wood, B.J.B. (Ed.), Microbiology of Fermented Foods. Elsevier Applied Science Publisher, UK.

Unites States Department of Agriculture, 2015a. National Agricultural Statistics Service. Vegetables 2014 Summary, USA.

United States Department of Agriculture, (2015b). Economics, statistics and market information system. <http://usda.mannlib.cornell.edu/MannUsda/viewDocumentInfo.do?documentID = 1397> (accessd January 2016).

Vaughn, R.H., Won, W.D., Spencer, F.B., Pappagranis, D., Foda, I.O., Krumperman, P.M., 1953.). *Lactobacillus plantarum*, the cause of yeast spots on olives. Appl. Microbiol. 1, 82–85.

Vaughn, R.H., King, A.D., Nagel, C.W., Ng, H., Levin, R.E., Macmilla, J.D., et al., 1969. Gram-negative bacteria associated with sloughing, a softening of Californian ripe olives. J. Food Sci. 34, 224–227.

Vaughn, R.H., Stevenson, K.E., Dave, B.A., Park, H.C., 1972. Fermenting yeast associated with softening and gas-pocket formation in olives. Appl. Microbiol. 23, 316–320.

Vega Leal-Sánchez, M., Ruiz Barba, J.L., Sánchez, A.H., Rejano, L., Jiménez Díaz, R., Garrido, A., 2003. Fermentation profile and optimization of green olive fermentation using *Lactobacillus plantarum* LPCO10 as a starter culture. Food Microbiol. 20, 421–430.

West, N.S., Gililland, J.R., Vaughn, R.H., 1941. Characteristics of coliform bacteria from olives. J. Bacteriol. 41, 341–353.

Wiese, S., Wubshet, S.G., Nielsen, J., Staerk, D., 2013. Coupling HPLC-SPE-NMR with a microplate-based high-resolution antioxidant assay for the efficient analysis of antioxidant in foods-validation and proof-of-concept study with caper buds. Food Chem. 141, 4010–4018.

Wilson, E.M., Johanningsmeier, S.D., Osborne, J.A., 2015. Consumer acceptability of cucumber pickles produced by fermentation in calcium chloride brines for reduced environmental impact. J. Food Sci. 80, S1360–S1367.

Yokotsouka, T., 1985. Fermented protein foods in the Orient, with emphasis on shoyu and miso in Japan. In: Wood, B.J.B. (Ed.), Microbiology *of Fermented Foods*. Elsevier Applied Science Publisher, UK.

Yoshida, H., Katsukaki, H., Ohta, R., Ishakawa, K., Fukuda, H., Fujino, T., et al., 1999. Antimicrobial activity of the thiosulfinates isolated from oil-macerated garlic extracts. Biosci. Biotechnol. Biochem. 63, 591–594.

Zhang, N., Huang, X., Zeng, Y., Wu, X., Peng, X., 2013. Study on prebiotic effectiveness of neutral garlic fructan *in vitro*. Food Sci. Hum. Wellness 2, 119–123.

Chapter 10

Wine Spoiling Phenomena

Daniela Campaniello and Milena Sinigaglia
University of Foggia, Foggia, FG, Italy

10.1 INTRODUCTION

Fermented alcoholic beverages (FABs) are produced by the fermentation of sugars or starches contained in fruits or grains by naturally occurring microorganisms, or by the addition of starter cultures; through biochemical reactions, these microorganisms change the substrates present in the raw materials into alcohol (Jeon et al., 2015).

Epidemiological studies have concluded that if moderately consumed, alcoholic beverages are able to confer various health benefits since the alcohol and antioxidants contained may help to prevent heart disease by increasing levels of high-density lipoprotein cholesterol (the good cholesterol) and protecting against artery damage. Thus, a reduction of the incidence of mortality from cardiovascular and cerebrovascular disease can occur (Wallerath et al., 2003).

Alcoholic beverages are generally carbonated (2.5−3 vol.), have a low level of oxygen, high carbon to nitrogen ratio, and contain up to 8.5% (alcohol by volume, abv). Due to their high ethanol content (> 4%) and low pH (2.4−4.5), FABs are generally recognized as microbiologically safe; nevertheless, some alcoholic beverages are considered potentially hazardous because of the abundant sugars and starches contained in raw materials which are attractive for microorganisms. Furthermore, the manufacturing processes are complex and, consequently, there are many potential sources of contamination which favor spoilage and pathogenic bacteria (Jeon et al., 2015). In particular, some fermentative yeasts and acid-tolerant bacteria, namely, lactic acid bacteria (LAB), and acetic acid bacteria (AAB), represent the major spoilage microbes.

In an interesting review, Loureiro and Malfeito-Ferreira (2003) reported that a univocal definition of "microbial spoilage" in fermented foods and beverages is not easy because the metabolites produced contribute to the flavor, aroma, and taste, affecting positively and/or negatively the quality of the final products. The authors highlighted a difference between what is perceived as spoilage or as benefic: the same compound can have spoilage or beneficial

The Microbiological Quality of Food. DOI: http://dx.doi.org/10.1016/B978-0-08-100502-6.00013-3

activity depending on the concentration in the alcoholic beverage. For example, they reported that the production of 4-ethylphenol (by *Brettanomyces/Dekkera* spp.) is considered as spoilage at levels higher than about 620 μg/L and as beneficial at levels less than 400 μg/L, because it contributes favorably by imparting aromatic notes of spices, leather, smoke, or game.

According to Loureiro and Malfeito-Ferreira (2003), few years later, Bartowsky (2009) reported that desirability of a flavor compound in wine is dependent on its concentration. For example, the buttery or butterscotch aromas of diacetyl (2,3-butanedione), an intermediate metabolite of citric acid metabolism in LAB, can add appreciable aromas and complexity to wine at concentrations below 4 mg/L; above this threshold value, diacetyl in wine becomes unpleasant with buttery notes.

Also *Saccharomyces cerevisiae* can exert both beneficial and detrimental activities as it is responsible for turning grape juice into wine but it may also exert undesirable effects (e.g., the production of hydrogen sulfide) (Malfeito-Ferreira, 2011).

Stratford (2006) proposed a new concept of spoilage as "the alteration of food recognized by consumer" and Malfeito-Ferreira (2011) reported that the presence of volatile phenols produced by *Dekkera bruxellensis* could be perceived of positively and negatively. Positively for some consumers which prefer wines tainted by these compounds and negatively for others which attributed to these volatile phenols (even at low concentrations) the reduction of the wine quality due to diminished flavor complexity.

In the following sections we will discuss the topic "alcoholic beverage" focusing on wine and microorganisms involved in the winemaking process with particular attention to these strains responsible for spoilage.

10.2 WINEMAKING PROCESSING AND MICROFLORA

Wine is an alcoholic beverage resulting from the fermentation of grape must through appropriate processes. Grape variety, soil composition, climate, geographical position, and vinification practices affect the diversity and quality of wine (Peñas et al., 2015). During enological process, grapes, must, and wine are susceptible to various defects due, in many cases, to potential microbial spoilage that significantly influences consumer acceptability (Peñas et al., 2015).

The "natural microflora" in wine includes yeasts (*S. cerevisiae* is predominant) and bacteria (LAB and AAB); more specifically, four genera of LAB (*Lactobacillus*, *Leuconostoc*, *Oenococcus*, and *Pediococcus*) and three genera of AAB (*Acetobacter*, *Gluconobacter*, *Gluconoacetobacter*) (Bartowsky and Henschke, 2008; Bartowsky, 2009).

Chemical compounds interact with yeasts, bacteria, and fungi to obtain the final product (du Toit and Pretorius, 2000): 95% of the sugar is converted to ethanol and carbon dioxide, 1% is converted to cellular material, and the remaining 4% is converted to other products.

The long-chain fatty acids, organic nitrogen-containing compounds, sulfur-containing compounds, and many others also concur to the fermentative process by penetration of the grape juice medium through the yeast cell wall membrane, thus participating in biochemical reactions and producing fermentative flavor. The reduced sulfur compounds, hydrogen sulfide, organic sulfides, and thiols are considered the most "negative" aroma compounds (Lambrechts and Pretorius, 2000).

Ethanol, glycerol, and carbon dioxide contribute to wine flavor but the main compounds that form the "fermentation bouquet" are the organic acids, higher alcohols, esters, and aldehydes. Some of these compounds (acetaldehyde, acetic acid, ethyl acetate, higher alcohols, and diacetyl) may also become undesirable when they are present in excessive concentrations.

The winemaking process has a long traditional history, but sometimes some problems may arise, affecting the final product.

10.2.1 First Stage

During the first stage of the winemaking process, the grapes are in contact with the winery equipment (du Toit and Pretorius, 2000). In wineries, the species frequently recovered in tank walls, crushers, presses, floor, winery walls, pipes, etc. are *S. cerevisiae*, *Rhodotorula* spp., *Cryptococcus* spp., *Debaryomyces hansenii*, *Aureobasidium pullulans*, *Pichia membranifaciens*, *Pichia anomala*, *Candida* spp., *Trichosporon cutaneum*, *Kloeckera apiculata*, *Metschnikowia pulcherrima*, and *Trichosporon rosei*.

Rhodotorula spp., *Cryptococcus* spp., *D. hansenii*, and *A. pullulans* are obligate aerobes, and thus have little or no ability to grow in wines, whilst *P. membranifaciens*, *P. anomala*, and *Candida* spp. cause film formation on the surface of bulk wines in unfilled containers and with sulfite levels insufficient to prevent their growth (Loureiro and Malfeito-Ferreira, 2003).

Concerning *Dekkera/Brettanomyces*, *Zygosaccharomyces bailii*, and *Saccharomycodes ludwigii* (known as dangerous yeasts), there are controversial opinions on their presence in wineries; literature suggests that they are present at low concentrations; in contrast, Loureiro and Malfeito-Ferreira (2003) and some predecessors (Van der Walt and van der Kerken, 1961) reported that these yeasts are common contaminants in wineries. They supposed that probably the utilization of inadequate culture media or analytical techniques, as well as short incubation periods favoring mold and fast growing species, contributed to the underestimation in wineries of these yeasts.

However, thanks to advances in technology and improvement in Good Manufacturing Practices (equipment design, sanitation procedures and use of preservatives) the growth of the yeasts is controlled and in many cases, contaminations and/or diseases were extinct (Loureiro and Malfeito-Ferreira, 2003). Nevertheless, yeasts remain the most feared contaminants leading to wine spoilage. In fact, if not properly sanitized, the equipment represents a source of contamination of the grape juice, and microorganisms have the possibility to enter the winemaking process, exerting an influence on the quality of the end product.

The healthy state of the grapes is also important: grape variety, the state of grapes at harvest, physical damage (due to birds, insects, harvesting and mold attack, temperature, rainfall), the use of insecticides and fungicides, and other viticultural practices are external conditions affecting the biodiversity of the microflora in the juice (du Toit and Pretorius, 2000).

Barata et al. (2008) found that damage represents the main factor determining the alteration of yeast ecology on grape surface: the most frequent species were *Issatchenkia* spp. and *Zygoascus hellenicus*, *Z. bailii*, and *Zygosaccharomyces bisporus* were also isolated in high amounts; thus, the authors demonstrated that *Z. bailii* was often associated with sour rot grapes and was able to survive during fermentation together with *S. cerevisiae*.

Furthermore, it must not be underestimated the importance of meteorological agents (hail, heavy rain, etc.), rainwater absorption or attack by *Drosophila* spp., honeybees, wasps, moths, and birds, on the damage on grapes. Attacks of phytopathogenic molds (causing powdery mildews, noble or gray rot) also affect their integrity and healthiness. In fact, grapes infected with powdery mildew harbor much higher microbial loads of microorganisms (yeasts, lactic, and acetic bacteria) and volatile compounds (ethanol, ethyl acetate, and acetic acid) than sound grapes. Among yeasts, *Dekkera* and *Kloeckera*, probably disseminated by insects attracted to the infected grapes by the volatile compounds given off by ripening berries, were mainly detected (Loureiro and Malfeito-Ferreira, 2003).

When *Botrytis cinerea* causes the well-known noble rot, the presence of *K. apiculata* and *Candida stellata* seems to be favored. *B. cinerea* frequently produces also gray rot which damages wine quality and causes serious economic losses. In addition, the presence of this rot is accompanied by an increase of fermentative yeasts (some of which are spoilers), molds, lactic, and acetic bacteria.

Both rots change the chemical composition of grape juice. In noble rot a significant increase in sugar concentration (generally higher than 300 g/L), a little increase in pH (usually about 0.3 pH units), production of gluconic acid, and a significant increase in acetic acid is observed. In gray rot the sugar concentration is not usually increased, but there is more glycerol, gluconic acid, and acetic acid (due to acetic bacteria). *Botrytis cinerea* also produces various antibiotic substances in grapes. In addition, if other molds are present like *Aspergillus niger*, *Penicillium* sp., and *Cladosporium* sp., grapes become bitter and with aromatic flavors, yielding wines with phenolic and iodine odors.

When grapes show browning and desegregation of the internal tissues, this is the result of being affected by a new type of rot, named "sour rot," where yeasts and acetic bacteria have a dominating role while molds are hardly detected (Loureiro and Malfeito-Ferreira, 2003).

10.2.2 Second Stage

During the second stage (fermentation), the composition of the grape juice (high sugar, acid content, and low pH) and the addition of sulfur dioxide (SO_2) are selective for the development of yeasts and bacteria during alcoholic fermentation. *Saccharomyces cerevisiae* is generally the dominant yeast during fermentation. Other yeast strains initially present (*Candida*, *Hanseniaspora*, *Kloeckera* and *Metschnikowia*, and less frequently *Kluyveromyces* and *Pichia*), named *non-Saccharomyces* yeasts, allow 10^6-10^7 CFU/mL and significantly influence the composition of the wine, but they are ethanol sensitive, thus when concentration starts to increase they die off (du Toit and Pretorius, 2000). The persistence of them during fermentation depends on the temperature of fermentation, nutrient availability, inoculum strength of *Saccharomyces*, use and levels of sulfur dioxide, and the quantity and identity of organisms initially present on the grapes (Lambrechts and Pretorius, 2000).

At pH >3.6 LAB and AAB found the ideal environment to grow, and in many cases, this may damage the quality of the wine.

10.2.3 Third Stage

During the third stage (postfermentation), the smell and taste of many wines improve, but when maturation and aging take place in wooden barrels, organoleptic alterations can occur due to undesirable microorganisms remaining in the pores of the wood after cleaning and sterilization. In fact, although cleaning methods are more sophisticated, the shape and microstructure of wooden barrels protect many undesirable organisms, encouraging their growth (Suárez et al., 2007). A good sanitation, exclusion of oxygen and the correct dosage of antimicrobial agents represent some critical factors in ensuring a stable product that will withstand attack from spoilage yeasts and bacteria (du Toit and Pretorius, 2000).

Wine bottling is a critical step since it is the last contamination source for wines (mainly for wines with residual sugar and for some dry white wines) before wine is marketed. The contaminating flora is represented by *Z. bailii*, *S. cerevisiae*, and *S. ludwigii*, species that are resistant to chemical preservatives and low water activity; thus high sulfite levels and sorbate, when used in sublethal doses, play an essential role in favoring highly resistant yeasts.

Finally, species of *Actinomyces* and *Streptomyces* could also be present in the corks or oak barrels (du Toit and Pretorius, 2000).

10.3 MICROORGANISMS RESPONSIBLE OF SPOILAGE

10.3.1 Yeasts

To assess the risk of potential spoilage, identification of yeasts is considered very important.

Two terms were introduced to include the yeasts that can bring about spoilage: "wine yeasts" and "wild yeasts."

- "wine yeasts" are considered those *Saccharomyces* yeasts, tolerant to high concentrations of ethanol and sugar, which perform a complete fermentation of grape juice without the production of off-flavors;
- "wild yeasts" are *non-Saccharomyces* yeasts which can perform a partial alcoholic fermentation, often by producing esters (du Toit and Pretorius, 2000). They are characterized by an anaerobic and aerobic growth and sometimes they persist during the fermentation, competing with *Saccharomyces* for nutrients, producing fatty acid esters, and other compounds affecting the fermentation bouquet of the wine (Lambrechts and Pretorius, 2000).

In Table 10.1 the main yeast strains responsible for wine spoilage are reported.

Malfeito-Ferreira (2011) reported another classification considering the role of the yeasts in hazards related to wine off-flavors. The author regrouped yeasts in:

- apiculate yeasts
- film-forming species
- fermenting yeasts
- and separately, discussed *Dekkera/Brettanomyces bruxellensis*

10.3.1.1 Apiculate Yeasts

Due to their lemon-shaped form, species of the genera *Kloeckera/Hanseniaspora* belong to this group and are found in grape surfaces or in juices, after grape crushing. They are inhibited during fermentation and produce metabolites such as ethyl acetate (causing a vinegar smell) in unwanted amounts, but adequate and preventive winemaking measures (low temperature, sulfur dioxide, hygiene) allow control of their growth, overcoming this problem.

10.3.1.2 Film-Forming Species

This denomination refers to microorganisms able to grow on the surface of wine, developing pellicles. Due to their aerobic nature and fast growth, *Pichia* and *Candida* are able to form film, as well as *S. cerevisiae*, *D. bruxellensis*, or *Z. bailii* may also be isolated from wine pellicles. Concerning *S. cerevisiae*, pellicles production is desirable for the race beticus, which is one of the agents of sherry-type wine production.

TABLE 10.1 Yeast Strains Involved in Wine Spoilage

Wine Yeasts	
S. cerevisiae	Is considered as a spoilage organism when causes re-fermentation with residual sugars.
Schizosaccharomyces pombe	Is associated with wine spoilage when growing in bottled wine it forms a sediment at the bottom.
Brettanomyces intermedius/ *Dekkera intermedia*	*Brettanomyces* is the nonsexual, nonsporulating form of *Dekkera*. These yeasts are most commonly found within the wood cooperage. In particular, *B. intermedius* produces volatile phenols, mousy off-flavor (resulting from isomers of tetrahydropyridines) and high levels of acetic acid.
Zygosaccharomyces (*Z. bailii*, *Z. bisporus*, *Z. rouxii*, and *Z. florentinus*)	*Zygosaccharomyces* species are osmophilic and grow at high sugar concentrations. In particular, *Z. bailii* re-ferments juice or wine during storage with large amounts of CO_2; it also causes turbidity and sediment. It is resistant to preservatives (SO_2 sorbic and benzoic acid used in grape juice and wine), and to high ethanol concentrations (>15%) and to low pH (<2.0).
Wild Yeasts	
Candida, *Debaryomyces*, *Hanseniaspora*, *Hansenula*, *Kloeckera*, *Pichia*, *Metschnikowia*, *Saccharomycodes*, *Torulaspora*	These yeasts are associated with the ester formation. Wild yeasts in spoilage at the early stages of alcoholic fermentation are correlated with large amounts of acetic acid (concentrations >200 mg/L of ethyl acetate and 0.6 mg/L of acetate cause the spoilage of wines). The genus *Candida* oxidizes ethanol, producing high concentrations of acetaldehyde, volatile acids, and esters. *Hanseniaspora uvarum* produces killer toxins.
	Hansenula anomala and *M. pulcherrima* grow as film layer and produce high levels of ethyl acetate, isoamyl acetate, methylbutyl acetate, and ethyl acetate, acetaldehyde, respectively.
	Kloeckera: during the initial phase of fermentation, if *Kloeckera* grows excessively, high levels of esters and volatile acids are produced.
	Saccharomycodes ludwigii produces high concentrations of acetaldehyde, flocculent masses settle as chunks and form a sliminess. It is highly tolerant to ethanol and resistant to preservative compounds (SO_2 and sorbate), in fact it is considered a dangerous spoiler yeast due to its sulfite tolerance (1000–1500 mg/L SO_2); furthermore, it forms "snowflake" particles and produces pungent apple and sweet butter off-flavor due to the acetaldehyde and acetoin, respectively (Juvonen et al., 2011). *Saccharomycodes ludwigii* as well as other sulfite-tolerant species (*Z. bailii*, *Zygosaccharomyces lentus*) can ferment residual sugars causing explosion of bottles due to the high pressure formed (up to 9 bars).

In bulk wines, *Pichia*, *Candida*, and *Metschnikowia* (which cover the surface when air is not removed from the top of storage vessels), produce acetic acid, acetate esters and acetaldehyde. This last one imparts an oxidized flavor (bruised apple) and sometimes may affect young red wines before malolactic fermentation (MLF) (when sulfite levels are low and tanks are not topped). The amount of acetaldehyde may increase during aging from oxidation of ethyl alcohol, aeration, or the activity of film yeasts (Lambrechts and Pretorius, 2000).

In bottled wines, when the cork does not prevent the diffusion of oxygen, or the level of free sulfur dioxide is too low, *Pichia* and *Candida* may produce a film or a ring of cells adherent to the bottleneck. Furthermore, when the initial load is high, these yeasts may cause sediments thus *Pichia* and *Candida* are regarded as indicators of poor Good Manufacturing Practices (GMP).

10.3.1.3 *Fermenting Yeasts*

Saccharomyces cerevisiae and *Saccharomyces bayanus* are considered desirable yeasts in wine fermentation; on the other hand, they may also be responsible for spoilage. During fermentation, nutritional imbalance in the grape juice can occur, thus reduced sulfur compounds can impart off-flavor production and these taints may persist. In addition, when environmental conditions are poor (caused, e.g., by lack of grape juice nutrients or inadequate temperature) yeasts can be inhibited and a stuck fermentation can occur.

10.3.1.4 Dekkera/Brettanomyces bruxellensis

In bulk or bottled wines *D. bruxellensis* produces a "horse sweat" taint which is a threat to wine quality, since it causes economic losses. The effect is direct, due to the production of volatile phenols, and indirect since the technological measures to control its activity may also concur to reduce wine quality.

Dekkera bruxellensis is an undesirable contaminant, due to acetic acid production and a "mousy" taint. It is a rather elusive yeast and due to its low growth rate, and it has been isolated only rarely from grapes (Guerzoni and Marchetti, 1987; Renouf and Lonvaud-Funel, 2007) and wineries (Connel et al., 2002). For this reason, selective media and long incubation periods are needed for its recovery. *Dekkera bruxellensis* is not tolerant to ethanol or preservatives but remains viable for long periods so that it can proliferate when conditions become less hard. Sometimes it was detected in sparkling wines thanks to its resistance to carbon dioxide or in still white wines, which causes cell death under normal amounts of ethanol and pH values.

In Table 10.2 are summarized the main compounds responsible of yeast spoilage.

TABLE 10.2 Compounds Responsible of Spoilage

Compounds	Characteristics
Sulfur compounds	Sulfur compounds, in high amounts, rend the quality of wine unacceptable, thus it is necessary to act preventively, selecting *S. cerevisiae* strains that produce limited amounts of hydrogen sulfide (H_2S).
	H_2S and volatile sulfur compounds are characterized by high volatility and reactivity and are responsible for off-flavors. During fermentation (through the sulfate reduction pathway) H_2S is produced by yeasts and by environmental factors (must composition, fermentation temperature, wine pH and the use of fungicides containing elemental sulfur). When H_2S reacts with other wine components (e.g., cystine, cysteine, glutathione, methionine, etc.) to produce mercaptans, thiols, and disulfides, they are perceived as skunky, onion, cabbage, rubber, and garlic off-flavors (du Toit and Pretorius, 2000).
	Tominaga et al. (1998) reported that the volatile thiols 4-mercapto-4-methylpentan-2-one, 4-mercapto-4-methylpentan-2-ol, and 3-mercaptohexan-l-ol, responsible for box tree, broom, passion fruit, guava, and conifer aromas in Sauvignon blanc wines, were formed by yeast during the alcoholic fermentation due to the degradation of the corresponding S-cysteine conjugated present in the grapes (du Toit and Pretorius, 2000).
Volatile phenol compounds and volatile acidity	Phenolic compounds such as vinylphenols (4-vinylguaiacol and 4-vinylphenol in white wines) and ethylphenols (4-ethylguaiacol and 4-ethyl phenol in red wines) are produced by decarboxylation (cinnamate decarboxylase) and reduction of hydroxycinnamic acids (p-coumaric- and ferulic acid). These compounds affect the taste, color, and odor of wines. They can give phenolic off aromas/odors often described as animal, stable, horse sweat, medical, "elastoplast" when present above their threshold levels (>750 and >440 μg/L for 4-vinylguaiacol and 4-vinylphenol; >100 and >600 μg/L for 4-ethylguaiacol and 4-ethyl phenol, respectively).
	Brettanomyces is most frequently identified in wines with phenolic off-odors described as mousy, barnyard-like, horsy, wet dog, tar, tobacco, creosote, leathery, and pharmaceutical.
	Contaminated wines are characterized by an increase in volatile acidity, due to the oxidation of acetaldehyde to acetic acid (instead of ethanol) and it is known that acetic acid is the major volatile acid in

(*Continued*)

TABLE 10.2 (Continued)

Compounds	Characteristics
	wine (>90%); its threshold value is of 0.7–1.1 g/L depending on type of wine. Yeasts involved in the acetification of wine above objectionable levels include *Brettanomyces* and its anamorph *Dekkera, P. anomala, K. apiculata,* and *Candida krusei.*
	Saccharomyces strains displayed variation in their production of acetate and this phenomenon is influenced by fermentation temperature, pH, juice composition (sugar and nitrogen levels), levels of acetyl-CoA synthetase enzyme and the presence of other microorganisms (du Toit and Pretorius, 2000).
Deacidification	The acidity of wine has a direct impact on the flavor of the wine and indirect on the pH, color, stability and quality of the wine. Grape juice and wine contain organic and inorganic acids (tartaric, malic, citric, acetic, lactic, and succinic). In cooler climatic regions (Europe, Canada, and the USA) the titratable acidity is high and the pH low, vice versa in warmer regions (South Africa, Australia, and South America); thus, the deacidification process become necessary in cooler climate regions to ensure a product that is balanced and not perceived as tart due to high levels of acidity and low pH.
	Yeast strains are also used to degrade malic acid: *S. cerevisiae* is rarely used as it is a weak utilizer of malic acid as species of the genus *Candida, Hansenula, Kloeckera,* and *Pichia,* whereas *S. pombe, Schizosaccharomyces malidevorans,* and *Z. bailii* can strongly degrade malate.
	In addition, deacidification causes the loss in red color due to an increase in pH: the color may change from full red to a bluish hue (du Toit and Pretorius, 2000).
Formation of ethyl carbamate	By using the arginase enzyme, *S. cerevisiae* degrades L-arginine (one of the major amino acids present in grape juice and wine) to ornithine, ammonia, and carbon dioxide.
	Urea, formed as an intermediate product, is secreted by certain yeasts into the wine, where the reaction between ethanol and urea produces ethyl carbamate (also known as urethane), which is considered to be carcinogen. *S. cerevisiae* can thus affect the quality of wine by providing precursors for the formation of ethyl carbamate; thus, it is preferable to select wine yeast strains that are low urea producers (du Toit and Pretorius, 2000).

10.3.2 Lactic Acid Bacteria (LAB)

LAB associated with wine are *Lactobacillus*, *Leuconostoc*, *Oenococcus*, and *Pediococcus*. They are Gram-positive, catalase-negative, nonmotile, nonspore-forming, rod- and coccus shaped.

They conduct MLF producing lactic acid but they affect also the quality of wine as spoilage microorganisms when they proliferate at the wrong time during the winemaking process. In addition, many secondary metabolites produced are volatile, thus potentially affecting the sensorial qualities of wine. Fortunately, most spoilage scenarios can be avoided with correct hygiene management during the vinification and maturation process (Bartowsky, 2009).

Initially, in healthy grapes, LAB are found in low numbers ($<10^3$ ufc/g); then fungi and/or AAB can stimulate their growth. When they survive to alcoholic fermentation, they can reach a population of 10^6-10^8 CFU/mL in the final product and, depending on the species, exert different detrimental effects on the quality of wine.

LAB are divided into three groups according to their metabolic activity:

- obligately homofermentative;
- obligately heterofermentative;
- facultatively heterofermentative.

Homofermentative and heterofermentative LAB produce D-lactic acid, which is associated with spoilage (as the L-lactic acid is produced during MLF); homofermentative LAB (lactobacilli and pediococci), through the Embden-Meyerhof-Parnas (glycolytic) pathway and heterofermentative lactobacilli (*Leuconostoc* and *Oenococcus* spp.) through the 6-phosphogluconate pathway. Heterofermentative LAB produce also acetic acid and concur to increase the acid content of wine. du Toit and Pretorius (2000) reported that acetic acid, implicated in stuck or sluggish fermentations as *Oenococcus oeni* and *Lactobacillus kunkeei*, can affect the rate of yeast-driven alcoholic fermentation. Practically, in the early stages of alcoholic fermentation, these LAB species (10^9 CFU/mL) produce acetic acid, which inhibits the growth and fermentation of *Saccharomyces*, influencing the rate at which the grape juice is converted to ethanol. Their presence is favored in wines without the addition of SO_2. Acid production problems caused by LAB can be removed if good winemaking practices are followed (du Toit and Pretorius, 2000).

Liu et al. (2006) and Zhang et al. (2005) introduced two new ethanol-tolerant pediococci, Pediococcus *cellicola* and Pediococcus *ethanolidurans*, respectively, as they tolerate 6.5% or more ethanol at pH 3.5.

In the following paragraphs we resumed the main causes of spoilage by LAB.

10.3.2.1 Re-Fermentation

Re-fermentation can occur in bottled wine with a $pH > 3.5$ in the presence of LAB and nutrients that enhance their growth. This problem can be corrected by the addition of tartaric acid (du Toit and Pretorius, 2000).

10.3.2.2 Mannitol

Heterofermentative lactobacilli are involved in the fructose metabolism, which can result in the formation of mannitol, a six-carbon sugar alcohol (Wisselink et al., 2002). Itself, mannitol is not considered dangerous or responsible for spoilage; the problem is the concomitant production of acetic acid, D-lactic acid, *n*-propanol, 2-butanol, and diacetyl, thus, the wine is perceived as viscous and sweetish. Spoiled wine can also be perceived of as having a slimy texture with a vinegar-estery aroma and slightly sweet taste (Bartowsky, 2009).

10.3.2.3 Ropiness

The genera *Leuconostoc* and *Pediococcus* are implicated in ropiness, which causes an increased viscosity and a slimy appearance. Viscosity is due to the production of extracellular polysaccharide, composed of D-glucan, in the presence of ethanol. *Pediococcus damnosus*, *P. pentosaceus*, and *P. claussenii* (Dobson et al., 2002) are the main species responsible of this spoilage and are resistant to ethanol, SO_2 and pH; thus ropiness could be controlled by lowering the pH to under 3.5 (du Toit and Pretorius, 2000; Bartowsky, 2009).

10.3.2.4 Mousiness

Similarly to the spoilage yeast due to *Brettanomyces*, heterofermentative lactobacilli (*Lactobacillus brevis*, *L. hilgardii*, and *L. cellobiosus*, now synonymous with *Lactobacillus fermentum*) are responsible for mousiness. Mousy wines result from the metabolism of ornithine and lysine, which lead to nitrogen-heterocyclic compounds (2-acetyltetrahydropyridine (ACTPY), 2-acetyl-1-pyrroline (ACPY), and 2-ethyltetrahydropyridine (ETPY)) perceived as unpleasant compounds (Costello et al., 2001; Costello and Henschke, 2002). More specifically, these compounds are perceived as a persistent aftertaste reminiscent of caged mice; an increase in pH renders the compounds volatile (Bartowsky, 2009).

10.3.2.5 Organic Acid Utilization

LAB metabolize citric, tartaric, and sorbic acids, affecting the wine quality to a degree such that the wine is considered spoiled (du Toit and Pretorius, 2000).

Citric acid catabolism is linked to malic acid degradation or MLF: wine acidity is reduced by the decarboxylation of L-malic acid to form L-lactic acid, thus reduced acidity together with an increase in pH impart a "softer mouth feel" in wine.

During citrate metabolism are produced some compounds that, when produced in elevated concentrations, contribute negatively to the complexity of the wine. The most important is the diketone ($CH_3-CO-CO-CH_3$), 2,3-butanedione or diacetyl (sometimes referred as biacetyl) (Bartowsky and Henschke, 2004). This compound at low concentrations (and combined with other compounds) is perceived as yeasty, nutty, and/or toasty while at high concentrations it has a characteristic buttery aroma and lactic character. In a range of 1−4 mg/L diacetyl confers a desirable "butterscotch or buttery" flavor to the wine but at exceeding concentrations (5−7 mg/L) it is considered as spoiler compound; its amount is higher in bottled red wine than white.

Diacetyl is produced by some species belonging to the genera *Streptococcus*, *Leuconostoc*, *Lactobacillus*, *Pediococcus*, and *Oenococcus*; in particular, pediococci strains cause haze and buttery and sour off-flavors from diacetyl and lactic acid formation (Juvonen et al., 2011).

Yeasts can also produce diacetyl from citrate, but the levels are not objectionable. In addition, ML bacteria may also influence aroma and flavor by various mechanisms, including the production of volatile secondary metabolites and the modification of grape and yeast-derived metabolites (Bartowsky and Henschke, 2004).

Lactobacillus plantarum and *L. brevis* are involved in the degradation of tartaric acid. *Lactobacillus plantarum* and *L. brevis* are facultatively and obligately heterofermentative, respectively, thus the degradation of tartaric acid leads to the obtainment of different products. In fact, through the tartrate dehydratase, tartaric acid is converted to oxaloacetic acid; then *L. plantarum* through its simple metabolism produces lactic acid, acetic acid, and CO_2 whilst *L. brevis* produces succinic acid, acetic acid, and CO_2. In both cases, acetic acid might increase the volatile acidity and wine becomes unacceptable (du Toit and Pretorius, 2000).

When sorbic acid (2,4-hexadienoic acid) is metabolized by LAB (in particular by *O. oeni*), the resulting is a "geranium tone," an off-odor typical of crushed geranium leaves. In detail sorbic acid is reduced to sorbinol through hydrogenation. Then, it will isomerize to form the alcohol 3,5-hexadiene-2-ol which reacts with ethanol to form 2-ethoxyhexa-3,5-diene, which has an odor reminiscent of crushed geranium leaves (*Pelargonium* spp.). Sorbic acid may be used as a chemical preservative to prevent yeast fermentation by *S. cerevisiae* in sweetened wines (but has no effect on LAB); in this case, care should be taken since auto-oxidation can take place resulting in products such as acrolein, crotonaldehyde, and formic acid (du Toit and Pretorius, 2000).

10.3.2.6 Amertume

This type of spoilage, amertume (bitter in French), is induced by acrolein, a chemical compound produced by *Lactobacillus*, *Leuconostoc*, *Oenococcus*, and *Pediococcus* (Bartowsky, 2009); they possess a dehydratase enzyme which converts glycerol into 3-hydroxypropionaldehyde and produces acrolein. The presence of this compound as a single component is not alarming, but when it reacts with the phenolic groups of anthocyanins, it produces bitterness in the wine (du Toit and Pretorius, 2000).

10.3.2.7 Biogenic Amines

Biogenic amines, deriving from their respective precursor amino acids, affect the hygienic quality or healthiness of the wine.

Histamine, tyramine, putrescine, cadaverine, and phenylethylamine are the best studied biogenic amines: at normal concentrations they are not considered dangerous, but the presence of ethanol and acetaldehyde might enhance their toxicity.

Certain wine compounds indicative of wine spoilage (higher alcohols, succinic acid, butyric acid, lactic acid, acetic acid, ethyl acetate, acetoin, and diethyl succinate) are correlated to high levels of biogenic amines.

Histamine can cause headaches, hypotension, and digestive problems, whereas tyramine and phenylethylamine are associated with migraines and hypertension if consumed in high concentrations (du Toit and Pretorius, 2000).

10.3.3 Acetic Acid Bacteria (AAB)

AAB belong to the family *Acetobacteriaceae* (Bartowsky and Henschke, 2008) and are Gram-negative, aerobic, catalase-positive microorganisms; from glucose AAB produce acetic acid. Their morphology may range from spherical, club-shaped, elongated, swollen, curved rods to filamentous. They are found in several habitats: flowers, fruit, and vegetables, wine and beer as spoilage microorganisms (because their metabolites result in unpleasant organoleptic characteristics) and vinegar as the primary fermenters.

Acetobacter, *Gluconobacter*, and *Gluconacetobacter* are the main genera associated with grape and wine spoilage. *Acetobacter* involved in winemaking are *A. aceti*, *A. hansenii*, *A. liquefaciens*, and *A. pasteurianus* (*A. hansenii* and *A. liquefaciens* have recently been reassigned as *Gluconacetobacter hansenii* and *G. liquefaciens*). *Gluconobacter* is represented by three species *G. asaii*, *G. frateurii*, and *G. oxydans*; this last one is the main specie found in association with grapes and grape must.

A recent taxonomy reports two new species named *Acetobacter oeni* and *A. tropicalis* isolated from spoiled red wine in Portugal (Silva et al., 2006) and from spontaneously fermenting Austrian wine, respectively (Silhavy and Mandl, 2006).

Due to their ethanol-tolerance, species of *Acetobacter* are often isolated from wine, whereas *Gluconobacter* prefer sugar-rich environments with low alcohol amounts, thus generally they are isolated from grape must. In addition, they are different for their oxidative ability: *Acetobacter* species oxidize ethanol to acetic acid and then to carbon dioxide and water, whereas *Gluconobacter* species cannot oxidize ethanol further than acetic acid as they do not have a complete citric acid cycle (Bartowsky and Henschke, 2008).

In unspoiled healthy grapes, AAB are generally present at 10^2-10^3 cells/g, and *G. oxydans* is the dominant species. When grapes are damaged, spoiled or infected, AAB reach a concentration of 10^6 cells/g, with *A. aceti* and *A. pasteurianus* as dominant species. The grapes assume an acetic smell, with the must containing levels as high as 3.9 g/L of acetic acid.

During bottling and storage of red wine *A. pasteurianus* can proliferate, thus an accurate management is necessary to avoid further spoilage (Bartowsky, 2009).

Acetic acid, acetaldehyde, and ethyl acetate are the main spoilage compounds produced by wine-associated AAB species.

Acetic acid confers a sour flavor and the legal limit is 1.2–1.4 g/L. At high concentrations, wine becomes objectionable and with a bitter, vinegar-like aroma.

It is known that acetic acid is generally produced in low concentrations by yeast (during alcoholic fermentation) and by LAB (during MLF) but through the metabolism of spoilage yeast, lactic, and AAB, it can happen that concentrations increase considerably.

Wines containing high amounts of AAB may contain significant amounts of acetaldehyde, an intermediate metabolite in the production of acetic acid from ethanol under low oxygen concentrations. Wines appear as "classic," nutty and sherry-like, being reminiscent of overripe bruised apples or nuts (aroma threshold 0.5 mg/L) (Francis and Newton, 2005). Acetaldehyde is able to bind SO_2, affecting the antimicrobial activity.

The ethyl ester of acetic acid, ethyl acetate, has a pungent aroma, solvent-like and reminiscent of nail polish remover (aroma threshold 7.5 mg/L) (Francis and Newton, 2005), and often reaches higher concentration.

Many of these aromas are particularly perceived upon opening a bottle of wine spoiled by AAB, due to the increase in acetic acid and ethyl acetate. In conjunction with aroma defects, often a visible ring of bacterial growth on the bottleneck is noted, at the junction between the surface of the wine and headspace created by the bottle closure.

Acetobacter aceti and *G. oxydans* convert glycerol (produced by yeast and molds) into dihydroxyacetone under aerobic conditions (ketogenesis). Dihydroxyacetone can affect the sensory quality of the wine with a sweet/

etherish property; it can also react with proline and produce a "crust-like" aroma. Dihydroxyacetone can affect the antimicrobial activity in the wine, due to its ability to bind SO_2 (Eschenbruch and Dittrich, 1986).

Acetobacter and *Gluconobacter* are able to oxidize lactic acid to acetoin when oxygen is present at low concentrations. Acetoin aroma is described as "butter-like," and the levels in wine ranged from 3 to 31.8 mg/L.

10.3.4 Endo-Spore-Forming Bacteria

Bacillus and *Clostridium* spp. have been rarely reported in microbiological spoilage of wines, nevertheless, when they found the optimal conditions for growth, they cause the reduction of the wine quality. du Toit and Pretorius (2000) reported that when *Bacillus subtilis*, *Bacillus circulans*, and *Bacillus coagulans* were inoculated into dessert wine, they were able to grow to 10^6-10^7 cells/mL, and an increased of volatile and total acidity was observed.

Clostridium spp. are implicated in low acid and high pH (>4.0) wine. They concur to the production of *n*-butyric acid, acetic acid, CO_2, hydrogen peroxide, butanol, acetone, and propanol.

10.3.5 Molds

Alternaria, *Aspergillus*, *Botrytis*, *Cladosporium*, *Mucor*, *Oidium*, *Penicillium*, *Plasmopara*, *Rhizopus*, and *Uncinula* are the most common molds isolated in grapes and are responsible for the spoilage of wine.

They cause:

1. Loss in juice yield
2. Prolonged pressing process caused by the slippery nature of infected grapes
3. Alteration of the chemical composition of wine as effect of the production of gluconic acid, higher levels of glycerol, oxidation of phenolic compounds
4. Production of ß-glucan that will negatively affect clarification
5. Production of off-flavors (such as acetic acid)
6. Stimulation of the growth of spoilage yeasts and bacteria

Molds are sensitive to ethanol, low pH, $SO_{2,}$ and anaerobiosis; although they are unable to survive in wine, molds secrete enzymes that alter the chemical composition of the grape juice. Molds produce antiyeast metabolites affecting alcoholic and MLFs and mycotoxins such as aflatoxins, patulin, and ochratoxin A (mainly produced by *Aspergillus* and *Penicillium* genera).

Finally, molds as *Aspergillus, Cladosporium, Monilia, Paecilomyces, Penicillium*, and *Trichoderma* are responsible of the contamination of corks and affect the quality of the wine by producing off-flavors perceived as a moldy, earthy or musty (du Toit and Pretorius, 2000).

10.4 CONCLUDING REMARKS AND FUTURE TRENDS

The winemaking process has been around for hundreds of years, but it is still threatened by some critical points that compromise its success, since contamination is always possible. Sometimes the solution of this problem is entrusted to the experience of winemakers without scientific background.

Chemical preservatives are generally used to control specific microorganisms that affect the quality of the end product. In addition, physical treatments (filtration, aseptic filling, thermal treatment) as well as nonthermal preservation techniques (high-pressure processing, high-pressure homogenization, pulsed electric field, hydrodynamic cavitation, ionizing irradiation, ultraviolet irradiation, and ultrasound) are used, but all of them offered efficient results only when the initial contamination is low. Challenge tests, mathematical models, and predictive microbiology are used as risk management tools to ensure microbiological safety and the stability of food products. Further efforts are necessary and the development of molecular biology techniques could be the future trend for identifying spoilers and assess the risks during vinification. On the other hand, a mutual exchange of information between researchers and winemakers could simplify and accelerate the process to ensure the stability of wine, but it will possible only if researchers and industries understand the importance of working more closely in the future.

REFERENCES

Barata, A., González, S., Malfeito-Ferreira, M., Querol, A., Loureiro, V., 2008. Sour-rot damaged grapes are sources of wine spoilage yeasts. FEMS Yeast Res. 8, 1008–1017.

Bartowsky, E.J., 2009. Bacterial spoilage of wine and approaches to minimize it. Lett. Appl. Microbiol. 48 (2), 149–156.

Bartowsky, E.J., Henschke, P.A., 2004. The 'buttery' attribute of wine – diacetyl – desirability, spoilage and beyond. Int. J. Food Microbiol. 96 (3), 235–252.

Bartowsky, E.J., Henschke, P.A., 2008. Acetic acid bacteria spoilage of bottled red wine-a review. Int. J. Food Microbiol. 125 (1), 60–70.

Connel, L., Stender, H., Edwards, C., 2002. Rapid detection and identification of *Brettanomyces* from winery air samples based on peptide nucleic acid analysis. Am. J. Enol. Vitic. 53, 322–324.

Costello, P., Lee, T.H., Henschke, P.A., 2001. Ability of lactic acid bacteria to produce *N*-heterocycles causing mousy off-flavor in wine. Aust. J. Grape Wine Res. 7 (3), 160–167.

Costello, P.J., Henschke, P.A., 2002. Mousy off-flavor of wine: precursors and biosynthesis of the causative *N*-heterocycles 2-ethyltetrahydropyridine, 2-acetyltetrahydropyridine, and 2-acetyl-1-pyrroline by *Lactobacillus hilgardii* DSM 20176. J. Agric. Food Chem. 50 (24), 7079–7087.

Dobson, C.M., Deneer, H., Lee, S., Hemmingsen, S., Glaze, S., Ziola, B., 2002. Phylogenetic analysis of the genus *Pediococcus*: *Pediococcus claussenii* sp. nov., a novel lactic acid bacterium isolated from beer. Int. J. Syst. Evol. Microbiol. 52 (6), 2003–2010.

Jeon, S.H., Kim, N.H., Shim, M.B., et al., 2015. Microbiological diversity and prevalence of spoilage and pathogenic bacteria in commercial fermented alcoholic beverages (Beer, Fruit Wine, Refined Rice Wine, and Yakju). J. Food Prot. 78 (4), 812–818.

du Toit, M., Pretorius, I.S., 2000. Microbial spoilage and preservation of wine: using weapons from nature's own arsenal-a review. South Af. J. Enol. Vitic. 21, 74–96.

Eschenbruch, B., Dittrich, H.H., 1986. Metabolism of acetic acid bacteria in relation to their importance to wine quality. Zentrablatt für Mikrobiologie 141 (4), 279–289.

Francis, I.L., Newton, J.L., 2005. Determining wine aroma from compositional data. Aust. J. Grape Wine Res. 11 (2), 114–126.

Guerzoni, E., Marchetti, R., 1987. Analysis of yeast flora associated with grape sour rot and of the chemical disease markers. Appl. Environ. Microbiol. 53 (3), 571–576.

Juvonen, R., Virkajärvi, V., Outi Priha, O., Laitila, A., 2011. Microbiological spoilage and safety risks in non-beer beverages. Espoo 2011. VTT Tiedotteita – Research Notes 2599. 1–107 + app. 4 p. <www.vtt.fi/inf/pdf/tiedotteet/2011/T2599.pdf> (accessed December 9, 2015).

Lambrechts, M.G., Pretorius, I.S., 2000. Yeast and its importance to wine aroma – a review. South Afr. J. Enol. Vitic. 21, 97–129.

Liu, L., Zhang, B., Tong, H., Dong, X., 2006. *Pediococcus ethanolidurans* sp. nov., isolated from the walls of a distilled-spirit-fermenting cellar. Int. J. Syst. Evol. Microbiol. 56 (10), 2405–2408.

Loureiro, V., Malfeito-Ferreira, M., 2003. Spoilage yeasts in the wine industry. Int. J. Food Microbiol. 86 (1–2), 23–50.

Malfeito-Ferreira, M., 2011. Yeasts and wine off-flavors: a technological perspective. Ann. Microbiol. 61, 95–102.

Peñas, E., di Lorenzo, C., Uberti, F., Restani, P., 2015. Allergenic proteins in enology: a review on technological applications and safety aspects. Molecules 20, 13144–13164 (accessed September 1, 2015).

Renouf, V., Lonvaud-Funel, A., 2007. Development of an enrichment medium to detect *Dekkera/Brettanomyces bruxellensis*, a spoilage wine yeast, on the surface of grape berries. Microbiol. Res. 162 (2), 154–157.

Silhavy, K., Mandl, K., 2006. *Acetobacter tropicalis* in spontaneously fermented wines with vinegar fermentation in Austria. Mitteilungen Klosterneuburg 56, 102–107.

Silva, L.R., Cleenwerck, I., Rivas, R., et al., 2006. *Acetobacter oeni* sp. nov., isolated from spoiled red wine. Int. J. Syst. Evol. Microbiol. 56 (1), 21–24.

Stratford, M., 2006. Food and beverage spoilage yeasts. In: Querol, A., Fleet, G. (Eds.), Yeasts in Food and Beverages. Springer, Berlin, pp. 335–379.

Suárez, R., Suárez-Lepe, J.A., Morata, A., Calderón, F., 2007. The production of ethylphenols in wine by yeasts of the genera *Brettanomyces* and *Dekkera*: a review. Food Chem. 102 (1), 10–21.

Tominaga, T., Peyrot Des Gachons, C., Dubourdieu, D., 1998. A new type of flavor precursors in *Vitis vinifera* L. cv. Sauvignon blanc: S-cysteine conjugates. J. Agric. Food Chem. 46, 5215–5219.

Van der Walt, J.P., van der Kerken, A.E., 1961. The wine yeasts of the cape: part V. Studies on the occurrence of *Brettanomyces intermedius* and *Brettanomyces schanderlii*. Antonie Van Leeuwenhoek 27, 81–90.

Wallerath, T., Poleo, D., Li, H., Förstermann, U., 2003. Red wine increases the expression of human endothelial nitric oxide synthase a mechanism that may contribute to its beneficial cardiovascular effects. J. Am. Coll. Cardiol. 41 (3), 471–478.

Wisselink, H.W., Weusthuis, R.A., Eggink, G., Hugenholtz, J., Grobben, G.J., 2002. Mannitol production by lactic acid bacteria: a review. Int. Dairy J. 12 (2–3), 151–161.

Zhang, B., Tong, H., Dong, X., 2005. *Pediococcus cellicola* sp. nov., a novel lactic acid coccus isolated from a distilled-spirit-fermenting cellar. Int. J. Syst. Evol. Microbiol. 55 (5), 2167–2170.

Part III

Future Trends in Food Spoilers

Chapter 11

The Impact of Biofilms on Food Spoilage

Barbara Speranza and Maria Rosaria Corbo
University of Foggia, Foggia, FG, Italy

11.1 INTRODUCTION

Biofilms are defined as aggregations of microorganisms attached to and growing on a surface (Costerton and Stewart, 2001): they are currently identified as an assemblage of surface-associated microbial cells that are enclosed in hydrated (from 85% to 95%) extracellular polymeric substances (EPS) mainly constituted of polysaccharides, proteins, phospholipids, teichoic, and even nucleic acids (Sauer et al., 2007).

The tendency to attach to surfaces is a spontaneous process and a prevalent mode of growth for microorganisms in nature. The first observation of microbial biofilms dates back to 1684 when, using a primitive light microscope, a microbial adhesion on tooth surfaces was observed and associated to a sessile community by Leeuwenhoek (1684). In 1920s the concept of bacterial biofilm was introduced by observing marine bacteria attached to ships' hulls; in 1980s, thanks to the development of new microscopy techniques, the bacteria were observed also on the solid surfaces of many ecological environments including waste water treatment systems and equipments used to produce some food products (Shi and Zhu, 2009).

A biofilm community may comprise single and/or multiple species of bacteria and form a single layer or three-dimensional structures. Any type of microorganisms, including those that cause spoilage and diseases, could form a biofilm (Parsek and Singh, 2003).

The first published report on food-borne bacterial biofilm is related to the adhesion of *Salmonella* to food surfaces, but the ability to form biofilms of *Listeria monocytogenes*, *Yersinia enterocolitica*, *Campylobacter jejuni*, *Escherichia coli* O157:H7, and other pathogens was also documented (Aarnisalo et al., 2007; Dykes et al., 2003; Lapidot et al., 2006; Waak et al., 2002).

The Microbiological Quality of Food. DOI: http://dx.doi.org/10.1016/B978-0-08-100502-6.00014-5

The big challenge linked to biofilm formation is that they represent a recalcitrant source for infections and over 80% of persistent bacterial infections were found to be associated with biofilms formed on industrial equipment surfaces in food processing plants (Janssens et al., 2008). Since biofilm inhabitants become more resistant to disinfectants and antimicrobials (Simões et al., 2010; Srey et al., 2013; Bridier et al., 2015), they have become problematic in a wide range of food industries, including dairy processing (Chmielewski and Frank, 2003), poultry and meat processing (Harvey et al., 2007; Sofos and Geornaras, 2010), and seafood processing (Shikongo-Nambabi, 2011). For example, *Bacillus cereus*, *E. coli*, *Shigella* spp., and *Staphylococcus aureus* have been often recovered in biofilms developing in the dairy and egg processing industries (Jan et al., 2011; Sharma and Anand, 2002; Shi and Zhu, 2009), whereas *Listeria* spp., *Staphylococcus* spp., and *Vibrio* spp. have been frequently isolated from industrial equipment surfaces in seafood processing plants (Bagge-Ravn et al., 2003; Gutierrez et al., 2012; Shi and Zhu, 2009).

Thus, it is clear that biofilm elimination from food processing facilities represents a big effort for food makers and researchers.

11.2 MECHANISMS OF MICROBIAL BIOFILM FORMATION

The biofilm formation is a stepwise and dynamical process consisting of (1) initial attachment, (2) irreversible attachment, (3) early development of biofilm architecture with maturation, and (4) biofilm dispersion (Fig. 11.1). The first step (1) includes four subprocesses which have been identified as: (i) preconditioning of the adhesion surface either by macromolecules present in the surrounding liquid or intentionally coated on the surface; (ii) transport of planktonic cells from the liquid to the surface; (iii) adsorption of cells at the surface; (iv) desorption of reversibly adsorbed cells. After the irreversible

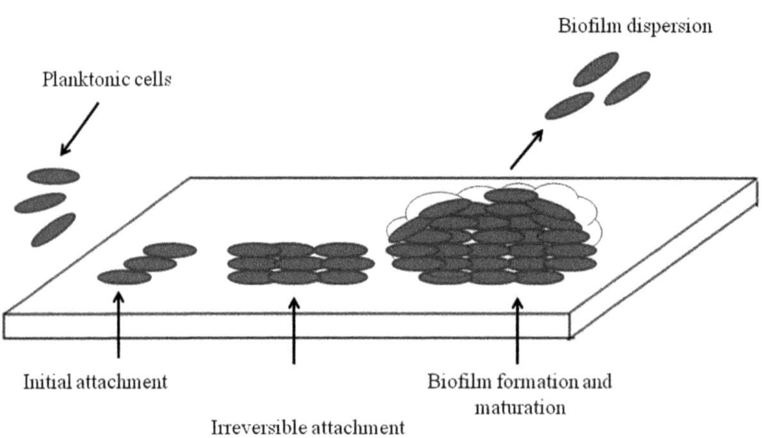

FIGURE 11.1 Steps of biofilm formation.

attachment of bacterial cells (2), other processes occur including the production of cell–cell communication molecules, transport of substrates to and within the biofilm and substrate metabolism by the biofilm cells. These processes are accompanied by cell growth, replication, and EPS production (3). The last step (4) consists into biofilm cells dispersion by detachment or sloughing (Breyers and Ratner, 2004).

In particular, the first initial attachment takes place in a very short time and it can be active or passive; in fact, it might be dependent on bacterial motility (active) or on the gravitational transportation of planktonic cells by diffusion or turbulence of the surrounding fluid (passive). At this stage, the adhesion is reversible since the attached microorganisms may detach from the surface and return to their planktonic forms at any moment (Stoodley et al., 2002a,b). The change from reversible to irreversible attachment is consequent to a shift from a weak interaction of the bacteria with the surface to a permanent bonding with the presence of EPS (Stoodley et al., 2002a,b). Simultaneously with EPS production, the accumulation and growth of adhered microorganisms leads to microcolony formation where the bond between the bacteria and the substratum was strengthened and the colony was stabilized (Donlan, 2002). This accumulation encourages the recruitment of planktonic cells from the surrounding medium as a result of cell-to-cell communication (Chmielewski and Frank, 2003).

The maturation is the step where the biofilm develops into an organized structure which can be flat or mushroom-shaped (Chmielewski and Frank, 2003): this step requires periods of 10 days or more (Stoodley et al., 2002a,b) during which different processes take place (i.e., alteration of genes encoding proteins involved in translation, metabolism, membrane transport and/or secretion, and gene regulation). A mature biofilm appears as a highly organized ecosystem where water channels are dispersed and able to provide passages for the exchange of nutrients, metabolites, and waste products (Sauer et al., 2007).

The last step in the biofilm formation cycle is dispersion where sessile cells might revert into their planktonic forms (Sauer et al., 2002). The potential causes of biofilm detachment are different: (1) external perturbation, such as increased fluid shear; (2) internal biofilm processes, such as endogenous enzymatic degradation; (3) release of EPS or surface-binding protein; (4) starvation and necessity to search for new nutrient-rich environments (Kaplan et al., 2003, 2004; O'Toole and Kaplan, 2000; Stoodley et al., 2002a,b). Regardless of the cause, the detachment may be defined as an active process allowing for the colonization of new niches (Sauer et al., 2002).

As it can be easily inferred, biofilm development is a very complex process, affected by several variables, such as environmental conditions (pH, nutrient availability, temperature, host proteins/adhesins, fluid dynamics), surface morphologies (chemistry, topography, physicochemistry), and microbiological factors (Gram-negative/positive, microbial shape, structure, molecular composition, species, physicochemistry, growth phase, age, presence of flagella, pili, capsules, or exopolymeric substances) (Guobjornsdottir et al., 2005). Fig. 11.2

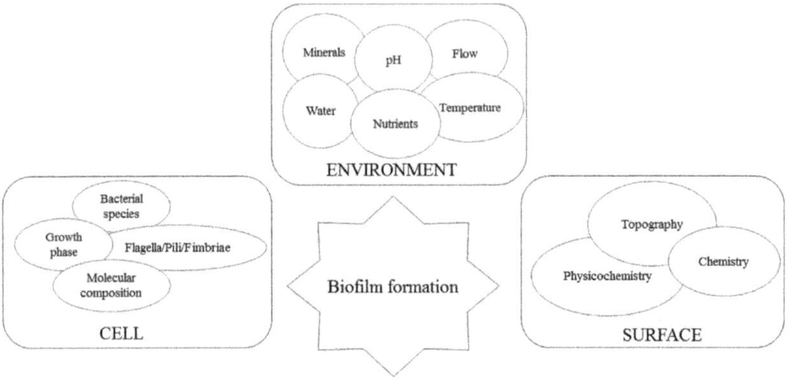

FIGURE 11.2 Main factors affecting biofilm formation.

summarizes the main factors affecting biofilm formation. Among these factors, environmental ones including pH, temperature, nutrient composition, and population characteristics of bacteria play an important role in the phenotypic change from planktonic cells to the sessile form. The effects of these variables on bacterial adhesion are not unique and may differ from species to species; for instance, it was demonstrated that maximum adhesion occurred at different pHs depending on the tested strain, different concentrations of phosphates or NaCl in the food matrix. The presence of other bacteria may influence the adhesion rate and biofilm maturation too. A good report covering all the factors affecting formation and architecture of microbial biofilm in the food industry is found in the review by Whitehead and Verran (2015).

The physical characteristics of solid surfaces are also one of the most important factors for biofilm formation because they influence initial cell attachment. High free energy and wet surfaces promote bacterial adhesion: hydrophilic surfaces (stainless steel, glass etc.), in fact, encourage cell adhesion more than hydrophobic surfaces (Buna-N rubber and other plastics). The extent of microbial attachment is correlated to the surface roughness. The conditioning of surface also plays a key role in the rate of bacterial attachment: the surfaces could be covered by a film of organic molecules such as proteins or polymeric substances produced by bacteria. Finally, the physiochemical properties of cell surfaces are an important aspect in active bacterial adhesion too: the surfaces of most bacterial cells are negatively charged and this keeps cells a short distance away from the surface due to an electrostatic repulsive force. The cell surface is able to reduce this repulsive force thanks to its hydrophobicity, which is due to fimbriae, flagella, and lipopolysaccharide (LPS) presence. Namely, fimbriae strongly help in attachment by overcoming the initial electrostatic repulsion barrier, even if many surface proteins are also reported to be involved in initial attachment (Latasa et al., 2005). The role of flagella is instead to bring bacteria to special adhesion sites rather than to act as adsorbents or adhesives, whereas LPS plays a role in initial surface attachment by conferring hydrophilic properties to the bacterial cell. Finally, the physiological status of cells also influences the hydrophobicity and the degree of bacterial adhesion; e.g., it is recognized that hydrophobicity of the cell surface increases with starvation and that spores possess higher hydrophobicity of their cell surfaces, thus adhering more quickly than vegetative cells to food-contact surfaces. Table 11.1 briefs the main variable shown to govern the attachment of the most studied pathogens.

11.3 BIOFILM IN FOOD ENVIRONMENTS, THEIR PERSISTENCE AND MAIN MICROORGANISMS INVOLVED

The formation of biofilms on industry surfaces can lead to serious hygiene problems and economic losses due to food spoilage and equipment impairment because pathogenic and spoilage biofilms could provide a reservoir of contamination increasing the risk for postprocessing contamination and leading to lowered shelf life of the food product and/or transmission of diseases (Bremer et al., 2006; Gram et al., 2007). Several reports revealed that biofilms are easily formed by various microorganisms on the equipment surfaces of the production line and those containing pathogens represent one of the major causes of contamination of food products (Guobjornsdottir et al., 2005; Shi and Zhu, 2009; Simões et al., 2010). The most common sources involved in biofilm accumulation are the floors, waste water pipes, bends in pipes, rubber seals, conveyor belts, stainless steel surfaces, etc. Buna-N and Teflon seals have also been implicated as important sites for biofilm formation (Chmielewski and Frank, 2003; Chmielewsky and Frank, 2006).

Biofilms not only present a considerable hygiene risk, but also cause serious economic losses through technical failures in water systems, cooling towers, heat exchangers, etc. For example, in the water distribution system, biofilms can increase energy consumption and mechanical blockage and accelerate corrosion of metal surfaces, leading to decrease in water velocity and carrying capacity, clogging of pipes, and decrease in efficiency of operations (Simões et al., 2010). Biofilm formation in heat exchangers and cooling towers has been a major problem for many years: the bacterial attachment, in fact, greatly reduces the heat transfer and operating efficiency of the processing equipment (Corbo et al., 2009). The microbial activity in biofilms, especially by the sulfate-reducing or acid-producing bacteria, causes corrosion of metal surfaces: these microorganisms catalyze chemical and biological reactions causing metal corrosion in pipelines and tanks, thus reducing the heat transfer efficacy if biofilms become sufficiently thick at plate heat exchangers and pipelines (Corbo et al., 2009). A significant number of reports underline the persistence of some food-borne pathogens on food-contact surfaces and biofilms able to seriously affect the quality and safety of the food products (Bridier et al., 2015; Srey et al., 2013).

TABLE 11.1 Main Variable Shown to Govern the Attachment of the Most Studied Pathogens

Microorganism	Main Variable Involved Into the Attachment	References
E. coli, Klebsiella pneumoniae, Salmonella enteritidis	Type 1 pili are the most common adhesion factor and their presence is critical to achieve a stable initial attachment.	Cookson et al. (2002), Schembri et al. (2005)
E. faecalis, Salmonella spp., *Vibrio parahaemolyticus* (VP), *S. aureus*	Surface proteins, such as Esp (enterococcal surface protein), BapA (biofilm-associated protein A), VP1443, and Bap (biofilm-associated protein) are involved in initial attachment or biofilm development on food-contact surfaces.	Latasa et al. (2005)
E. coli, Y. enterocolitica, B. cereus	Flagella have been shown to be critical to the initial contact and structuring of biofilms.	Kim et al. (2008), Houry et al. (2010), Van Houdt and Michiels (2005)
Gram-positive bacteria	A ubiquitous protein cell wall anchoring system mediated by transpeptidases is able to allow the covalent linkage of cell surface proteins harboring an LPXTG (Leu-Pro-any-Thr-Gly) aminoacid signature to the cell wall.	Guiton et al. (2009)
Gram-positive bacteria, particularly *Enterococcus* spp.	Pili have been definitely assigned as important players into the initial adhesion of bacteria to surfaces, in bacterial aggregation as well as in biofilm formation.	Garsin and Willems (2010), Mandlik et al. (2008), Nallapareddy et al. (2006), Sillanpaa et al. (2010)
L. monocytogenes	The surface roughness was reported as the most significant variable influencing on the biofilm forming ability.	Chaturongkasumrit et al. (2011)
L. monocytogenes	The electrostatic charge of bacterial cell walls (conferred by peptidoglycan anionic teichoic acids) and cell surface hydrophobicity (enhanced by the presence of lactic acid) have been shown to govern its attachment.	Briandet et al. (1999a,b)
Salmonella enterica	The BapA surface protein is directly involved in biofilm formation.	Latasa et al. (2005)
Salmonella spp.	The surface roughness has no correlation with adhesion.	Chia et al. (2009)
Salmonella spp., *L. monocytogenes*	Hydrophobic surfaces encourage the adhesion more than the hydrophilic ones.	Sinde and Carballo (2000)

As seen before, due to the wide range of contributing factors (surface type, availability of nutrients and oxygen, microbial species, etc.) biofilms can be quite diverse; in essence it can be said that each biofilm is different and the time available for its formation depends on numerous factors.

11.3.1 Dairy Industry

In dairy industry, biofilm formation depends on many factors: milk is a very perishable product and is truly vulnerable to contamination by various microorganisms. The major sources of contaminated milk and milk products are usually considered to come from improperly cleaned and sanitized equipments (Jessen and Lammert, 2003). Food-borne pathogens can enter the milk processing equipment by direct contact with contaminants in the farm environment (e.g., fecal contamination) and/or also through the water used in the milking machines (Oliver et al., 2005). These contaminating microorganisms are able to form biofilms that are difficult to eradicate and can act as a harbor and/or substrate for other microorganisms less inclined to form biofilm, increasing their probability of survival and subsequent contamination during food processing (Lapidot et al., 2006; Lehner et al., 2005; Lomander et al., 2004; Møretrø et al., 2003; Møretrø and Langsrud, 2004).

In dairy environments, the most commonly encountered bacteria belong to the genus *Enterobacter*, *Lactobacillus*, *Listeria*, *Micrococcus*, *Streptococcus*, *Bacillus*, and *Pseudomonas* (Salo et al., 2006; Sharma and Anand, 2002; Waak et al., 2002).

Outbreaks of pathogens associated with biofilms have been related to the presence of *L. monocytogenes*, *Y. enterocolitica*, *C. jejuni*, *Salmonella* spp., *Staphylococcus* spp., and *E. coli* O157:H7 (Aarnela et al., 2007; Dykes et al., 2003; Lapidot et al., 2006; Sharma and Anand, 2002; Waak et al., 2002). Table 11.2 reports on the microbial species most encountered in the dairy industry and other food environments.

11.3.2 Poultry and Meat Industry

Organic residues in poultry and meat processing could be a niche for microorganism accumulation and biofilm formation becoming a potential source of cross-contamination (Brooks and Flint, 2008; McLandsborough et al., 2006; Simões et al., 2010). Approximately 50% of the strains isolated on poultry farms were able to produce biofilms (Marin et al., 2009). In particular, *Salmonella* strains are the most common pathogens found on meat/poultry processing equipments since they are able to adhere to different materials (Teflon, stainless steel, rubber, and polyurethane) (Chia et al., 2009). Besides *Salmonella* spp., *Campylobacter* spp., and *E. coli* O157:H7 are also commonly found on contact surfaces of meat, poultry, and ready-to-eat meat products (Sanders et al., 2007; Silagyi et al., 2009), as shown in Table 11.2.

TABLE 11.2 Some Evidences About Biofilm Producers Food Industries (Dairy, Poultry and Meat, Seafood)

Dairy Industry		
Pseudomonas spp.	*Pseudomonas* spp. biofilms are the most important sessile bacteria causing spoilage of conventionally pasteurized milk products, acting in two different ways. First, they produce lipolytic and proteolytic enzymes that can survive pasteurization and even ultra-high temperature treatments, thus reducing the sensory quality and shelf life of the processed milk products. Second, *Pseudomonas* spp. can act in the postpasteurization step, causing spoilage of product during refrigerated storage.	Dogan and Boor (2003), Wiedmann et al. (2000)
Lactobacillus curvatus, *Lactobacillus fermentum*	They persisted on milk residues in cheese processing plants even after repeated cleaning.	Wong (1998)
Bacillus spp., particularly *B. cereus*	In dairy plants *B. cereus* represent more than 12% of the biofilms constitutive microflora. Contamination with this species is almost inevitable. In addition, *B. cereus* spores are both highly resistant and very hydrophobic, able to adhere easily to food processing equipments.	Lindsay et al. (2006), Sharma and Anand (2002)
Listeria spp.	This bacterium has been found in different parts of dairy plants and it is considered as a major food safety challenge in the dairy industry. It may also survive for a long time in dairy processing facilities, even for 7 years.	Vilar et al. (2007), Waak et al. (2002)
Poultry and Meat Industry		
C. jejuni	In poultry processing, *C. jejuni* in biofilm can survive longer than planktonic cells; in addition, it is well recognized that its attachment on surfaces is facilitated by a preexisting biofilm.	Hanning et al. (2008)
E. coli O157:H7	At low temperatures, this microorganism is able to adhere and survive on meat contact surfaces, and even increase in population as time progressed. An important factor influencing the attachment of *E. coli* O157:H7 is the presence of other microorganisms on the surfaces.	Castonguay et al. (2006), Dourou et al. (2011), Kim et al. (2006), Klayman et al. (2009), Marouani-Gadri et al. (2009)
Acinetobacter calcoaceticus	This species is able to form well-organized biofilms able to increase the opportunities for pathogens to grow in the food industry.	Burmølle et al. (2006), Habimana et al. (2010), Stewart and Franklin (2008)

Salmonella spp.	Biofilms of this pathogen are widespread in the poultry and meat industry. One of the most serious outbreaks associated to this pathogen took place in Spain and resulted in 2138 cases due to consumption of precooked chicken contaminated by the biofilm of *Salmonella* Hadar formed on a bend of a pipe transporting gravy in the processing site. *Salmonella* biofilms are very resistant to environmental stresses, antibiotics and disinfectants, and consequently they are extremely difficult to eradicate in the food industry.	Hoiby et al. (2010)
Seafood		
Aeromonas hydrophila	*A. hydrophila* easily produces a thin biofilm that can cover 40–50% of the surface of stainless steel and polystyrene acting as a severe source of contamination in processing facilities of seafood products.	Jahid et al. (2014)
V. parahaemolyticus	*V. parahaemolyticus* can form biofilms on various surfaces representing a serious problem especially for shrimp factories.	Aagesen et al. (2013), Frischkorn et al. (2013), Vezzulli et al. (2008)
V. cholerae (toxigenic or nontoxigenic)	*V. cholerae* forms biofilms by attaching to the different surfaces. Some authors suggested that biofilm formation is a major feature of *V. cholerae* virulence.	Donlan and Costerton (2002)
V. vulnificus	Pathogenic *V. vulnificus* are biofilm producers on the surfaces of polystyrene, glass tubes, and glass covers' lips. Mixed species of exogenous bacteria were recently reported to be capable of enhancing *V. vulnificus* biofilm formation.	Joseph and Wright (2004)
Salmonella spp.	Contamination by *Salmonella* spp. biofilms is a big challenge in seafood industry. Some authors reported a strong correlation between persistence of *Salmonella* spp. in the fish-processing industry and their biofilm formation ability.	Vestby et al. (2009)
L. monocytogenes	This organism has been shown to contaminate food processing apparatuses, raw materials, employees, and products, by forming biofilm. These biofilms are highly resistant to UV light, desiccation, and the sanitizing chemicals that are typically used for sterilization of processing equipment and surfaces, providing *L. monocytogenes* with the opportunity to spread to fish products.	Cruz and Fletcher (2011), Keskinen et al. (2009), Mai and Conner (2007), Stepanovic et al. (2004), Takahashi et al. (2009)

11.3.3 Seafood Processing

In the fish processing industry, both equipment and water quality are considered to be major concerns. Many types of fish-contaminated-bacteria are found to be biofilm-forming, including *Vibrio cholerae* (Faruque et al., 2006), *Vibrio parahaemolyticus* (Enos-Berlage et al., 2005), *Vibrio vulnificus* (Joseph and Wright, 2004), and *Vibrio alginolyticus* (Mizan et al., 2015). Many genera other than *Vibrio*, such as *L. monocytogenes*, *Salmonella* spp., *Bacillus* spp., *Aeromonas*, and *Pseudomonas* spp., are also known to be biofilm forming in fish and seafood processing (Rajkowski, 2009).

In a study by Guðbjörnsdóttir et al. (2005) it was shown that adhered bacteria are found frequently on the seafood processing lines, despite the fact that thorough cleaning and disinfection are carried out regularly. Gram-negative rods, particularly *Pseudomonas* spp., *Aeromonas* spp., *Enterobacteriaceae*, and yeasts, were isolated especially from shrimp processing plants, whereas *Pseudomonas* spp. and *Enterobacteriaceae* were found mainly in fish processing plants. *Pseudomonas putida* and *Pseudomonas fluorescens* were the main species of *Pseudomonas* spp. isolated from the shrimp factories. Genera such as *Pseudomonas* spp. and *Aeromonas* spp. are found to be significant biofilm producers able to enhance the biofilm formation of other genera. *Salmonella agona* and *Salmonella montevideo* were found to be good biofilm producers and their persistence in the factory environment depends significantly on the strain's biofilm forming ability (Vestby et al., 2009). Table 11.2 also summarizes some evidences about biofilm forming bacteria in seafood industries.

11.4 CONTROL OF BIOFILM FORMATION USING TRADITIONAL AND NOVEL METHODS

Since biofilms are a great concern in the food sectors, many studies have been done in order to gain a better understanding of their development and spread. The first and most important thing to do is to prevent biofilm formation by regularly cleaning and disinfecting in order to avoid allowing the cells to firmly attach to contact surfaces (Midelet and Carpentier, 2004; Simões et al., 2006).

Meyer (2003) suggested three different strategies to prevent biofilm formation: (1) disinfection "in time," before biofilm develops, (2) disinfection of biofilms using harsh disinfectants, and (3) inhibition of attachment of microbes by selecting surface materials not promoting attachment. More recently some researchers have suggested the incorporation of antimicrobial products into the surface materials themselves (Knetsch and Koole, 2011; Park et al., 2011) by coating surfaces with antimicrobials (Thouvenin et al., 2003) or by modifying the surfaces' physicochemical properties (Chandra et al., 2005; Rosmaninho et al., 2007). For instance, some authors proposed silicone rubber implants with covalently coupled quaternary ammonium coatings (Gottenbos et al., 2001), surfaces coated with silver (Hashimoto, 2001; Klueh et al., 2000) or surfaces preconditioned with surfactants (Cloete and Jacobs, 2001).

Apart from the aforementioned innovative techniques, good and regular cleaning remains the most recommended measure to prevent this troublesome problem (Simões et al., 2010). Many different chemical products may be used in cleaning, including surfactants, acid or alkaline products able to dissolve the EPS matrix and gain access to the bacteria cells (Simões et al., 2006). The use of high temperatures in cleaning, as well as the application of water turbulence and/or scrubbing, are effective means of reducing the physical force required (Maukonen et al., 2003).

It is worth noting that the application of cleaning procedures alone does not allow the complete removal of biofilm, but approximately 90% of the bacteria are detached from the surfaces and not killed: consequently, microorganisms might later reattach to other surfaces and form a biofilm in another site. Therefore a disinfection treatment is indispensable to completely eliminate sessile cells in order to prevent microbial growth on surfaces before a new production restarts (Gram et al., 2007).

A good disinfectant must be effective, safe, easy to use, quickly rinsed off from surfaces, leaving no toxic residues that could affect the health properties and sensory values of the final products. There is no one strategy able to control absolutely biofilm formation. The chemicals currently used in disinfection processes belong to the following types: acidic compounds, aldehyde-based biocides, caustic products; chlorine, hydrogen peroxide, iodine, isothiazolinones, ozone, peracetic acid, phenols, biguanidines, surfactants (Bremer et al., 2006; Chmielewski and Frank, 2007; Dosti et al., 2005; Simões et al., 2006). Their effectiveness can be influenced by several factors, such as the presence of organic material (including fat, carbohydrates, and protein-based materials), pH, temperature, water hardness, chemical inhibitors, concentration, and contact time (Bremer et al., 2002; Kuda et al., 2008). Among disinfectants, sodium hypochlorite (NaClO) is one of the most effective for biofilm inactivation (Ozdemir et al., 2010; da Silva et al., 2011); it is more powerful in low pH than alkaline pH environments (Araújo et al., 2011) and it was reported to be a potential biofilm antimicrobial against *S. aureus* (Toté et al., 2010), *Prevotella intermedia*, *Peptostreptococcus micros*, *Streptococcus intermedius*, *Fusobacterium nucleatum*, and *Enterococcus faecalis* (Spratt et al., 2001).

One widely used disinfectant is hydrogen peroxide (H_2O_2); it is very efficient against biofilms because of its highly oxidizing capacity based on the production of free radicals affecting the biofilm matrix (de Carvalho, 2007; de Carvalho and da Fonseca, 2007; Toté et al., 2010). In addition, this compound is safe and does not cause allergic reactions (Rideout et al., 2005), thus it can be used at a high concentration without negatively affecting the product quality.

Ozone is also a potent antimicrobial agent which can be used against bacteria, fungi, viruses, protozoa, and bacterial and fungal spores (Khardre et al., 2001): the microorganisms are eradicated by the disruption or of the cell membrane, which in turn leads to the leakage of the cell contents (Dosti et al., 2005; Lagrange et al., 2004).

An ideal antimicrobial agent due to its extreme oxidizing capacity is peracetic acid, which is used in water purification and as a disinfectant. Since this compound decomposes into safe and environmentally friendly residues in food (acetic acid and hydrogen peroxide), it can be applied without rinsing (Simões et al., 2010).

However, even if regularly applied, cleaning and disinfection procedures are not fully effective on biofilm structures and can induce the selection of resistant phenotypes (Simões et al., 2010). It was frequently demonstrated that microorganisms could develop a certain degree of resistance to such strategies or acquire it later through mutation or genetic exchange. These processes enable the bacteria to survive and proliferate also under higher concentration of disinfectants (Gilbert and Mcbain, 2003; McBain et al., 2002).

Consequently, new approaches to controlling biofilms in the food industry have been recently proposed. Simões et al. (2010) reviewed the most innovative green strategies proposed during the last decade, including those using enzymes, phages, microbial interactions, or metabolite molecules.

During the last 10 years, the international scientific community has been focusing its efforts on developing active "not-chemical" biofilm-targeted treatments able to eradicate biofilms, such as biocides generated on-site (ozone or acidic electrolyzed water) or natural compounds extracted from bacterial cultures or aromatic plants (e.g., surfactins or oils from oregano and thyme) (Ashraf et al., 2014).

Bacteriophages have been also identified as potential candidates to attack pathogens in bacterial biofilms because they can diffuse through the biofilm matrix to perform their antibacterial efficacy (Briandet et al., 2008; Donlan, 2009). In this regard, natural and engineered bacteriophages were proposed as an effective means against food-borne pathogen biofilms due to their ability to infect bacteria and provide a natural, highly specific, nontoxic, feasible approach to controlling several microorganisms involved in biofilm formation. When phages come into contact with biofilms, if the phage also possesses polysaccharide-degrading enzymes, or if considerable cell lysis is caused by the phage, the integrity of the biofilm may rapidly be destroyed (Simões et al., 2010). Despite these evidences, this technology has not yet been successfully developed and relatively little information is actually available on the action of bacteriophages on biofilms (Sutherland et al., 2004).

An additional tool to improve disinfection processes is the use of enzyme-based detergents. Enzymes can target and loosen the matrix triggering cell release from biofilms and allowing conventional disinfection agents to reach their bacterial targets. Depending on the composition of the biofilm matrix, different enzymes may be preferred, including proteases, cellulases, polysaccharide depolymerases, alginate lyases, dispersin B or DNAses (Xavier et al., 2005; Orgaz et al., 2007; Jabbouri and Sadovskaya, 2010; Lequette et al., 2010). The use of enzymes in biofilm control is still limited due to the low prices of the chemicals used today compared with the costs of the enzymes. In fact, the technology and production of these enzymes and the enzyme-based detergents are mostly patent-protected.

Another innovative approach to removing biofilms is the use of hyper-swimming tunneling bacteria. A recent study demonstrated that planktonic bacilli propelled by flagella were able to tunnel deep into a biofilm structure: the transient pores created in the matrix by these bacteria facilitated the penetration and action of disinfectants from the environment, enhancing the killing of biofilm cells (Houry et al., 2012).

Another promising way of preventing biofilm formation is the use of molecules which are able to interfere with bacterial communication, in order to prevent biofilm implementation or maturation: an example is found in brominated furanones which were shown to successfully reduce biofilm formation (Ni et al., 2009; Sintim et al., 2010).

According to the concept of guided microbial ecology, protective biofilms have been proposed as barrier microflora against the potential establishment of food-borne pathogens: some studies in this area have focused on the ecology of wooden shelves used for cheese ripening for the inhibition of the pathogen *L. monocytogenes* (Mariani et al., 2007; Speranza et al., 2009).

In addition to biological and/or chemical alternatives, some physical processes are also promising options to remove biofilms; e.g., ultrasonication was reported to be used as an efficient biofilm removal method (Oulahal-Lagsir et al., 2000a,b).

Table 11.3 focuses on some innovative techniques recently proposed to control biofilm formation.

11.5 FUTURE TRENDS

Spoilage and pathogenic microorganisms in biofilms formed in different food industries settings are a serious source of food contamination and clinical infection. As highlighted, biofilm formation depends on properties of the substratum and bacterial cells as well as on environmental factors including pH, temperature, and nutrient components. A better understanding of the mechanisms in cell adhesion and biofilm formation should facilitate the development of new strategies for cleaning, and inhibit biofilm formation, thus reducing disease outbreaks and economic losses.

Even though up to now the conventional control strategies are still used and still being developed, due to the increased resistance of biofilms to these conventional processes, novel means for their control are required. The search for new biofilm control strategies appears to be a necessity in

TABLE 11.3 Some Innovative Techniques Recently Proposed to Control Biofilm Formation

Treatment	Biofilm Type	References
Application of oregano oil, thymol, and carvacrol	Control of *Staphylococcus* biofilm formation	Nostro et al. (2007)
Application of essential oils, including thymol, carvacrol, and eugenol and their combinations	Control of industrial paper mill biofilms	Neyret et al. (2014)
Use of engineered phages	Control of *E. coli* and *L. monocytogenes* biofilm formation	Lu and Collins (2007)
Application of ultrasonication	Removal of dairy biofilms on stainless steel and polypropylene surfaces	Oulahal-Lagsir et al. (2000a,b)
Application of ultrasonication combined with ethylenediamine-tetraacetic acid (EDTA) or enzymes	Control of *S. aureus* and *E. coli* biofilm formation	Oulahal et al. (2007)
Application of ozonation and sonication	Removal of *L. monocytogenes* biofilm	Baumann et al. (2009)
Use of protease enzymes	Removal of *P. fluorescens* biofilm	Molobela et al. (2010)
Use of pectin esterase, pectin lyase, and cellulose	Removal of mature *P. fluorescens* biofilm	Orgaz et al. (2007)
Use of a combination of NaClO and hydrogen peroxide	Killing and removing Pseudomonas *aeruginosa* biofilm	DeQueiroz and Day (2007)
Application of a combined treatment using H_2O_2 and UV	Treatment of biofilms in a pilot-scale system	Vankerckhoven et al. (2011)
Use of a bacteriophage with an antibiotic compound	Removal of old *K. pneumoniae* biofilm	Verma et al. (2010)
Application of ultrasonic waves in presence of proteolytic and glycolytic enzymes	Removal of *E. coli* biofilm	Oulahal-Lagsir et al. (2003)
Synergistic use of an alkaline cleaner and a bacteriophage	Removal of *E. coli* O157:H7 biofilm formed on stainless steel	Sharma et al. (2005)
Application of biosurfactants produced by *Lactococcus lactis* 53	Control of biofilm formation on silicone rubber	Rodrigues et al. (2004)
Application of surfactin released by *Bacillus subtilis*	Control of biofilm formation by *S. enterica, E. coli,* and *Proteus mirabilis*	Mireles et al. (2001)
Application of a soluble polysaccharide released by *E. coli*	Control of biofilm formation by a wide range of Gram-positive and Gram-negative bacteria	Valle et al. (2006)
Application of cis-2decenoic acid released by *P. aeruginosa*	Control of biofilm formation by *B. subtilis, E. coli, S. aureus, K. pneumoniae, P. aeruginosa, P. mirabilis, Streptococcus pyogenes,* and *Candida albicans*	Davies and Marques (2009)

overcoming the biofilm resistance. Since each biofilm is different, until now there is no unique system which is able to remove all biofilms: consequently, any biofilm problem should be analyzed in detail to determine its nature in order to define and implement an effective cleaning and disinfection operation. In addition, further investigations should be performed to clarify some important aspects still lacking, such as the impact of antimicrobial products on microbial biofilms, their recovery responses to damage, how microorganisms can develop resistance and subsequently survive previously effective control procedures, and other mysterious aspects still connected with this phenomenon.

REFERENCES

Aagesen, A.M., Phuvasate, S., Su, Y.C., Hase, C.C., 2013. Persistence of *Vibrio parahaemolyticus* in the Pacific oyster *Crassostrea gigas* is a multifactorial process involving pili and flagella but not type III secretion systems or phase variation. Appl. Environ. Microbiol. 79, 3303–3305.

Aarnela, K., Lunden, J., Korkeala, H., Wirtanen, G., 2007. Susceptibility of *Listeria monocytogenes* strains to disinfectants and chlorinated alkaline cleaners at cold temperatures. LWT - Food Sci. Technol. 40, 1041–1048.

Aarnisalo, K., Lundén, J., Korkeala, H., Wirtanen, G., 2007. Susceptibility of *Listeria monocytogenes* strains to disinfectants and chlorinated alkaline cleaners at cold temperatures. LWT - Food Sci. Technol. 40 (6), 1041–1048.

Araújo, P., Lemos, M., Mergulhão, F., Melo, L., Simões, M., 2011. Antimicrobial resistance to disinfectants in biofilms. In: Mendéz-Vilas, A. (Ed.), Science against microbial pathogens: communicating current research and technological advances. Formatex Research Center, pp. 826–834.

Ashraf, M.A., Ullah, S., Ahmad, I., Qureshi, A.K., Balkhair, K.S., Abdur Rehman, M., 2014. Green biocides, a promising technology: current and future applications to industry and industrial processes. J. Sci. Food Agric. 94, 388–403.

Bagge-Ravn, D., Ng, Y., Hjelm, M., Christiansen, J.N., Johansen, C., Gram, L., 2003. The microbial ecology of processing equipment in different fish industries- analysis of the microflora during processing and following cleaning and disinfection. Int. J. Food Microbiol. 87 (3), 239–250.

Baumann, A.R., Martin, S.E., Feng, H., 2009. Removal of *Listeria monocytogenes* biofilms from stainless steel by use of ultrasound and ozone. J. Food Prot. 72 (6), 1306–1309.

Bremer, P.J., Monk, I., Butler, R., 2002. Inactivation of *Listeria monocytogenes/Flavobacterium* spp. biofilms using chlorine: impact of substrate, pH, time and concentration. Lett. Appl. Microbiol. 35 (4), 321–325.

Bremer, P.J., Fillery, S., McQuillan, A.J., 2006. Laboratory scale clean-in-place (CIP) studies on the effectiveness of different caustic and acid wash steps on the removal of dairy biofilms. Int. J. Food Microbiol. 106 (3), 254–262.

Breyers, J.D., Ratner, J.P., 2004. Bioinspired implant materials befuddle bacteria. ASM News 70, 232–237.

Briandet, R., Leriche, V., Carpentier, B., Bellon-Fontaine, M.N., 1999a. Effects of the growth procedure on the surface hydrophobicity of *Listeria monocytogenes* cells and their adhesion to stainless steel. J. Food Prot. 62, 994–998.

Briandet, R., Meylheuc, T., Maher, C., Bellon-Fontaine, M.N., 1999b. *Listeria monocytogenes* Scott A: cell surface charge, hydrophobicity and electron donor and acceptor characteristics under different enviromental growth conditions. Appl. Environ. Microbiol. 65, 5328–5333.

Briandet, R., Lacroix-Gueu, P., Renault, M., Lecart, S., Meylheuc, T., Bidnenko, E., et al., 2008. Fluorescence correlation spectroscopy to study diffusion and reaction of bacteriophages inside biofilms. Appl. Environ. Microbiol. 74 (7), 2135–2143.

Bridier, A., Sanchez-Vizuete, P., Guilbaud, M., Piard, J.C., Naïtali, M., Briandet, R., 2015. Biofilm-associated persistence of food-borne pathogens. Food Microbiol. 45, 167–178.

Brooks, J.D., Flint, S.H., 2008. Biofilms in the food industry: problems and potential solutions. Int. J. Food Sci. Technol. 43 (12), 2163–2176.

Burmølle, M., Webb, J.S., Rao, D., Hansen, L.H., Sørensen, S.J., Kjelleberg, S., 2006. Enhanced biofilm formation and increased resistance to antimicrobial agents and bacterial invasion are caused by synergistic interactions in multispecies biofilms. Appl. Environ. Microbiol. 72 (6), 3916–3923.

Castonguay, M.H., van der Schaaf, S., Koester, W., Krooneman, J., van der Meer, W., Harmsen, H., et al., 2006. Biofilm formation by *Escherichia coli* is stimulated by synergistic interactions and co-adhesion mechanisms with adherence proficient bacteria. Res. Microbiol. 157 (5), 471–478.

Chandra, J., Patel, J.D., Li, J., Zhou, G., Mukherjee, P.K., McCormick, T.S., et al., 2005. Modification of surface properties of biomaterials influences the ability of *Candida albicans* to form biofilms. Appl. Environ. Microbiol. 71 (12), 8795–8801.

Chaturongkasumrit, Y., Takahashi, H., Keeratipibul, S., Kuda, T., Kimura, B., 2011. The effect of polyesterurethane belt surface roughness on *Listeria monocytogenes* biofilm formation and its cleaning efficiency. Food Control. 22 (12), 1893–1899.

Chia, T.W.R., Goulter, R.M., McMeekin, T., Dykes, G.A., Fegan, N., 2009. Attachment of different *Salmonella* serovars to materials commonly used in a poultry processing plant. Food Microbiol. 26 (8), 853–859.

Chmielewski, R.A.N., Frank, J.F., 2003. Biofilm formation and control in food processing facilities. Compr. Rev. Food Sci. Food Saf. 2 (1), 22–32.

Chmielewsky, R.A.N., Frank, J.F., 2006. A predictive model for heat inactivation of *Listeria monocytogenes* biofilm on buna-N rubber. LWT - Food Sci. Technol. 39, 11–19.

Chmielewski, R.A.N., Frank, J.F., 2007. Inactivation of *Listeria monocytogenes* biofilms using chemical sanitizers and heat. In: Blaschek, H.P., Wang, H.H., Agle, M.E. (Eds.), Biofilms in the Food Environment. Blackwell Publishing, Ames, IA, pp. 73–104.

Cloete, T.E., Jacobs, L., 2001. Surfactants and the attachment of *Pseudomonas aeruginosa* to 3CR12 stainless steel and glass. Water SA 27, 21–26.

Cookson, A.I., Cooley, W.A., Woodward, M.J., 2002. The role of type 1 and curli fimbriae of Shiga toxin-producing *Escherichia coli* in adherence to abiotic surfaces. Int. J. Med. Microbiol. 292, 195–205.

Corbo, M.R., Speranza, B., Sinigaglia, M., 2009. Microbial biofilms in the food industry: only negative implications? In: Bailey, W.C. (Ed.), Biofilms: Formation, Development and Properties. Nova Science Publishers. pp. 585–594.

Costerton, J.W., Stewart, P.S., 2001. Battling biofilms. Sci. Am. 285 (1), 74–81.

Cruz, C.D., Fletcher, G.C., 2011. Prevalence and biofilm-forming ability of *Listeria monocytogenes* in New Zealand mussel (*Perna canaliculus*) processing plants. Food Microbiol. 28, 1387–1393.

da Silva, P.M.B., Acosta, E.J.T.R., de Rezende Pinto, L., Graeff, M., Spolidorio, D.M.P., Almeida, R.S., et al., 2011. Microscopical analysis of *Candida albicans* biofilms on heat-polymerised acrylic resin after chlorhexidine gluconate and sodium hypochlorite treatments. Mycoses 54 (6), 712–717.

Davies, D.G., Marques, C.N., 2009. A fatty acid is responsible for inducing dispersion in microbial biofilms. J. Bacteriol. 191, 1393–1403.

de Carvalho, C.C.C.R., 2007. Biofilms: recent developments on an old battle. Recent Patents Biotechnol. 1 (1), 49–57.

de Carvalho, C.C.C.R., da Fonseca, M.M., 2007. Assessment of three-dimensional biofilm structure using an optical microscope. Biotechniques 42 (5), 616–620.

DeQueiroz, G.A., Day, D.F., 2007. Antimicrobial activity and effectiveness of a combination of sodium hypochlorite and hydrogen peroxide in killing and removing *Pseudomonas aeruginosa* biofilms from surfaces. J. Appl. Microbiol. 103, 794–802.

Dogan, B., Boor, K.J., 2003. Genetic diversity and spoilage potential among *Pseudomonas* spp. isolated from fluid milk products and dairy processing plants. Appl. Environ. Microbiol. 69, 130–138.

Donlan, R., Costerton, J.W., 2002. Biofilms: survival mechanisms of clinically relevant microorganisms. Clin. Microbiol. Rev. 15, 167–193.

Donlan, R.M., 2002. Biofilms: microbial life on surfaces. Emerg. Infect. Dis. 8 (9), 881–890.

Donlan, R.M., 2009. Preventing biofilms of clinically relevant organisms using bacteriophage. Trends Microbiol 17, 66–72.

Dosti, B., Guzel-Seydim, Z., Greene, A.K., 2005. Effectiveness of ozone, heat and chlorine for destroying common food spoilage bacteria in synthetic media and biofilms. Int. J. Dairy Technol. 58 (1), 19–24.

Dourou, D., Beauchamp, C.S., Yoon, Y., Geornaras, I., Belk, K.E., Smith, G.C., et al., 2011. Attachment and biofilm formation by *Escherichia coli* O157:H7 at different temperatures, on various food-contact surfaces encountered in beef processing. Int. J. Food Microbiol. 149 (3), 262–268.

Dykes, G.A., Sampathkumar, B., Korber, D.R., 2003. Planktonic or biofilm growth affects survival, hydrophobicity and protein expression patterns of a pathogenic *Campylobacter jejuni* strain. Int. J. Food Microbiol. 89 (1), 1–10.

Enos-Berlage, J.L., Guvener, Z.T., Keenan, C.E., McCarter, L.L., 2005. Genetic determinants of biofilm development of opaque and translucent *Vibrio parahaemolyticus*. Mol. Microbiol. 55 (4), 1160–1182.

Faruque, S.M., Biswas, K., Udden, S.M.N., Ahmad, Q.S., Sack, D.A., Nair, G.B., et al., 2006. Transmissibility of cholera: in vivo-formed biofilms and their relationship to infectivity and persistence in the environment. Proc. Natl. Acad. Sci. USA. 103 (16), 6350–6355.

Frischkorn, K.R., Stojanovski, A., Paranjpye, R., 2013. *Vibrio parahaemolyticus* type IV pili mediate interactions with diatom-derived chitin and point to an unexplored mechanism of environmental persistence. Environ. Microbiol. 15, 1416–1427.

Garsin, D.A., Willems, R.J., 2010. Insights into the biofilm lifestyle of enterococci. Virulence 1, 219–221.

Gilbert, P., Mcbain, A.J., 2003. Potential impact of increased use of biocides in consumer products on prevalence of antibiotic resistance. Clin. Microbiol. Rev. 16 (2), 189–208.

Gottenbos, B., van der Mei, H.C., Klatter, F., Nieuwenhuis, P., Busscher, H.J., 2001. In vitro and in vivo antimicrobial activity of covalently coupled quaternary ammonium silane coatings on silicone rubber. Biomaterials 23, 1417–1423.

Gram, L., Bagge-Ravn, D., Ng, Y.Y., Gymoese, P., Vogel, B.F., 2007. Influence of food soiling matrix on cleaning and disinfection efficiency on surface attached *Listeria monocytogenes*. Food Control 18 (10), 1165–1171.

Guðbjörnsdóttir, B., Einarsson, H., Thorkelsson, G., 2005. Microbial adhesion to processing lines for fish fillets and cooked shrimp: influence of stainless steel surface finish and presence of gram-negative bacteria on the attachment of *Listeria monocytogenes*. Food Technol. Biotechnol. 43 (1), 55–61.

Guiton, P.S., Hung, C.S., Kline, K.A., Roth, R., Kau, A.L., Hayes, E., et al., 2009. Contribution of autolysin and sortase A during *Enterococcus faecalis* DNA-dependent biofilm development. Infect. Immun. 77, 3626–3638.

Gutierrez, D., Delgado, S., Vazquez-Sanchez, D., Martinez, B., Cabo, M.L., Rodriguez, A., et al., 2012. Incidence of *Staphylococcus aureus* and analysis of associated bacterial communities on food industry surfaces. Appl. Environ. Microbiol. 78, 8547–8554.

Habimana, O., Heir, E., Langsrud, S., Asli, A.W., Møretrø, T., 2010. Enhanced surface colonization by *Escherichia coli* O157:H7 in biofilms formed by an *Acinetobacter calcoaceticus* isolate from meat-processing environments. Appl. Environ. Microbiol. 76 (13), 4557–4559.

Hanning, I., Jarquin, R., Slavik, M., 2008. *Campylobacter jejuni* as a secondary colonizer of poultry biofilms. J. Appl. Microbiol. 105 (4), 1199–1208.

Harvey, J., Keenan, K.P., Gilmour, A., 2007. Assessing biofilm formation by *Listeria monocytogenes* strains. Food Microbiol. 24 (4), 380–392.

Hashimoto, H., 2001. Evaluation of the anti-biofilm effect of a new anti-bacterial silver citrate/lecithin coating in an in-vitro experimental system using a modified Robbins device. Kansenshogaku Zasshi 75 (8), 678–685.

Hoiby, N., Bjarnsholt, T., Givskov, M., Molin, S., Ciofu, O., 2010. Antibiotic resistance of bacterial biofilms. Int. J. Antimicrob. Agents 35, 322–332.

Houry, A., Briandet, R., Aymerich, S., Gohar, M., 2010. Involvement of motility and flagella in *Bacillus cereus* biofilm formation. Microbiology 156, 1009–1018.

Houry, A., Gohar, M., Deschamps, J., Tischenko, E., Aymerich, S., Gruss, A., et al., 2012. Bacterial swimmers that infiltrate and take over the biofilm matrix. Proc. Natl. Acad. Sci. USA. 109, 13088–13093.

Jabbouri, S., Sadovskaya, I., 2010. Characteristics of the biofilm matrix and its role as a possible target for the detection and eradication of *Staphylococcus epidermidis* associated with medical implant infections. FEMS Immunol. Med. Microbiol. 59, 280–291.

Jahid, I.K., Han, N., Ha, S.D., 2014. Inactivation kinetics of cold oxygen plasma depends on incubation conditions of *Aeromonas hydrophila* biofilm on lettuce. Food Res. Int. 55, 181–189.

Jan, S., Brunet, N., Techer, C., Le Marechal, C., Kone, A.Z., Grosset, N., et al., 2011. Biodiversity of psychrotrophic bacteria of the *Bacillus cereus* group collected on farm and in egg product industry. Food Microbiol. 28, 261–265.

Janssens, J.C.A., Steenackers, H., Robijns, S., Gellens, E., Levin, J., Zhao, H., et al., 2008. Brominated furanones inhibit biofilm formation by *Salmonella enterica* serovar *typhimurium*. Appl. Environ. Microbiol. 74 (21), 6639–6648.

Jessen, B., Lammert, L., 2003. Biofilm and disinfection in meat processing plants. Int. Biodeterior. Biodegradation 51 (4), 265–269.

Joseph, L.A., Wright, A.C., 2004. Expression of *Vibrio vulnificus* capsular polysaccharide inhibits biofilm formation. J. Bacteriol. 186 (3), 889–893.

Kaplan, J.B., Ragunath, C., Ramasubbu, N., Fine, D.H., 2003. Detachment of *Actinobacillus actinomycetemcomitans* biofilm cells by an endogenous b-hexosaminidase activity. J. Bacteriol. 185 (16), 4693–4698.

Kaplan, J.B., Ragunath, C., Velliyagounder, K., Fine, D.H., Ramasubbu, N., 2004. Enzymatic detachment of *Staphylococcus epidermidis* biofilms. Antimicrob. Agents Chemother 48 (7), 2633–2636.

Keskinen, L.A., Burke, A., Annous, B.A., 2009. Efficacy of chlorine, acidic electrolyzed water and aqueous chlorine dioxide solutions to decontaminate *Escherichia coli* O157:H7 from lettuce leaves. Int. J. Food. Microbiol. 132 (2–3), 134–140.

Khardre, M.A., Yousef, A.E., Kim, J.G., 2001. Microbiological aspects of ozone applications in food: a review. J. Food. Sci. 66 (9), 1242–1252.

Kim, H., Ryu, J.H., Beuchat, L.R., 2006. Attachment of biofilm formation by *Enterobacter sakazakii* on stainless steel and enteral feeding tubes. Appl. Environ. Microbiol. 72 (9), 5846–5856.

Kim, T.J., Young, B.M., Young, G.M., 2008. Effect of flagellar mutations on *Yersinia enterocolitica* biofilm formation. Appl. Environ. Microbiol. 74, 5466–5474.

Klayman, B.J., Volden, P.A., Stewart, P.S., Camper, A.K., 2009. *Escherichia coli* O157:H7 requires colonizing partner to adhere and persist in a capillary flow cell. Environ. Sci. Technol. 43 (6), 2105–2111.

Klueh, U., Wagner, V., Kelly, S., Johnson, A., Bryers, J.D., 2000. Efficacy of silver coated fabric to prevent bacterial colonization and subsequent device-based biofilm formation. J. Biomed. Mater. Res. 53 (6), 621–631.

Knetsch, M.L.W., Koole, L.H., 2011. New strategies in the development of antimicrobial coatings: the example of increasing usage of silver and silver nanoparticles. Polymers 3 (1), 340–366.

Kuda, T., Yano, T., Kuda, M.T., 2008. Resistances to benzalkonium chloride of bacteria dried with food elements on stainless steel surface. LWT - Food Sci. Technol. 41 (6), 988–993.

Lagrange, F., Reiprich, W., Hoffmann, M., 2004. CIP-cleaning and disinfection with ozone water. Fleischwirtschaft 84 (2), 112–114.

Lapidot, A., Romling, U., Yaron, S., 2006. Biofilm formation and the survival of *Salmonella typhimurium* on parsley. Int. J. Food Microbiol. 109 (3), 229–233.

Latasa, C., Roux, A., Toledo-Arana, A., Ghigo, J.M., Gamazo, C., Penades, J.R., et al., 2005. BapA, a large secreted protein required for biofilm formation and host colonization of *Salmonella enterica* serovar *enteritidis*. Mol. Microbiol. 58, 1322–1339.

Leeuwenhoek, A.V., 1684. Some microscopical observations about animals in the scurf of the teeth. Phil. Trans. 14, 568–574.

Lehner, A., Riedel, K., Eberl, L., Breeuwer, P., Diep, B., Stephan, R., 2005. Biofilm formation, extracellular polysaccharide production, and cell-to-cell signalling in various *Enterobacter sakazakii* strains: aspects promoting environmental persistence. J. Food Prot. 68, 2287–2294.

Lequette, Y., Boels, G., Clarisse, M., Faille, C., 2010. Using enzymes to remove biofilms of bacterial isolates sampled in the food -industry. Biofouling 26, 421–431.

Lindsay, D., Brozel, V.S., von Holy, A., 2006. Biofilm-spore response in *Bacillus cereus* and *Bacillus subtilis* during nutrient limitation. J. Food Prot. 69, 1168–1172.

Lomander, A., Schreuders, P., Russek-Cohen, E., Ali, L., 2004. Evaluation of chlorines' impact on biofilms on scratched stainless steel surfaces. Bioresour. Technol. 94 (3), 275–283.

Lu, T.K., Collins, J.J., 2007. Dispersing biofilms with engineered enzymatic bacteriophage. Proceedings of the National Academy of Sciences of the United States of America 104, 11197–11202.

Mai, T.L., Conner, D.E., 2007. Effect of temperature and growth media on the attachment of *Listeria monocytogenes* to stainless steel. Int. J. Food Microbiol. 120, 282–286.

Mandlik, A., Swierczynski, A., Das, A., Ton-That, H., 2008. Pili in Gram-positive bacteria: assembly, involvement in colonization and biofilm development. Trends Microbiol. 16, 33–40.

Mariani, C., Briandet, R., Chamba, J.F., Notz, E., Carnet-Pantiez, A., Eyoug, R.N., et al., 2007. Biofilm ecology of wooden shelves used in ripening the French raw milk smear cheese Reblochon de Savoie. J. Dairy Sci. 90, 1653–1661.

Marin, C., Hernandiz, A., Lainez, M., 2009. Biofilm development capacity of *Salmonella* strains isolated in poultry risk factors and their resistance against disinfectants. Poult. Sci. 88 (2), 424–431.

Marouani-Gadri, N., Augier, G., Carpentier, B., 2009. Characterization of bacterial strains isolated from a beef-processing plant following cleaning and disinfection and influence of isolated strains on biofilm formation by *Sakai*¨and EDL 933 *E. coli* O157:H7. Int. J. Food Microbiol. 133 (1–2), 62–67.

Maukonen, J., Mättö, J., Wirtanen, G., Raaska, L., Mattila-Sandholm, T., Saarela, M., 2003. Methodologies for the characterization of microbes in industrial environments: a review. J. Ind. Microbiol. Biotechnol. 30 (6), 327–356.

McBain, A.J., Rickard, A.H., Gilbert, P., 2002. Possible implications of biocide accumulation in the environment on the prevalence of bacterial antibiotic resistance. J. Ind. Microbiol. Biotechnol. 29 (6), 326–330.

McLandsborough, L., Rodriguez, A., Pérez-Conesa, D., Weiss, J., 2006. Biofilms: at the interface between biophysics and microbiology. Food Biophys. 1 (2), 94–114.

Meyer, B., 2003. Approaches to prevention, removal and killing of biofilms. Int. Biodeter. Biodegr. 51 (4), 249–253.

Midelet, G., Carpentier, B., 2004. Impact of cleaning and disinfection agents on biofilm structure and on microbial transfer to a solid model food. J. Appl. Microbiol. 97 (2), 262–270.

Mireles, J.R., Toguchi, A., Harshey, R.M., 2001. *Salmonella enterica* serovar Typhimurium swarming mutants with altered biofilm-forming abilities: surfactin inhibits biofilm formation. J. Bacteriol. 183, 5848–5854.

Mizan, F.R., Jahid, I.K., Ha, S., 2015. Microbial biofilms in seafood: a food-hygiene challenge. Food Microbiol. 49, 41–55.

Molobela, I.P., Cloete, T.E., Beukes, M., 2010. Protease and amylase enzymes for biofilm removal and degradation of extracellular polymeric substances (EPS) produced by *Pseudomonas fluorescens* bacteria. J. Microbiol. 4 (14), 1515–1524.

Møretrø, T., Hermansen, L., Holck, A.L., Sidhu, M.S., Rudi, K., Langsrud, S., 2003. Biofilm formation and the presence of the intercellular adhesion *Locus ica* among *Staphylococci* from food and food processing environments. Appl. Environ. Microbiol. 69 (9), 5648–5655.

Møretrø, T., Langsrud, S., 2004. *Listeria monocytogenes*: biofilm formation and persistence in food processing environments. Biofilms 1, 107–121.

Nallapareddy, S.R., Singh, K.V., Sillanpaa, J., Garsin, D.A., Hook, M., Erlandsen, S.I., et al., 2006. Endocarditis and biofilm-associated pili of *Enterococcus faecalis*. J. Clin. Investig. 116, 2799–2807.

Neyret, C., Herry, J.M., Meylheuc, T., Dubois-Brissonnet, F., 2014. Plant-derived compounds as natural antimicrobials to control paper mill biofilms. J. Ind. Microbiol. Biotechnol. 41, 87–96.

Ni, N., Li, M., Wang, J., Wang, B., 2009. Inhibitors and antagonists of bacterial quorun sensing. Med. Res. Rev. 29, 65–124.

Nostro, A., Sudano Roccaro, A., Bisignano, G., Marino, A., Cannatelli, M.A., Pizzimenti, F.C., et al., 2007. Effects of oregano, carvacrol and thymol on *Staphylococcus aureus* and *Staphylococcus epidermidis* biofilms. J. Med. Microbiol. 56, 519–523.

O'Toole, G.A., Kaplan, H.B., 2000. Biofilm formation as microbial development. Annu. Rev. Microbiol.49–79.

Oliver, S.P., Jayarao, B.M., Almeida, R.A., 2005. Food borne pathogens in milk and the dairy farm environment: food safety and public health implications. Foodborne. Pathog. Dis. 2, 115–129.

Orgaz, B., Neufeld, R.J., Sanjose, C., 2007. Single-step biofilm removal with delayed release encapsulated pronase mixed with soluble enzymes. Enzyme 40, 1045–1051.

Oulahal, N., Martial-Gros, A., Bonneau, M., Blum, L.J., 2007. Removal of meat biofilms from surfaces by ultrasounds combined with enzymes and/or a chelating agent. Innov. Food Sci. Emerg. Technol. 8 (2), 192–196.

Oulahal-Lagsir, N., Martial-Gros, A., Boistier, E., Blum, L.J., Bonneau, M., 2000a. The development of an ultrasonic apparatus for the non-invasive and repeatable removal of fouling in food processing equipment. Lett. Appl. Microbiol. 30 (1), 47–52.

Oulahal-Lagsir, N., Martial-Gros, A., Bonneau, M., Blum, L.J., 2000b. Ultrasonic methodology coupled to ATP bioluminescence for the non-invasive detection of fouling in food processing equipment e validation and application to a dairy factory. J. Appl. Microbiol. 89 (3), 433–441.

Oulahal-Lagsir, N., Martial-Gros, A., Bonneau, M., Blum, L.J., 2003. "*Escherichia coli*-milk" biofilm removal from stainless steel surfaces: synergism between ultrasonic waves and enzymes. Biofouling 19, 159–168.

Ozdemir, H.O., Buzoglu, H.D., Calt, S., Stabholz, A., Steinberg, D., 2010. Effect of ethylenediaminetetraacetic acid and sodium hypochlorite irrigation on *Enterococcus faecalis* biofilm colonization in young and old human root canal dentin: in vitro study. J. Endod. 36 (5), 842–846.

Park, S.H., Jarquin, R., Hanning, I., Almeida, G., Ricke, S.C., 2011. Detection of *Salmonella* spp. survival and virulence in poultry feed by targeting the hilA gene. J. Appl. Microbiol. 111 (2), 426–432.

Parsek, M.R., Singh, P.K., 2003. Bacterial biofilms: an emerging link to disease pathogenesis. Annu. Rev. Microbiol. 57, 677–701.

Rajkowski, K.T., 2009. Biofilms in fish processing. In: Fratamico, P.M., Annous, B.A., Gunther, N.W. (Eds.), Biofilms in the Food and Beverage Industries. Woodhead Publishing, Oxford, pp. 499–516.

Rideout, K., Teschke, K., Dimich-Ward, H., Kennedy, S.M., 2005. Considering risks to healthcare workers from glutaraldehyde alternatives in high-level disinfection. J. Hosp. Infect. 59 (1), 4–11.

Rodrigues, L.R., van der Mei, H.C., Teixeira, J.A., Oliveira, R., 2004. Biosurfactant from *Lactococcus lactis* 53 inhibits microbial adhesion on silicone rubber. Appl. Microbiol. Biotechnol. 66, 306–311.

Rosmaninho, R., Santos, O., Nylander, T., Paulsson, M., Beuf, M., Benezech, T., et al., 2007. Modified stainless steel surfaces targeted to reduce fouling – evaluation of fouling by milk components. J. Food Eng. 80 (4), 1176–1187.

Salo, S., Ehavald, H., Raaska, L., Vokk, R., Wirtanen, G., 2006. Microbial surveys in Estonian dairies. LWT - Food Sci. Technol. 39 (5), 460–471.

Sanders, S.Q., Boothe, D.H., Frank, J.F., Arnold, J.W., 2007. Culture and detection of *Campylobacter jejuni* within mixed microbial populations of biofilms on stainless steel. J. Food Prot. 70 (6), 1379–1385.

Sauer, K., Camper, A.K., Ehrlich, G.D., Costerton, J.W., Davies, D.G., 2002. *Pseudomonas aeruginosa* displays multiple phenotypes during development as a biofilm. Bacteriology 184 (4), 1140–1154.

Sauer, K., Rickard, A.H., Davies, D.G., 2007. Biofilms and biocomplexity. Am. Soc. Microbiol. 2, 347–353.

Schembri, M.A., Blom, J., Krogfelt, K.A., Klemm, P., 2005. Capsule and fimbria interaction in *Klebsiella pneumonia*. Infect. Immunol. 73, 4626–4633.

Sharma, M., Anand, S.K., 2002. Characterization of constitutive microflora of biofilms in dairy processing lines. Food Microbiol. 19, 627–636.

Sharma, M., Ryu, J.H., Beuchat, L.R., 2005. Inactivation of *Escherichia coli* O157:H7 in biofilm on stainless steel by treatment with an alkaline cleaner and a bacteriophage. J. Appl. Microbiol. 99, 449–459.

Shia, X., Zhua, X., 2009. Biofilm formation and food safety in food industries. Trends Food Sci. Technol. 20, 407–413.

Shikongo-Nambabi, M., 2011. Control of bacterial contamination during marine fish processing. J. Biol. 3 (1), 1–17.

Silagyi, K., Kim, S.H., Lo, Y.M., Wei, C., 2009. Production of biofilm and quorum sensing by *Escherichia coli* O157:H7 and its transfer from contact surfaces to meat, poultry, ready-to-eat deli, and produce products. Food Microbiol. 26 (5), 514–519.

Sillanpaa, J., Nallapareddy, S.R., Singh, K.V., Prakash, V.P., Fothergill, T., Ton-That, H., et al., 2010. Characterization of the ebp(fm) pilus-encoding operon of *Enterococcus faecium* and its role in biofilm formation and virulence in a murine model of urinary tract infection. Virulence 1, 236–246.

Simões, M., Simões, L.C., Machado, I., Pereira, M.O., Vieira, M.J., 2006. Control of flow-generated biofilms with surfactants: evidence of resistance and recovery. Food Bioprod. 84 (4), 338–345.

Simões, M., Simões, L.C., Vieira, M.J., 2010. A review of current and emergent biofilm control strategies. LWT - Food Sci. Technol. 43 (4), 573–583.

Sinde, E., Carballo, J., 2000. Attachment of *Salmonella* spp. and *Listeria monocytogenes* to stainless steel, rubber and polytetrafluorethylene: the influence of free energy and the effect of commercial sanitizers. Food Microbiol. 17, 439–447.

Sintim, H.O., Smith, J.A., Wang, J., Nakayama, S., Yan, L., 2010. Paradigm shift in discovering next-generation anti-infective agents: targeting quorum sensing, c-di-GMP signaling and biofilm formation in bacteria with small molecules. Future Med. Chem. 2, 1005–1035.

Sofos, J.N., Geornaras, I., 2010. Overview of current meat hygiene and safety risks and summary of recent studies on biofilms, and control of *Escherichia coli* O157:H7 in nonintact, and *Listeria monocytogenes* in ready-to-eat, meat products. Meat Sci. 86 (1), 2–14.

Speranza, B., Sinigaglia, M., Corbo, M.R., 2009. Non starter lactic acid bacteria biofilms: a means to control the growth of *Listeria monocytogenes* in soft cheese. Food Control 20, 1063–1067.

Spratt, D.A., Pratten, J., Wilson, M., Gulabivala, K., 2001. An in vitro evaluation of the antimicrobial efficacy of irrigants on biofilms of root canal isolates. Int. Endod. J. 34 (4), 300–307.

Srey, S., Jahid, I.K., Ha, S., 2013. Biofilm formation in food industries: a food safety concern. Food Control 31, 572–585.

Stepanovic, S., Cirkovic, I., Ranin, L., Svabic-Vlahovic, M., 2004. Biofilm formation by *Salmonella* spp. and *Listeria monocytogenes* on plastic surface. Lett. Appl. Microbiol. 38 (5), 428–432.

Stewart, P.S., Franklin, M.J., 2008. Physiological heterogeneity in biofilms. Nat. Rev. Microbiol. 6 (3), 199–210.

Stoodley, P., Cargo, R., Rupp, C.J., Wilson, S., Klapper, I., 2002a. Biofilm material properties as related to shear-induced deformation and detachment phenomena. J. Ind. Microbiol. Biotechnol. 29 (6), 361–367.

Stoodley, P., Sauer, K., Davies, D.G., Costerton, J.W., 2002b. Biofilms as complex differentiated communities. Annu. Rev. Microbiol. 56, 187–209.

Sutherland, I.W., Hughes, K.A., Skillman, L.C., Tait, K., 2004. The interaction of phage and biofilms. FEMS Microbiol. Lett. 232, 1–6.

Takahashi, H., Miya, S., Igarashi, K., Suda, T., Kuramoto, S., Kimura, B., 2009. Biofilm formation ability of *Listeria monocytogenes* isolates from raw ready-to-eat seafood. J. Food Prot. 72, 1476–1480.

Thouvenin, M., Langlois, V., Briandet, R., Langlois, J.Y., Guerin, P.H., Peron, J.J., et al., 2003. Study of erodable paint properties involved in antifouling activity. Biofouling 19 (3), 177–186.

Toté, K., Horemans, T., Vanden Berghe, D., Maes, L., Cos, P., 2010. Inhibitory effect of biocides on the viable masses and matrices of *Staphylococcus aureus* and *Pseudomonas aeruginosa* biofilms. Appl. Environ. Microbiol. 76 (10), 3135–3142.

Valle, J., Re, D.S., Henry, N., Fontaine, T., Balestrino, D., Latour-Lambert, P., et al., 2006. Broad-spectrum biofilm inhibition by a secreted bacterial polysaccharide. Proc. Natl. Acad. Sci. USA. 103, 12558–12563.

Van Houdt, R., Michiels, C.W., 2005. Role of bacterial cell surface structures in *Escherichia coli* biofilm formation. Res. Microbiol. 156 (5–6), 626–633.

Vankerckhoven, E., Verbessem, B., Crauwels, S., Declerck, P., Muylaert, K., Willems, K.A., et al., 2011. Exploring the potential synergistic effects of chemical disinfectants and UV on the inactivation of free-living bacteria and treatment of biofilms in a pilot-scale system. Water Sci. Technol. 64 (6), 1247–1253.

Verma, V., Harjai, K., Chhibber, S., 2010. Structural changes induced by a lytic bacteriophage make ciprofloxacin effective against older biofilm of *Klebsiella pneumoniae*. Biofouling 26 (6), 729–737.

Vestby, L.K., Møretrø, T., Langsrud, S., Heir, E., Nesse, L.L., 2009. Biofilm forming abilities of *Salmonella* are correlated with persistence in fish meal- and feed factories. BMC Vet. Res. 5, 20.

Vezzulli, L., Pezzati, E., Repetto, B., Stauder, M., 2008. A general role for surface membrane proteins in attachment to chitin particles and copepods of enviromental and clinical vibrios. Lett. Appl. Microbiol. 46, 119–125.

Vilar, M.J., Yus, E., Sanjuan, M.L., Dieguez, F.J., Rodriguez-Otero, J.L., 2007. Prevalence of and risk factors for *Listeria* species on dairy farms. J. Dairy Sci. 90, 5083–5088.

Waak, E., Tham, W., Danielsson-Tham, M.L., 2002. Prevalence and fingerprinting of *Listeria monocytogenes* strains isolated from raw whole milk in farm bulk tanks and in dairy plant receiving tanks. Appl. Environ. Microbiol. 68 (7), 3366–3370.

Whitehead, K.A., Verran, J., 2015. Formation, architecture and functionality of microbial biofilms in the food industry. Curr. Opin. Food Sci. 2, 84–91.

Wiedmann, M., Weilmeier, D., Dineen, S.S., Ralyea, R., Boor, K.J., 2000. Molecular and phenotypic characterization of *Pseudomonas* spp. isolated from milk. Appl. Environ. Microbiol. 66 (5), 2085–2095.

Wong, A.C.L., 1998. Biofilms in food processing environments. J. Dairy Sci. 81 (10), 2765–2770.

Xavier, J.B., Picioreanu, C., van Loosdrecht, M.C., 2005. A framework for multidimensional modelling of activity and structure of multispecies biofilms. Environ. Microbiol. 7, 1085–1103.

Chapter 12

Food Spoilage and Food Safety: Is There a Link?

Altieri Clelia
University of Foggia, Foggia, FG, Italy

12.1 MICROBIAL FOOD SAFETY

The microbiological aspects of food safety have been studied for long time and all over the world, and in the last 20 years, at least in the industrialized world, food-borne diseases caused by bacteria, parasites, viruses, and prions have fed significantly into political strategies and generated media attention.

Public health efforts have been mainly directed toward the well-known pathogens in food and food-borne diseases, and improvements in the microbiological safety of foods have been driven by public demand in response to disease outbreaks. Throughout the 1990s and until today, three major food-borne bacterial targets (*Salmonella* spp., *Campylobacter* spp., and *Escherichia coli*) have persisted, commanding the most research and surveillance attention from government agencies and, to a large extent, the most awareness from the food industry. More recently there has been a growing concern about *Listeria monocytogenes*. These bacterial pathogens together constitute the greatest burden of food-borne illness for which etiology is known. Not surprisingly therefore these diseases command the majority of public health interest and policy maker awareness in intestinal infectious diseases. They also provide clear examples of the persistence of bacterial food-borne pathogens, despite considerable efforts aimed at prevention and control. Nowadays, major elements in microbial food safety are (1) the possible presence of emerging and not yet well-known pathogens, (2) the possible presence of opportunistic pathogens, (3) possible physiological aspects performed by a variety of microorganisms, as well as interactions among species/strains and overall antibiotic resistance.

12.1.1 Safety Assessment of Food

Despite all the efforts from scientists, governments, and industry it appears that food-borne disease will continue as a major public health problem

The Microbiological Quality of Food. DOI: http://dx.doi.org/10.1016/B978-0-08-100502-6.00015-7

worldwide, with enormous implications for both the social welfare of populations and national economies.

Food-borne outbreaks appear to be on the rise again in some industrialized countries, not only with foods from animal origin, but also with dry products and ingredients, fresh foods (Anonymous, 2008), and shellfish (Pontrelli et al., 2008). New risks are being encountered because of changes in characteristics of the relevant microorganisms, in production methodologies, in the environment and ecology, and also because of an increase in the global trade of foodstuffs. In addition, demands on food safety increase steadily. Due to the nature of microbes and the modern food chain, measures to ensure food safety have to be implemented through a global approach, in fact management of microbial food safety has to be considered a balancing act involving several different factors. All over the world a lot of food laws are issued and e.g., EU (Anonymous, 2012) law in fact states that a high level of protection of human life and health should be assured in the pursuit of community policies.

Monitoring of contamination in the food chain, combined with surveillance of human illness and epidemiological investigations of outbreaks and sporadic cases, continue to be important sources of information in order to evaluate the success of current food safety management systems and to identify new hazards. Surveillance is defined as "the ongoing and systematic collection, analysis, and interpretation of data about a disease or health condition; used in planning, implementing, and evaluating public health programs" (Anonymous, 2000, 2007a,b). Surveillance can be aimed at outbreaks or sporadic cases of food-borne disease, and continues to be a cornerstone of food safety management.

The EU is the world's largest exporter and the world's second largest importer of food and drink products and a major trader in agricultural commodities. Global trade highlights the need for scientists and risk assessors to cooperate at global level. In this context, it is extremely important to ensure that the most robust scientific knowledge is applied to protect public health, thereby sustaining confidence in the food systems. Nevertheless the terms "risk manager" and "risk management" are still not unequivocal and are used to indicate several functions and persons.

Formally, the risk manager in all countries is the government's minister of public health and/or his colleague of agriculture, as they set the standards to which food producers must adhere. Often, however, the person responsible for compliance to procedures within a food manufacturing company is called a "risk manager," with this term describing very well her/his day-to-day activities. Although these functions are less clearly defined in smaller operations and in primary production, the role should be fulfilled in any food operation. Risk assessors provide the risk manager with science-based advice on the magnitude of risks and cost-effective ways to reduce these. This advice enables the risk manager to take decisions on measures to control risks, by setting standards, by implementing in-plant control measures and/or enforcing existing regulations. Researchers, in their role as risk assessors, need to provide the regulators, in their role as risk managers, with advice that can be implemented in a practical manner.

Food safety starts at primary production. To reduce the risk of food-borne gastroenteritis, especially for foods to be eaten raw, such as fresh produce and shellfish, knowledge about contamination routes, and preventative measures is of great importance. In certain cases interventions could be effective early in the primary production phase, including the production environment. In order to stop an ongoing outbreak, such outbreak investigations should be carried out quickly, which relies on close collaboration between different scientific disciplines such as microbiology, epidemiology, wildlife control, risk communication, etc., often represented by different organizations or institutes, which may hamper the investigation (Havelaar et al., 2010).

These examples clearly show the benefits of inter- and intrascientific collaboration, developing a benefit system based on inter- and intra-scientific collaboration, able to give a wide-ranging view on such a complex problem as food safety is.

12.2 FUTURE CHALLENGES TO FOOD MICROBIAL SAFETY

The continuing and rapid globalization of food production and distribution is the first important challenge because it could mean the potential for international public health problems, involving food contaminated with pathogens or chemicals.

New food production technologies also present "new challenges." As examples, we may list as follows: nanotechnology, genomics, cloning, and proteomics just to name a few. But a challenge is also to upgrade existing technologies and design new risk assessment practices to keep up with the pace of science and innovation.

Such a problem cannot be approached without regard to a correct surveillance system, able to highlight the etiologic agents, foods, and settings involved most often in food-borne disease outbreaks and able to help to identify food commodities and preparation settings in which interventions might be most effective. Analysis of data collected over several years of surveillance provides a means to assess changes in the food commodities associated most frequently with outbreaks that might occur following improvements in food safety or changes in consumption patterns or food preparation practices. Prevention of food-borne disease depends on targeted interventions at appropriate points from food production to food preparation.

But the new challenge to food microbial safety is to understand that the effective way to face this problem is through developing a system approach, able to analyze food safety as well as a complex system of factors, with management of microbial food safety as a balancing act involving disparate factors. In particular, according to Havelaar et al. (2010) a risk management has to consider (1) food safety demands, (2) clear separation between risk management and risk assessment, (3) a scientific basis for the approach used, by an interaction between scientists and regulators.

In order to analyze future challenges to food safety, it must be considered that the correct approach is to investigate this problem on large scales of space and time, as well as on small scales; in fact certain aspects slowly change at a global scale (i.e., climate changes), while others occur on very short time scales (i.e., genetic or molecular mutation). Also the spatial scale is very important as well, as changes occur in microorganisms or humans or in an environment. To better explore the complexity of such a system, mathematical models may be useful, because they are able to represent the essential aspects of an existing or constructed system. A mathematical model describes a system by a set of numerical variables and a set of equations, representing relationships between the considered variables.

Some variables are not easily predictable and the use of mathematical models can be helpful in order to obtain a reasonable forecast of the future development of them in the system.

Nowadays some of the most interesting variables to be considered as hardly predictable factors in evaluating food chain safety are the following: (1) the possible significance of spoilers as opportunistic pathogens, (2) the dissemination and maintenance of antibiotic resistance within the system.

An opportunistic pathogen is normally a commensal microorganism, not harming its host, but it can cause disease when the host's resistance is low because of several different factors.

Several microbial genera and species play a spoiling role in food, but is it possible that some spoilers opportunistically play a different role, performing pathogenic characteristics because of different factors?

Among the spoilers described above (*Pseudomonas* and related species, lactic acid bacteria (LAB), specific spoilage organisms, spore-forming bacteria) are there any species able to harm the host?

In the light of the real significance of the term "opportunistic pathogen," the answer is: may be. In fact nowadays, when technology is correctly applied, it is hard to evaluate the real probability for a spoiler to play a pathogenic role without evaluating the single host, whose characteristics, as well as a particularly low resistance, represent the most important risk factor.

These microorganisms perform different roles, other than the role of spoilers, and these involve safety and consumer health. Different physiological phenomena in microbial life could harm consumers, or could influence food safety, as well as antibiotic resistance.

12.2.1 Dissemination and Maintenance of Antibiotic Resistance

Antibiotics are manufactured on a very large scale and have been used for human therapy and other purposes since the mid-20th century. The impact of this practice on the ecology of bacteria populations has been very profound, in particular more and more microbial strains in any given species have become resistant to antibiotics and they are able to release resistance factors in the environment.

The use of antibiotics as growth promoters in livestock and their role in selecting antibiotic resistant bacteria have been extensively reviewed (Butaye et al., 2003; Kazimierczak et al., 2006; Landers et al., 2012). In particular antibiotics have been used for decades in animal farming at subtherapeutic doses as growth promoters, until this practice was recognized in the early 1980s as a major determinant contributing to selection of antibiotic resistant strains in the guts of treated livestock, and to their subsequent release into the environment. Food products derived from animals colonized by antibiotic resistant bacteria are considered therefore as potential transmission vehicles of antibiotic resistance to humans (Gonzalez-Zorn and Escudero, 2012).

12.2.1.1 Development of Antibiotic Resistance

Development of antibiotic resistance in bacteria is mainly based on two factors: (1) the presence of resistance genes and (2) the external selective pressure. In fact resistance to a given antibiotic can be intrinsic to a bacterial species or genus (inherent or natural resistance) that results in an organism's ability to thrive in the presence of an antimicrobial agent due to an inherent characteristic of the organism. Intrinsic resistance is not horizontally transferable, and poses no risk in nonpathogenic bacteria. On the other hand, acquired resistance is present in some strains within a species usually susceptible to the antibiotic under consideration, and might be horizontally spread among bacteria. Acquired resistance to antimicrobial agents can arise either from mutations in the bacterial genome or through the acquisition of additional genes, coding for a resistance mechanism.

Genes that can confer resistant phenotypes are collectively termed the resistome (Forsberg et al., 2012), which is a useful organizing concept, but is not a natural or functional grouping, as it comprises diverse genes whose functions may have simply been coopted to confer resistance (Dantas and Sommer, 2012; Gillings et al., 2015). Nevertheless, the resistome concept helps to highlight the point that resistance genes in pathogens often originate from unrelated environmental organisms, and that the resistance genes of medical concern are just a small proportion of the resistome. Culture independent sampling of human microbiota reveals a vast reservoir of potential resistance genes. However, the majority of these genes have not been identified in clinical contexts, and are evolutionarily distant from known resistance determinants.

In contrast, if aerobic proteobacteria are cultured from fecal samples, over half of the resistome elements are identical to known resistance genes (Gillings, 2013). These resistance determinants are readily exchanged with pathogens, and spread rapidly between individuals. This suggests that proteobacteria from environmental sources might act as a conduit for clinically-relevant resistance genes to make their way into the human microbiome and thence into pathogens.

12.2.1.2 Antibiotic Resistance in Environment

Antibiotic resistance spreading is an emerging worldwide concern, developing also in different environments considered apparently safe from this point of view. For instance the presence of antibiotic resistance determinants and virulence factors associated with mobile genetic elements (integrons, plasmids, and transposons) compounds the risks associated with the introduction of human bacterial strains to apparently safe environments, including even the Antarctic, where, for instance, the rapid emergence of antibiotic resistance has demonstrated the importance of horizontal gene transfer to bacterial evolution and led the idea of "gene pollution" (Boto, 2010; Power et al., 2016). In general genetic pollution has generated a pool of genes in the natural environment that can be sequestered by diverse microbial species including opportunistic pathogens, and those that already have antibiotic or other resistance traits (Gillings, 2014).

Antimicrobial molecules and resistance mechanisms are abundant in any bacterial community, and play an important evolutionary and regulatory role, and the development of antibiotic resistance in natural environments may be considered to be a result of competition among different bacterial taxa (Bhullar et al., 2012) and intra-species competition is sufficient to drive the acquisition of resistance. For instance growth of methicillin-resistant *Staphylococcus aureus* (MRSA) has been shown to induce the rapid generation of a mutated strain that attempts to outcompete the wild-type strain(O) through secretion of surfactant molecules and the bacteriocin Bsa (Koch et al., 2014; Becattini et al., 2016).

12.2.1.3 Antibiotic Resistance in Food

Antimicrobial resistance is a very particular issue also in the food chain. In fact, for more than 50 years, antimicrobial agents have been an essential component of infectious disease treatment, in both human and veterinary medicine, and the use of such agents has resulted in the development and spread of antimicrobial resistance, due to the selective pressure that drives rapid development of resistant strains also along the food chain, as previously described.

The public health consequences of this resistance can involve increased frequency of treatment failures and severity of infection, including prolonged duration of illness, progression to systemic infections, increased hospitalization, and increased mortality.

In addition, horizontal resistance gene transfer may occur between different bacterial species in the guts of animals and humans (e.g., from *Salmonella* to *E. coli*) (Balis et al., 1996) so antibiotic use for growth promotion has been prohibited in the EU since 2003 by Regulation 1831/2003 of the European Parliament and Council, but still occurs in many countries elsewhere.

In the case of enteric zoonotic bacteria (e.g., *Salmonella*, *Campylobacter*), there is clear evidence of an unfavorable impact on human health. More complex is the case of indicator organisms (i.e., enterococci and *E. coli* strains that cause no disease in animals but can cause disease in humans). For example, avoparcin (a glycopeptide like vancomycin), was used in Europe as a growth promoter for animals until 1997 and epidemiological and molecular studies have linked avoparcin use to the carriage and dissemination of vancomycin resistant enterococci (VRE) from animals to humans in close contact with these animals (Graham et al., 2009). Owing to public health concerns over resistance to glycopeptides, avoparcin use was banned in the EU in 1997 and a subsequent reduction in VRE prevalence in humans and animals was observed (Wegener, 2003).

Indistinguishable strains of *E. coli* and MRSA have also been found in animals and humans in close contact with these animals, and interspecies transfer has been documented (Khanna et al., 2008; van Loo et al., 2007) even if evidence suggests that animal strains do not persist in the human gut for long periods of time.

The magnitude of the problem is significantly increased by the possibility for bacteria to transfer resistance determinants horizontally and by the mounting increase in the use (overuse and misuse) of antibiotics, which has created an enormous selective pressure toward resistant bacteria (Levy, 1992).

Gene transfer occurs widely in vivo between gastrointestinal tract bacteria, and between gastrointestinal tract bacteria and pathogenic bacteria, as identical resistance genes are present in diverse bacterial species from different hosts. In fact we face the frightening probability that most pathogenic bacteria that threaten human health will soon be resistant to all known antibiotics.

For several decades studies in the selection and dissemination of antibiotic resistance have focused mainly on clinically relevant bacterial species, but the evolution of antibiotic resistant food-borne pathogens has been amply documented (Witte, 1998; Ridley and Threlfall, 1998; Teuber, 1999; Teuber and Perreten, 2000; Threlfall et al., 2000; White et al., 2002) and many investigators have speculated that commensal bacteria may act as reservoirs of antibiotic resistance genes similar to those found in human pathogens; thus it seems to be very important in our understanding of how antibiotic resistance genes are maintained and spread through bacterial populations (Levy and Salyers, 2002).

The main threat associated with these bacteria is that they can transfer resistance genes to pathogenic bacteria, but resistance genes could be similarly transferred also among microbiota genera, performing possible perturbations to the composition and functions of the microbiota, that may have significant effects on health (Cho and Blaser, 2012).

Reservoir organisms could possibly be found in various foods and food products containing high densities of nonpathogenic bacteria as a result of their natural production process, and in addition they could be found in human and animal microbiota, and the food chain could be considered as the main route of transmission of antibiotic resistant bacteria between the animal and human population.

In the light of these considerations, increasing focus has been given to food as vehicles of antibiotic resistance genes (Klein et al., 2000; Munsch-Alatossava et al., 2012).

12.2.1.4 Spoilers Role in Spreading Antibiotic Resistance

Gram-Negatives

In taking into consideration the species which have been detailed in previous chapters as spoilers, it is interesting to underline their possible role in spreading antibiotic resistances. In fact, extensively drug resistant (XDR) and pandrug resistant (PDR) *Pseudomonas aeruginosa* and *Acinetobacter* spp. are a problem in many parts of the world (Magiorakos et al., 2011; Zavascki et al., 2010; Morrow et al., 2013). *P. aeruginosa* has become increasingly difficult to treat in an impressively short time, as practically all known mechanisms of antimicrobial resistance have been described in this pathogen (Breidenstein et al., 2011). These mechanisms are often present simultaneously, thereby conferring XDR phenotypes and limiting empirical treatment options.

The results of a Chinese surveillance study provide a glimpse into this problem: 7.5% of *P. aeruginosa* and 14.1% of Acinetobacter *baumannii* in the study were PDR and thus without an active antibacterial treatment option (Wang et al., 2010).

European data from 2008 (EARS-Net 2008) highlight that 17% of *P. aeruginosa* isolates were found to be resistant to three or more antipseudomonal antibiotics (piperacillin–tazobactam, ceftazidime, fluoroquinolones, aminoglycosides, and carbapenems), with 6% demonstrating resistance to all five classes of tested antipseudomonal antibiotics (XDR).

Data from the United States are incomplete, with a pooled mean carbapenem resistance of 12% (Hidron et al., 2008) and an overall rate of multidrug resistance rate of c. 15% based on 2722 isolates sampled across the United States and Puerto Rico (Morrow et al., 2013).

In the Asia-Pacific region, carbapenem resistance is prevalent and estimated to be c. 40% (Hsueh, 2012). Latin American countries have a similar resistance rate of 30–50% (Gales et al., 2012; Amabile-Cuevas, 2010).

Acinetobacter spp. are emerging as PDR/XDR nosocomial and community-acquired pathogens in some settings (Slama, 2008). Although not as prevalent as other Gram-negative organisms, *Acinetobacter* spp. demonstrates an incredible ability to escape antibacterial activity and to develop resistance to multiple classes of antibiotics, both of which have been important factors in the increasing recognition of these organisms' clinical significance (Peleg et al., 2008; O'Shea, 2012).

Large-scale comparable surveillance data on combined resistance in *Acinetobacter* spp. are quite lacking, but they should have be performed, in order to have a more complete picture of this bacterium.

Infections with multiresistant *E. coli* are also an important public health problem. Commensal *E. coli* can be a source of resistance genes for pathogenic *E. coli* strains.

The animal to human spread of resistance genes from *E. coli*, and from commensal to pathogenic Enterobacteriaceae, has been well documented, but the role of food in such transfer is not yet known and needs further investigation. In particular, the transfer of resistance genes from *E. coli* to *Salmonella* has been demonstrated experimentally in the poultry intestinal tract (Poppe et al., 2005) and there are some reports indicating the acquisition of resistance plasmids by *E. coli* and *Salmonella* in the human gut. Moreover, the exchange of resistance genes between bacterial clones has been demonstrated experimentally in water, soil, on kitchen towels, on cutting boards, and on the surface of foods (Walsh et al., 2010).

Gram-Positives

On the whole, while pathogenic species, mostly Gram-negative, represent an immediate threat to human/animal health due to their difficult eradication when carrying antibiotic resistance determinants, antibiotic resistant Gram-positives consist of mostly nonpathogenic species (or opportunistic pathogens). Among them, several LAB species are present in the raw materials employed in fermented food production (milk, meat, vegetables), where they play a crucial role in leading fermentations by acting as natural starters of environmental origin. Therefore LAB are integral components of fermented foods, where they carry out primary and secondary fermentations, up to the final products (Rattanachaikunsopon and Phumkhachorn, 2010). Their long history of safe use in food production earned most LAB species the generally regarded as safe status by the US Food and Drug Administration and the qualified presumption of safety classification by the European Food Safety Authority. In the past decade the increasing interest in probiotics has led to extensive characterization of the nutritional and health-associated features of LAB, especially of lactobacilli (and of bifidobacteria) (Gonzalez-Rodriguez et al., 2013) which are highly represented in the human gut, mainly to substantiate health claims of commercial probiotic products.

Both lactobacilli and lactococci are mostly composed of nonpathogenic species, but rather display probiotic features and were never considered a threat for humans; nevertheless antibiotic resistant strains have been increasingly described, demonstrating LAB as reservoir of antibiotic resistance genes, horizontally transmissible to pathogens through the food chain (Devirgiliis et al., 2013). A growing body of literature is now available on this issue in isolates deriving from various environmental sources, but also LAB strains widely used as probiotics or in starter cultures have the potential to serve as a host of antibiotic resistance genes with the risk of transferring the genes in many LAB and other species of pathogenic bacteria.

Most data exist on opportunistic pathogenic enterococci, while the number of reports on lactococci and lactobacilli is limited. As described before, VRE have emerged as a frequent cause of nosocomial infections. Of considerable concern is the possibility that VRE, selected and enriched by the use of avoparcin (with cross resistance to vancomycin) as a growth promoter in animal husbandry, are spread via the food chain (Klein et al., 2000; Pavia et al., 2000).

Enterococcal food isolates (mainly *Enterococcus faecalis* and *Enterococcus faecium*) were analyzed for resistances to a broader range of different antibiotics using phenotypic susceptibility testing, both in raw meat (Klein et al., 1998; Robrido et al., 2000) and fermented milk and meat products (Teuber and Perreten, 2000; Franz et al., 2001; Giraffa, 2002). Their data suggest a high prevalence of (multiple) antibiotic resistant enterococci in foods, which nevertheless were mostly susceptible to the clinically-relevant antibiotics ampicillin and vancomycin. Butaye et al. (2000) studied the in vitro susceptibility and resistance of *E. faecium* strains isolated from raw poultry meat, cheese, raw pork, and preparations of cheese and raw pork to growth promoting antibacterials used in animals and antibiotics used therapeutically in humans. Resistance against bacitracin, virginiamycin, narasin, and tylosin, was found to be high among strains from poultry meat.

Although enterococci are generally regarded as being intrinsically resistant to low levels of gentamicin, a high-level gentamicin resistance was detected in many dairy isolates. Donabedian et al. (2003) evaluated the molecular mechanisms of gentamicin resistance in *Enterococcus* spp. isolated from animals, foods, and humans, suggesting that there are similarities in gentamicin resistance in enterococci isolated from humans, retail food, and farm animals from geographically diverse areas and there is evidence of the spread of gentamicin resistant enterococci from animals to humans through the food supply.

The tetracycline resistance could be linked to the presence of *tet*(M) genes in Enterococcal isolates, all of which also harbored a member of the Tn916–Tn1545 conjugative transposon family. Cataloluk and Gogebaken (2004) reported the prevalence of *tet*(M) and *erm*(B) genes in the majority (61.961%) of lactobacilli of human and dairy origin isolated from Turkey.

The resistant strains belonged to *Lactobacillus acidophilus*, *Lactobacillus crispatus*, *Lactobacillus gasseri*, and *Lactobacillus plantarum*. It shows that drug resistance may be acquired in the intestinal tract during passage and spread to dairy products by the hands of workers during production (Ouoba et al., 2008).

An overview of antibiotic resistances can be summarized by stating that only a limited number of papers have reported the prevalence of antibiotic resistance in *Lactobacillus* spp. isolated from raw meat and fermented food products. A few studies have reported an overall susceptibility to antimicrobial agents (with the exception of intrinsic resistances) in strains used as meat starter cultures (Holley and Blaszyk, 1997) or dairy starter cultures (Katla et al., 2001).

Nevertheless, a higher and higher number of food-borne *Lactobacillus* species have been reported to carry one or more antibiotic resistance genes, although the association of such genes with mobile elements as well as their possible horizontal transfer were not always investigated among these bacteria. The most common antibiotics employed for selection of antibiotic resistant lactobacilli in these studies were tetracycline and erythromycin, followed by chloramphenicol, streptomycin, ampicillin, vancomycin, and clindamycin.

Ammor et al. (2008) found two tetracycline resistance genes, *tet*(M) and *tet*(L), coexisting in a food-borne strain of *Lactobacillus sakei*, and they were shown to reside within a transposon-like element and a plasmid, respectively. A *tet*(M) gene carried by a tetracycline-resistant strain of Lactobacillus *paracasei* of dairy origin was associated with the broad host range Tn*916* transposon, which could be transferred to *E. faecalis* in filter-mating assays, although with low frequency (Devirgiliis et al., 2009); nucleotide sequence of the erythromycin resistance plasmid pLFE1 from *L. plantarum* strain M345, isolated from raw-milk cheese, revealed genes involved in conjugal transfer. Filter-mating experiments demonstrated pLFE1 can be transferred to Lactobacillus *rhamnosus*, *Lactococcus lactis*, *Listeria innocua*, *E. faecalis*, and *L. monocytogenes*, suggesting a broad host range (Feld et al., 2009).

Like other LAB, lactococci can acquire antibiotic resistance under selective pressure, can survive antimicrobial treatments and consequently act as reservoir for antibiotic resistance genes transmissible to other bacteria. Several studies reported the susceptibility of *L. lactis* to Gram-positive spectrum antibiotics (erythromycin, lincomycin, vancomycin, novobiomicin, teicoplanin), to beta-lactams, and to some broad-spectrum antibiotics (rifampicin, chloramphenicol, spectinomycin). On the other hand, most lactococcal species display intrinsic resistance to metronidazole, trimethoprim, and cefoxitin, and to the aminoglycosides gentamicin and kanamycin (Katla et al., 2001; Ammor et al., 2008; Florez et al., 2012).

Finally, the ability of *Lactococcus* to act as recipient in conjugal transfer experiments using a *Lactobacillus* donor, was demonstrated by using a *tet*(M) determinant characterized in *L. plantarum*, which was successfully transferred to *L. lactis* strain BU-2-60 (Toomey et al., 2010). While *L. lactis* subsp. *lactis* and *L. lactis* subp. *cremoris* are nonpathogenic and used in starter cultures for dairy products, *Lactococcus garvieae* is a serious fish pathogen, and also causes mastitis in cows (Pitkala et al., 2004). However, *L. garvieae* strains from dairy sources have been shown to be free of virulence determinants (i.e., hemolysin and gelatinase), suggesting that *L. garvieae* of dairy origin are unconnected to the pathogenic strains (Foschino et al., 2008). This species was isolated from raw milk as well as from artisanal cheese (Casalta and Montel, 2008). The activity of *L. garvieae* strains in dairy fermentations seems to contribute to the final sensory features (Fernandez et al., 2010) and no evidence was ever reported of an association between raw-milk cheese consumption and human disease. Walther et al. (2008) reported the presence of antibiotic resistance genes in *L. lactis* and *L. garvieae* strains isolated from raw-milk, tested for susceptibility to 17 antibiotics. Most of the *L. garvieae* strains showed phenotypic resistance to tetracycline and harbored *tet*(S)and *tet*(M) determinants. The authors also report phenotypic resistance to clindamycin, erythromycin, streptomycin, and nitrofurantoin. In particular, all erythromycin resistant isolates were shown to harbor the *erm*(B) gene. The multidrug transporter *mdt*(A) was also detected in this work for the first time in *L. garvieae*. The transporter *mdt*(A) confers resistance to macrolides, lincosamides, streptogramins, and tetracycline and it had previously been described only in *L. lactis* (Walther et al., 2008). A previous safety investigation by Fortina et al. (2007) revealed strains of *L. garvieae* of dairy origin moderately resistant to kanamycin, as well as some tetracycline-resistant biotypes harboring *tet*(M) and *tet*(S).

Further testing in dairy strains of *L. garvieae* against 14 antibiotics was carried out by Fernandez et al. (2010) who found five isolates showing phenotypic resistance associated to the presence of *tet*(M).

Antibiotic Resistance as a Risk Factor

Nowadays, it is possible to observe and verify that the success of antibiotics for treating infections has a downside: emergence and rapid dissemination of antibiotic resistant bacteria, meaning a substantial risk for human health. Antibiotic resistance has become an increasingly serious problem with global human deaths due to antibiotic resistant infections predicted to reach 10 million by 2050, more than the current death toll associated with different forms of cancer (O'Neill, 2014).

In addition, the number of approvals of new antibiotics are significantly and steadily decreasing (Alanis, 2005).

For this reason, there is an urgent need to develop novel antimicrobial agents, including alterative drugs based on antimicrobial peptides (Lata et al., 2007) that have already been described as evolutionarily ancient weapons against microbial infections.

Future Perspective

A possible future perspective could be the study and use of antimicrobial peptides, for their interesting properties, such as broad-spectrum (antibacterial, antifungal, and antiviral) effects and low levels of induced resistance. They could be considered as promising alternatives to conventional antibiotics (Wang et al., 2016). Antimicrobial peptides are small biological molecules (<10 kDa, 12–50 amino acids) which usually function through a defined high-affinity antimicrobial target and which can exert multiple antimicrobial activities that might provide a strategy to prevent bacteria from developing resistance.

Unlike conventional antibiotics, which usually function through a defined high-affinity antimicrobial target and which can induce resistance in microorganisms, these molecules exert multiple antimicrobial activities that include responding to microbial infections by acting on host targets rather than microbial targets which would be an advantage over the traditional antibiotics.

These peptides have a strong potential for application as feed additives in different animal species, in fact these antimicrobial peptides have been demonstrated to improve growth performance, promote nutrient digestibility and gut health, positively alter intestinal microbiota, and enhance immune function in pigs and broilers. The effects of these natural molecules on growth performance are probably due to their antimicrobial and immunomodulating activity, thereby promoting nutrient digestibility and health, however their fate in vivo is still poorly understood. Consequently there is a need for research evaluating the pharmacokinetics of these antimicrobial small peptides.

REFERENCES

Alanis, A.J., 2005. Resistance to antibiotics: are we in the post-antibiotic era? Arch. Med. Res. 36, 697–705.

Amabile-Cuevas, C., 2010. Antibiotic resistance in Mexico: a brief overview of the current status and its causes. J. Infect. Dev. Countries 4, 126–131.

Ammor, M.S., Flórez, A.B., van Hoek, A.H.A.M., de los Reyes-Gavilán, C.G., Aarts, H.J.M., Margolles, A., et al., 2008. Molecular characterization of intrinsic and acquired antibiotic resistance in lactic acid bacteria and bifidobacteria. J. Mol. Microbiol. Biotechnol. 14, 6–15.

Anonymous, 2000. Dorland's Illustrated Medical Dictionary, twentyninth ed. WB Saunders Company, Philadelphia, PA.

Anonymous, 2007a. Opinion of the Scientific Panel on Biological Hazards on microbiological criteria and targets based on risk analysis. EFSA J. 462, 1–29.

Anonymous, 2007b. EFSA Scientific Committee, introduction of a Qualified Presumption of Safety (QPS) approach for assessment of selected microorganisms referred to EFSA. EFSA J. 587, 1–16.

Anonymous, 2008. Outbreak of Salmonella serotype saintpaul infections associated with multiple raw produce items—United States, 2008. MMWR Morb. Mortal. Wkly. Rep. 57, 929–934.

Anonymous, 2012. EFSA Panel on Biological Hazards (BIOHAZ). Scientific Opinion on Public health risks represented by certain composite products containing food of animal origin. EFSA J. 10 (5), 2662.

Balis, E., Vatopoulos, A.C., Kanelopoulou, M., Mainas, E., Hatzoudis, G., Kontogianni, V., et al., 1996. Indications of in vivo transfer of an epidemic R plasmid from *Salmonella* enteritidis to *Escherichia coli* of the normal human gut flora. J. Clin. Microbiol. 4, 977–988.

Becattini, S., Taur, Y., Pamer, E.G., 2016. Antibiotic-induced changes in the intestinal microbiota and disease. Trends Mol. Med. 22, 458–478. Available from: http://dx.doi.org/10.1016/j.molmed.2016.04.003.

Bhullar, K., Waglechner, N., Pawlowski, A., Koteva, K., Banks, E.D., Johnston, M.D., et al., 2012. Antibiotic resistance is prevalent in an isolated cave microbiome. PLoS ONE 7, e34953.

Boto, L., 2010. Horizontal gene transfer in evolution: facts and challenges. Proc. Biol. Sci. 277, 819–827.

Breidenstein, E.B.M., de la Fuente-Núñez, C., Hancock, R.E.W., 2011. *Pseudomonas aeruginosa*: all roads lead to resistance. Trends Microbiol. 19, 419–426.

Butaye, P., Damme, K.V., Devriese, L.A., Damme, L.V., Baele, M., Lauwers, S., et al., 2000. In vitro susceptibility of *Enterococcus faecium* isolated from food to growth-promoting and therapeutic antibiotics. Int. J. Food Microbiol. 54, 181–187.

Butaye, P., Devriese, L.A., Haesebrouck, F., 2003. Antimicrobial growth promoters used in animal feed: effects of less well known antibiotics on Gram-positive bacteria. Clin. Microbiol. Rev. 16, 175–188. Available from: http://dx.doi.org/10.1128/CMR.16.2.175-188.2003.

Casalta, E., Montel, M.C., 2008. Safety assessment of dairy microorganisms: the *Lactococcus* genus. Int. J. Food Microbiol. 126, 271–273. Available from: http://dx.doi.org/10.1016/j.ijfoodmicro.2007.08.013.

Cataloluk, O., Gogebaken, B., 2004. Presence of drug resistance in intestinal lactobacilli of dairy and human origin. FEMS Microbiol. Lett. 236, 7–12.

Cho, I., Blaser, M.J., 2012. The human microbiome: at the interface of health and disease. Nat. Rev. Genet. 13, 260–270.

Dantas, G., Sommer, M.O.A., 2012. Context matters—the complex interplay between resistome genotypes and resistance phenotypes. Curr. Opin. Microbiol. 15, 577–582.

Devirgiliis, C., Coppola, D., Barile, S., Colonna, B., Perozzi, G., 2009. Characterization of the Tn916 conjugative transposon in a food-borne strain of *Lactobacillus paracasei*. Appl. Environ. Microbiol. 75, 3866–3871. Available from: http://dx.doi.org/10.1128/AEM.00589-09.

Devirgiliis, C., Zinno, P., Perozzi, G., 2013. Update on antibiotic resistance in foodborne *Lactobacillus* and *Lactococcus* species. Front Microbiol. 4, 1–13.

Donabedian, S.M., Thal, L.A., Hershberger, E., Perri, M.B., Chow, J.W., Bartlett, et al., 2003. Molecular characterization of gentamicin resistant enterococci in the United States: evidence of spread from animals to humans through food. J. Clin. Microbiol. 41 (3), 1109–1113.

Feld, L., Bielak, E., Hammer, K., Wilcks, A., 2009. Characterization of a small erythromycin resistance plasmid pLFE1 from the food-isolate *Lactobacillus plantarum* M345. Plasmid 61, 159–170. Available from: http://dx.doi.org/10.1016/j.plasmid.2009.01.002.

Fernandez, E., Alegria, A., Delgado, S., Mayo, B., 2010. Phenotypic, genetic and technological characterization of *Lactococcus garvieae* strains isolated from a raw milk cheese. Int. Dairy J. 20, 142–148. Available from: http://dx.doi.org/10.1016/j.idairyj.2009.11.004.

Florez, A.B., Reimundo, P., Delgado, S., Fernandez, E., Alegria, A., Guijarro, J.A., et al., 2012. Genome sequence of *Lactococcus garvieae* IPLA 31405, a bacteriocin-producing, tetracycline-resistant strain isolated from a raw-milk cheese. J. Bacteriol. 194, 5118–5119. Available from: http://dx.doi.org/10.1128/JB.00975-12.

Forsberg, K.J., Reyes, A., Wang, B., Selleck, E.M., Sommer, M.O.A., Dantas, G., 2012. The shared antibiotic resistome of soil bacteria and human pathogens. Science 337, 1107–1111.

Fortina, M.G., Ricci, G., Foschino, R., Picozzi, C., Dolci, P., Zeppa, G., et al., 2007. Phenotypic typing, technological properties and safety aspects of *Lactococcus garvieae* strains from dairy environments. J. Appl. Microbiol. 103, 445–453. Available from: http://dx.doi.org/10.1111/j.1365-2672.2006.03265.x.

Foschino, R., Nucera, D., Volponi, G., Picozzi, C., Ortoffi, M., Bottero, M.T., 2008. Comparison of *Lactococcus garvieae* strains isolated in northern Italy from dairy products and fishes through molecular typing. J. Appl. Microbiol. 105, 652–662. Available from: http://dx.doi.org/10.1111/j.1365-2672.2008.03780.x.

Franz, C.M., Muscholl-Silberhorn, A.B., Yousif, N.M.K., Vancanneyt, M., Swings, J., Holzapfel, W.H., 2001. Incidence of virulence factors and antibiotic resistance among enterococci isolated from food. Appl. Environ. Microbiol. 67, 4385–4389.

Gales, A.C., Castanheira, M., Jones, R.N., Sader, H.S., 2012. Antimicrobial resistance among Gram-negative bacilli isolated from Latin America: results from SENTRY Antimicrobial Surveillance Program (Latin America, 2008–2010). Diagn. Microbiol. Infect. Dis. 73, 354–360.

Gillings, M.R., 2013. Evolutionary consequences of antibiotic use for the resistome, mobilome and microbial pangenome. Front Microbiol. 4, 4. Available from: http://dx.doi.org/10.3389/fmicb.2013.00004.

Gillings, M.R., 2014. Integrons: past, present, and future. Microbiol. Mol. Biol. Rev. 78, 257–277.

Gillings, M.R., Paulsen, I.T., Tetu, S.G., 2015. Ecology and evolution of the human microbiota: fire, farming and antibiotics. Genes 6, 841–857. Available from: http://dx.doi.org/10.3390/genes6030841.

Giraffa, G., 2002. Enterococci from foods. FEMS Microbiol. Rev. 26, 163–171.

Gonzalez-Rodriguez, I., Ruiz, L., Gueimonde, M., Margolles, A., Sanchez, B., 2013. Factors involved in the colonization and survival of bifidobacteria in the gastrointestinal tract. FEMS Microbiol. Lett. 340, 1–10. Available from: http://dx.doi.org/10.1111/1574-6968.12056.

Gonzalez-Zorn, B., Escudero, J.A., 2012. Ecology of antimicrobial resistance: humans, animals, food and environment. Int. Microbiol. 15, 101–109.

Graham, J.P., Price, L.B., Evans, S.L., Graczyk, T.K., Silbergeld, E.K., 2009. Antibiotic resistant enterococci and staphylococci isolated from flies collected near confined poultry feeding operations. Sci. Total Environ. 407, 2701–2710. Available from: http://dx.doi.org/10.1016/j.scitotenv.2008.11.056.

Havelaar, A.H., Brul, S., de Jong, A., de Jong, R., Zwietering, M.H., ter Kuile, B.H., 2010. Future challenges to microbial food safety. Int. J. Food Microbiol. 139, S79–S94.

Hidron, A.I., Edwards, J.R., Patel, J., Horan, T.C., Sievert, D.M., Pollock, D.A., et al., 2008. Participating National Healthcare Safety Network Facilities NHSN annual update: antimicrobial-resistant pathogens associated with healthcare-associated infections: annual summary of data reported to the National Healthcare Safety Network at the Centers for Disease Control and Prevention, 2006–2007. Infect. Control Hosp. Epidemiol. 29, 996–1011.

Holley, R.A., Blaszyk, M., 1997. Antibiotic challenge of meat starter cultures and effects upon fermentations. Food Res. Int. 30, 513–522.

Hsueh, P.R., 2012. Study for monitoring antimicrobial resistance trends (SMART) in the Asia-Pacific region, 2002–2010. Int. J. Antimicrob. Agents 40, S1–S3.

Katla, A.K., Kruse, H., Johnsen, G., Herikstad, H., 2001. Antimicrobial susceptibility of starter culture bacteria used in Norwegian dairy products. Int. J. Food Microbiol. 67, 147–152.

Kazimierczak, K.A., Flint, H.J., Scott, K.P., 2006. Comparative analysis of sequences flanking tet(W) resistance genes in multiple species of gut bacteria. Antimicrob. Agents Chemother. 50, 2632–2639. Available from: http://dx.doi.org/10.1128/AAC.01587-05.

Khanna, T., Friendship, R., Dewey, C., Weese, J., 2008. Methicillin resistant *Staphylococcus aureus* colonization in pigs and pig farmers. Vet. Microbiol. 128, 298–303.

Klein, G., Pack, A., Reuter, G., 1998. Antibiotic resistance patterns of enterococci and occurrence of vancomycin-resistant enterococci in raw minced beef and pork in Germany. Appl. Environ. Microbiol. 64, 1825–1830.

Klein, G., Hallman, C., Casas, I.A., Abad, J., Louwers, J., Reuter, G., 2000. Exclusion of vanA, vanB and vanC type glycopeptideresistance in strains of *Lactobacillus reuteri* and *Lactobacillus rhamnosus* used as probiotics by polymerase chain reaction and hybridization methods. J. Appl. Microbiol. 89, 814–815.

Koch, G., Yepes, A., Förstner, K.U., Wermser, C., Stengel, S.T., Modamio, J., et al., 2014. Evolution of resistance to a last-resort antibiotic in *Staphylococcus aureus* via bacterial competition. Cell 158 (5), 1060–1071. Available from: http://dx.doi.org/10.1016/j.cell.2014.06.046.

Landers, T.F., Cohen, B., Wittum, T.E., Larson, E.L., 2012. A review of antibiotic use in food animals: perspective, policy and potential. Public Health Rep. 127, 4–22.

Lata, S., Sharma, B.K., Raghava, G.P., 2007. Analysis and prediction of antibacterial peptides. BMC Bioinformatics 8, 263.

Levy, S.B., 1992. The Antibiotic Paradox: How Miracle Drugs are Destroying the Miracle. Plenum Press, New York, NY.

Levy, S.B., Salyers, A.A., 2002. Reservoirs of antibiotic resistance (ROAR) network. <http://www.healthsci.tufts.edu/apua/Roar/roarhome.htm>.

Magiorakos, A.P., Srinivasan, A., Carey, R.B., Carmeli, Y., Falagas, M.E., Giske, C.G., et al., 2011. Multidrug-resistant, extensively drug-resistant and pandrug-resistant bacteria: an international expert proposal for interim standard definitions for acquired resistance. Clin. Microbiol. Infect. 18, 268–281.

Morrow, B.J., Pillar, C.M., Deane, J., Sahm, D.F., Lynch, A.S., Flamm, R.K., et al., 2013. Activities of carbapenem and comparator agents against contemporary US *Pseudomonas aeruginosa* isolates from the CAPITAL surveillance program. Diagn. Microbiol. Infect. Dis. 75, 412–416.

Munsch-Alatossava, P., Gauchi, J., Chamlagain, B., Alatossava, T., 2012. Trends of Antibiotic Resistance in Mesophilic and Psychrotrophic Bacterial Populations during Cold Storage of Raw Milk. International Scholarly Research Network ISRN Microbiology Volume, Article ID 918208, 13 p. <http://dx.doi.org/10.5402/2012/918208>.

O'Neill, J., 2014. Antimicrobial peptides: Tackling of a crisis for the health and wealth of nations. The review on Antimicrobial Resistance. Available from: <http://amrreview.org/sites/fefault/files/AMR%20Review%20paper%20%20Tackling%20a%20crisis%20for%20the%20health%20and%20wealth%20of%20nations_1.pdf> (accessed December 2014).

O'Shea, M.K., 2012. *Acinetobacter* in modern warfare. Int. J. Antimicrob. Agents 39, 363–375. Available from: http://dx.doi.org/10.1016/j.ijantimicag.2012.01.018. Epub 2012 Mar 28.

Ouoba, L.I.I., Lei, V., Jensen, L.B., 2008. Resistance of potential probiotic lactic acid bacteria and bifidobacteria of African and European origin to antimicrobials: determination and transferability of the resistance genes to other bacteria. Int. J. Food Microbiol. 121, 217–224.

Pavia, M., Nobile, C.G.A., Salpietro, L., Angelillo, I.F., 2000. Vancomycin resistance and antibiotic susceptibility of enterococci in raw meat. J. Food Prot. 63, 912–915.

Peleg, A.Y., Seifert, H., Paterson, D.L., 2008. *Acinetobacter baumannii*: emergence of a successful pathogen. Clin. Microbiol. Rev. 21, 538–582.

Pitkala, A., Haveri, M., Pyorala, S., Myllys, V., Honkanen-Buzalski, T., 2004. Bovine mastitis in Finland 2001–prevalence, distribution of bacteria, and antimicrobial resistance. J. Dairy Sci. 87, 2433–2441. Available from: http://dx.doi.org/10.3168/jds.S0022-0302(04)73366-4.

Pontrelli, G., Boccia, D., Di Renzi, M., Massari, M., Giugliano, F., Celentano, et al., 2008. Epidemiological and virological characterization of a large community-wide outbreak of hepatitis A in Southern Italy. Epidemiol. Infect. 136, 1027–1034.

Poppe, C., Martin, L.C., Gyles, C.L., Reid-Smith, R., Boerlin, P., McEwen, S.A., et al., 2005. Acquisition of resistance to extended-spectrum cephalosporins by *Salmonella enterica* subsp. enterica Serovar Newport and *Escherichia coli* in the turkey poult intestinal tract. Appl. Environ. Microbiol. 71, 1184–1192.

Power, M.L., Samuel, A., Smith, J.J., Stark, J.S., Gillings, M.R., Gordon, D.M., 2016. *Escherichia coli* out in the cold: dissemination of human-derived bacteria into the Antarctic microbiome. Environ. Pollut. 215, 58–65.

Rattanachaikunsopon, P., Phumkhachorn, P., 2010. Lactic acid bacteria: their antimicrobial compounds and their uses in food production. Ann. Biol. Res. 1, 218–228.

Regulation (EC) No. 1831/2003 of the European Parliament and of The Council of 22 September 2003 on additives for use in animal nutrition. Off. J. Eur. Union L, 268, 2003, 29–43.

Ridley, A., Threlfall, E.J., 1998. Molecular epidemiology of antibiotic resistance genes in multiresistant epidemic *Salmonella typhimurium* DT 104. Microb. Drug Resist. 4, 113–118.

Robrido, B., Singh, K.V., Baquero, F., Murray, B.E., Torres, C., 2000. Vancomycin-resistant enterococci isolated from animals and food. Int. J. Food Microbiol. 54, 197–204.

Slama, T., 2008. Gram-negative antibiotic resistance: there is a price to pay. Crit. Care 12 (Suppl. 4), S4.

Teuber, M., 1999. Spread of antibiotic resistance with food-borne pathogens. Cell. Mol. Life Sci. 56, 755–763.

Teuber, M., Perreten, V., 2000. Role of milk and meat products as vehicles for antibiotic resistant bacteria. Acta Vet. Scand. Suppl. 93, 75–87.

Threlfall, E.J., Ward, L.R., Frost, J.A., Willshaw, G.A., 2000. The emergence and spread of antibiotic resistance in food-borne bacteria. Int. J. Food Microbiol. 62, 1–5.

Toomey, N., Bolton, D., Fanning, S., 2010. Characterisation and transferability of antibiotic resistance genes from lactic acid bacteria isolated from Irish pork and beef abattoirs. Res. Microbiol. 161, 127–135. Available from: http://dx.doi.org/10.1016/j.resmic.2009.12.010.

van Loo, I., Huijsdens, X., Tiemersma, E., de Neeling, A., van de Sande-Bruinsma, N., Beaujean, D., et al., 2007. Emergence of methicillin-resistant *Staphylococcus aureus* of animal origin in humans. Emerg. Infect. Dis. 13, 1834–1839.

Walsh, F., Cooke, N.M., Smith, S.G., Moran, G.P., Cooke, F.J., Ivens, A., et al., 2010. Comparison of two DNA microarrays for detection of plasmid-mediated antimicrobial resistance and virulence factor genes in clinical isolates of Enterobacteriaceae and non-Enterobacteriaceae. Int. J. Antimicrob. Agents 35, 593–598.

Walther, C., Rossano, A., Thomann, A., Perreten, V., 2008. Antibiotic resistance in *Lactococcus* species from bovine milk: presence of a mutated multidrug transporter mdt(A) gene in susceptible *Lactococcus garvieae* strains. Vet. Microbiol. 131, 348–357. Available from: http://dx.doi.org/10.1016/j.vetmic.2008.03.008.

Wang, H., Chen, M., Ni, Y., Liu, Y., Sun, H., Yu, Y., et al., 2010. Antimicrobial resistance among clinical isolates from the Chinese Meropenem Surveillance Study (CMSS), 2003–2008. Int. J. Antimicrob. Agents 35, 227–234.

Wang, S., Zeng, X., Yang, Q., Qiao, S., 2016. Antimicrobial peptides as potential alternatives to antibiotics in food animal industry. Int. J. Mol. Sci. 17, 603. Available from: http://dx.doi.org/10.3390/ijms17050603.

Wegener, H.C., 2003. Antibiotics in animal feed and their role in resistance development. Curr. Opin. Microbiol. 6, 439–445.

White, D.G., Zhao, S., Simjee, S., Wagner, D.D., McDermott, P.F., 2002. Antimicrobial resistance of foodborne pathogens. Microb. Infect. 4, 405–412.

Witte, W., 1998. Medical consequences of antibiotic use in agriculture. Science 279, 996–997.

Zavascki, A.P., Carvalhaes, C.G., Picão, R.C., Gales, A.C., 2010. Multidrug-resistant *Pseudomonas aeruginosa* and *Acinetobacter baumannii*: resistance mechanisms and implications for therapy. Expert Rev. Anti. Infect. Ther. 8, 71–93.

Index

Note: Page numbers followed by "*f*" and "*t*" refer to figures and tables, respectively.

E

F

K

L

M

T

CPI Antony Rowe
Chippenham, UK
2017-02-01 18:52